Lecture Notes in Electrical Engineering
Volume 110

For further volumes:
http://www.springer.com/series/7818

Sio Iong Ao • Oscar Castillo • Xu Huang
Editors

Intelligent Control and Innovative Computing

Springer

Editors
Sio Iong Ao
International Association of Engineers
Hung To Road
Unit 1, 1/F, 37-39
Hong Kong
China, People's Republic
siao@graduate.hku.hk

Dr. Oscar Castillo
Tijuana Institute of Technology
Fracc. Tomas Aquino CP 22379
Tijuana
Baja California, Mexico
ocastillo@tectijuana.mx

Xu Huang
Professor and Head of Engineering
Faculty of Information Sciences
and Engineering
University of Canberra
ACT Canberra, 2601
Australia
xu.huang@canberra.edu.au

ISSN 1876-1100
ISSN 1876-1119
ISBN 978-1-4614-1694-4
e-ISBN 978-1-4614-1695-1
DOI 10.1007/978-1-4614-1695-1
Springer New York Dordrecht Heidelberg London

Library of Congress Control Number: 2011943339

Printed on acid-free paper

Springer is part of Springer Science+Business Media (www.springer.com)

Preface

A large international conference on Advances in Intelligent Control and Innovative Computing was held in Hong Kong, March 16–18, 2011, under the auspices of the International MultiConference of Engineers and Computer Scientists (IMECS 2011). The IMECS is organized by the International Association of Engineers (IAENG). IAENG is a non-profit international association for the engineers and the computer scientists, which was founded in 1968 and has been undergoing rapid expansions in recent years. The IMECS conferences have served as excellent venues for the engineering community to meet with each other and to exchange ideas. Moreover, IMECS continues to strike a balance between theoretical and application development. The conference committees have been formed with over 250 members who are mainly research center heads, deans, department heads (chairs), professors, and research scientists from over 30 countries. The conference participants are also truly international with a high level of representation from many countries. The responses for the conference have been excellent. In 2011, we received more than 800 manuscripts, and after a thorough peer review process 54.31% of the papers were accepted (http://www.iaeng.org/IMECS2011).

This volume contains 32 revised and extended research articles written by prominent researchers participating in the conference. Topics covered include artificial intelligence, decision supporting systems, automated planning, automation systems, control engineering, systems identification, modeling and simulation, communication systems, signal processing, and industrial applications. The book offers the state of the art of tremendous advances in intelligent control and innovative computing and also serves as an excellent reference text for researchers and graduate students, working on intelligent control and innovative computing.

Sio Iong Ao
Oscar Castillo
Xu Huang

Contents

1 **Reciprocally Convex Approach for the Stability of Networked Control Systems** 1
Jeong Wan Ko and PooGyeon Park

2 **Space Robot Control for Unknown Target Handling** 11
Shinichi Tsuda and Takuro Kobayashi

3 **Multi-Sensor Architectures** 25
D.M. Akbar Hussain, Zaki Ahmed, and M.Z. Khan

4 **Five-Axis Curve Machining for Computer Control Machine Tools** 39
Rong-Shine Lin and Shyh-Leh Chen

5 **Approximate Linear Quadratic Regulator Problem and Its Application to Optimal Control in Discrete-Time LTI Systems** 53
Kazuyoshi Mori

6 **Structured Robust Control for MIMO Systems Using Artificial Intelligent Techniques** 65
Somyot Kaitwanidvilai, Piyapong Olranthichachat, and Natthaphob Nimpitiwan

7 **Fuzzy Robust Trajectory Tracking Control of WMRs** 77
Jafar Keighobadi and Yaser Mohamadi

8 **Balanced Performance/Robustness PID Design** 91
Orlando Arrieta, Ramon Vilanova, and Víctor M. Alfaro

9 **Hybrid Genetic Algorithm Optimisation of Distribution Networks—A Comparative Study** 109
Romeo Marin Marian, Lee Luong, and Son Duy Dao

10 Cooking-Step Scheduling Algorithm for Simultaneous Cooking of Multiple Dishes 123
Nobuo Funabiki and Yukiko Matsushima

11 Estimation of Distributed Generation Using Complex-Valued Network Inversion with Regularization 137
Takehiko Ogawa, Kyosuke Nakamura, and Hajime Kanada

12 A New Method of Cooperative PSO: Multiple Particle Swarm Optimizers with Inertia Weight with Diversive Curiosity 149
Hong Zhang

13 A New Design Method for Train Marshaling Evaluating the Transfer Distance of Locomotive 163
Yoichi Hirashima

14 Cost Adjusted MIQ: A New Tool for Measuring Intelligence Value of Systems 177
Tarik Ozkul and Ismail H. Genc

15 Harmony Search Algorithm with Various Evolutionary Elements for Fuzzy Aggregate Production Planning 189
Pasura Aungkulanon, Busaba Phruksaphanrat, and Pongchanun Luangpaiboon

16 A Compression Technique for XML Element Retrieval 203
Tanakorn Wichaiwong and Chuleerat Jaruskulchai

17 Words-of-Wisdom Search Based on User's Sentiment 217
Akiyo Nadamoto and Kouichi Takaoka

18 Configurable Application Designed for Mining XML Document Collections 233
Mihaela Juganaru Mathieu, Cristal Karina Galindo-Durán, and Héctor Javier Vázquez

19 Integrated Approach for Privacy Preserving Itemset Mining 247
Barış Yıldız and Belgin Ergenç

20 Mining General Fuzzy Sequences Based on Fuzzy Ontology 261
Mehdi Gholizadeh, Mir Mohsen Pedram, and Jamshid Shanbezadeh

21 A Comparative Study of Density-based Clustering Algorithms on Data Streams: Micro-clustering Approaches 275
Amineh Amini and Teh Ying Wah

22 A New Clustering Algorithm for Noisy Image Retrieval 289
Morteza Analoui and Maedeh Beheshti

23 **Page as a Meeting Place: Web Search Augmented with Social Communication** 303
Yuya Matsui, Yukiko Kawai, and Jianwei Zhang

24 **The Game of n-Player White-Black Cutthroat and Its Complexity** 319
Alessandro Cincotti

25 **Combinatorial Problems With Closure Structures** 325
Stefan Porschen

26 **Decidable Containment Problems of Rational Word Relations** 337
Wojciech Fraczak and Stéphane Hassen

27 **Pattern Matching Algorithm Using a Succinct Data Structure for Tree-Structured Patterns** 349
Yuko Itokawa, Masanobu Wada, Toshimitsu Ishii, and Tomoyuki Uchida

28 **Network Traffic Screening Using Frequent Sequential Patterns** 363
Hisashi Tsuruta, Takayoshi Shoudai, and Jun'ichi Takeuchi

29 **Informative Gene Discovery in DNA Microarray Data Using Statistical Approach** 377
Kentaro Fukuta and Yoshifumi Okada

30 **High Capacity Image Steganography Based on Genetic Algorithm and Wavelet Transform** 395
Elham Ghasemi, Jamshid Shanbehzadeh, and Nima Fassihi

31 **Reduce Measurement Uncertainty of Ball Crater Method by Using Image Processing Techniques** 405
Jungang Huang, Panos Liatsis, and Kevin Cooke

32 **Decision Level Fusion Framework for Face Authentication System** 413
V. Vaidehi, Teena Mary Treesa, N.T. Naresh Babu, A. Annis Fathima, S. Vasuhi, P. Balamurali, and Girish Chandra

Index 429

Chapter 1
Reciprocally Convex Approach for the Stability of Networked Control Systems

Jeong Wan Ko and PooGyeon Park

Abstract This chapter deals with the problem of stability analysis for networked control systems via the time-delayed system approach. The network-induced delays are modeled as two additive time-varying delays in the closed-loop system. To check the stability of such particular featured systems, an appropriate Lyapunov–Krasovskii functional is proposed and the Jensen inequality lemma is applied to the integral terms that are derived from the derivative of the Lyapunov–Krasovskii functional. Here, the cascaded structure of the delays in the system enables one to partition the domain of the integral terms into three parts, which produces a linear combination of positive functions weighted by inverses of convex parameters. This is handled efficiently by the authors' lower bounds lemma that handles the so-called reciprocally convex combination.

Keywords Reciprocally convex combination • Delay systems • Stability • Networked control systems

1 Introduction

It is well known that the presence of delay elements can bring about system instability and performance degradation, leading to design flaws and incorrect analysis conclusions (see [5]). Hence, a lot of attention has been paid to the time-delayed systems in recent years.

J.W. Ko
Department of Electrical Engineering, Pohang University of Science and Technology, Pohang, Republic of Korea
e-mail: hrp26@postech.ac.kr

P.G. Park (✉)
Division of IT Convergence Engineering, Pohang University of Science and Technology, Pohang, Republic of Korea
e-mail: ppg@postech.ac.kr

S.I. Ao et al. (eds.), *Intelligent Control and Innovative Computing*, Lecture Notes in Electrical Engineering 110, DOI 10.1007/978-1-4614-1695-1_1,

Among the relevant topics in this field, networked control systems have emerged as one of the most attractive issues in line with the rapidly growing network environments (see [1,4,7,8]).

In the networked control systems, interpreting in the time-delayed system perspective, signals transmitted from one point to another may encounter two network-induced delays: $d_s(t)$ from the sampler (sensor) to the controller and $d_h(t)$ from the controller to the holder (actuator). This causes an introduction of delay elements in the closed-loop system. For example, when the ordinary plant is considered, $\dot{x}(t) = Ax(t) + Bu(t)$, the closed-loop system becomes

$$\dot{x}(t) = Ax(t) + BK(x - d_s(t) - d_h(t)), \tag{1.1}$$

that is, systems with two additive time-varying delays.

As concerns about it, a conventional approach is to assemble the induced delays as a single one, $d(t) = d_s(t) + d_h(t)$, and to consider (1.1) as a single delayed system with $d(t)$ as the system delay (see [9, 11]). Recently, [4, 7] have also shown the possible reduction of conservatism in the analysis and synthesis problems by treating $d_s(t)$ and $d_h(t)$ separately in constructing the Lyapunov–Krasovskii functional. However, to take into account the relationship between the two delays, they have to introduce slightly excessive free weighting matrices.

As a way of reducing the number of decision variables, this chapter focuses on the Jensen inequality lemma (see [5]). It is well known that relaxations based on the Jensen inequality lemma in delayed systems produce a special type of function combinations, a linear combination of positive functions weighted by inverses of convex parameters, *say a reciprocally convex combination*. Here, the cascaded structure of delays in the system enables one to partition the domain of the integral terms that are derived from the derivative of the Lyapunov–Krasovskii functional into three parts, which produces a reciprocally convex combination having three convex parameters as the weights. This can be handled efficiently by Park's [10] lower bounds lemma that can be applied for all finite reciprocally convex combinations.

The chapter is organized as follows. Section 2 will explain the structure of the networked control systems and develop the corresponding stability criterion. Section 3 will show simple examples for verification of the criterion.

2 Main Results

2.1 System Description

Let us consider the system:

$$\dot{x}(t) = Ax(t) + Bu(t), \tag{1.2}$$

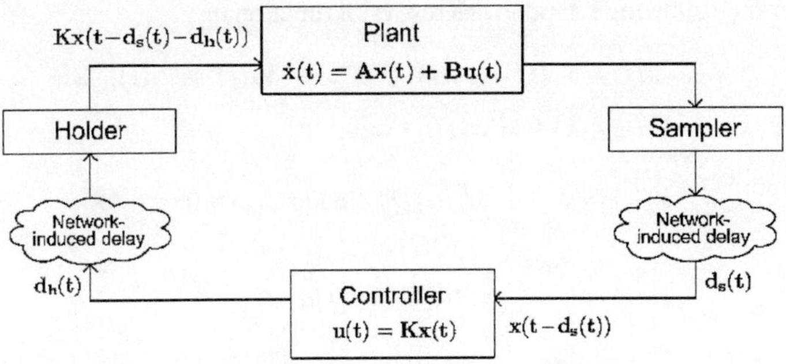

Fig. 1.1 Networked control systems

which, with the introduction of the two network-induced delays: $d_s(t)$ from the sampler (sensor) to the controller and $d_h(t)$ from the controller to the holder (actuator), produces a special type of closed-loop system as:

$$\dot{x}(t) = Ax(t) + BKx(t - d_s(t) - d_h(t)), \tag{1.3}$$

that is, systems with two additive time-varying delays. Checking the stability of such particular featured systems is the focus of the forthcoming section.

2.2 *Robust Stability Analysis*

Let us consider the following delayed system with model uncertainties:

$$\begin{aligned} \dot{x}(t) &= Ax(t) + A_d x(t - d_1(t) - d_2(t)) + Dp(t), \quad t \geq 0, \\ q(t) &= Ex(t) + E_d x(t - d_1(t) - d_2(t)), \quad t \geq 0, \\ x(t) &= \phi(t), \quad -\bar{d}_1 - \bar{d}_2 \leq t \leq 0, \end{aligned} \tag{1.4}$$

where $0 \leq d_1(t) \leq \bar{d}_1$, $0 \leq d_2(t) \leq \bar{d}_2$, $\dot{d}_1(t) \leq \bar{\tau}_1$, $\dot{d}_2(t) \leq \bar{\tau}_2$, $p(t) = \Delta(t)q(t)$, $\Delta^T(t)\Delta(t) \leq \gamma^{-2}I$, and $\phi(t) \in \mathcal{C}^1(\bar{d}_1 + \bar{d}_2)$, the set of continuously differentiable functions in the domain $[-2(\bar{d}_1 + \bar{d}_2), 0]$. Let us define $\bar{d} = \bar{d}_1 + \bar{d}_2$, $\bar{\tau} = \bar{\tau}_1 + \bar{\tau}_2$, $d(t) = d_1(t) + d_2(t)$, $\chi(t) = \mathrm{col}\{x(t), x(t - d_1(t)), x(t - d(t)), x(t - \bar{d}), p(t)\}$ and the corresponding block entry matrices as:

$$e_1 = [I\,0\,0\,0\,0]^T,\ e_2 = [0\,I\,0\,0\,0]^T, \ldots,\ e_5 = [0\,0\,0\,0\,I]^T,\ e_6 = \left(Ae_1^T + A_d e_3^T + De_5^T\right)^T,$$

so that the system can be written as $\dot{x}(t) = e_6^T \chi(t)$.

Consider the following Lyapunov–Krasovskii functional:

$$V(t) = V_1(t) + V_2(t) + V_3(t) + V_4(t) + V_5(t), \tag{1.5}$$

$$V_1(t) = x^T(t)Px(t),\ P > 0, \tag{1.6}$$

$$V_2(t) = \int_{t-d_1(t)}^{t} x^T(\alpha)Q_1x(\alpha)\mathrm{d}\alpha,\ Q_1 > 0, \tag{1.7}$$

$$V_3(t) = \int_{t-d(t)}^{t} x^T(\alpha)Q_2x(\alpha)\mathrm{d}\alpha,\ Q_2 > 0, \tag{1.8}$$

$$V_4(t) = \int_{t-\bar{d}}^{t} x^T(\alpha)Q_3x(\alpha)\mathrm{d}\alpha,\ Q_3 > 0, \tag{1.9}$$

$$V_5(t) = \bar{d}\int_{-\bar{d}}^{0}\int_{t+\beta}^{t} \dot{x}^T(\alpha)R\dot{x}(\alpha)\mathrm{d}\alpha\mathrm{d}\beta,\ R > 0. \tag{1.10}$$

Then, the time derivatives of $V_i(t)$ become

$$\dot{V}_1(t) = 2\dot{x}^T(t)Px(t) = 2\chi^T(t)e_6Pe_1^T\chi(t), \tag{1.11}$$

$$\dot{V}_2(t) = \chi^T(t)\left\{e_1Q_1e_1^T - \left(1-\dot{d}_1(t)\right)e_2Q_1e_2^T\right\}\chi(t), \tag{1.12}$$

$$\dot{V}_3(t) = \chi^T(t)\left\{e_1Q_2e_1^T - \left(1-\dot{d}(t)\right)e_3Q_2e_3^T\right\}\chi(t), \tag{1.13}$$

$$\dot{V}_4(t) = \chi^T(t)\left\{e_1Q_3e_1^T - e_4Q_3e_4^T\right\}\chi(t), \tag{1.14}$$

$$\dot{V}_5(t) = \bar{d}^2\chi^T(t)e_6Re_6^T\chi(t) - \bar{d}\int_{t-\bar{d}}^{t}\dot{x}^T(\beta)R\dot{x}(\beta)\mathrm{d}\beta, \tag{1.15}$$

so that $\dot{V}(t)$ can be upper-bounded by the following quantity:

$$\begin{aligned}\dot{V}(t) \le\ & \chi^T(t)\Omega_1\chi(t)\\ & -\bar{d}\int_{t-d_1(t)}^{t}\dot{x}^T(\beta)R\dot{x}(\beta)\mathrm{d}\beta\\ & -\bar{d}\int_{t-d(t)}^{t-d_1(t)}\dot{x}^T(\beta)R\dot{x}(\beta)\mathrm{d}\beta\\ & -\bar{d}\int_{t-\bar{d}}^{t-d(t)}\dot{x}^T(\beta)R\dot{x}(\beta)\mathrm{d}\beta \end{aligned} \tag{1.16}$$

$$\le \chi^T(t)\Bigg\{\Omega_1 \tag{1.17}$$

$$-\frac{1}{\alpha}(e_1-e_2)R(e_1-e_2)^T$$

$$-\frac{1}{\beta}(e_2-e_3)R(e_2-e_3)^T$$

$$\left.-\frac{1}{\gamma}(e_3-e_4)R(e_3-e_4)^T\right\}\chi(t) \tag{1.18}$$

$$\leq \chi^T(t)(\Omega_1+\Omega_2)\chi(t), \tag{1.19}$$

where

$$\Omega_1 = e_6 P e_1^T + e_1 P e_6^T + e_1(Q_1+Q_2+Q_3)e_1^T - e_4 Q_3 e_4^T$$
$$-(1-\bar{\tau}_1)e_2 Q_1 e_2^T - (1-\bar{\tau})e_3 Q_2 e_3^T + \bar{d}^2 e_6 R e_6^T, \tag{1.20}$$

$$\Omega_2 = -\begin{bmatrix}(e_1-e_2)^T\\(e_2-e_3)^T\\(e_3-e_4)^T\end{bmatrix}^T \begin{bmatrix} R & S_{1,2} & S_{1,3}\\ * & R & S_{2,3}\\ * & * & R\end{bmatrix}\begin{bmatrix}(e_1-e_2)^T\\(e_2-e_3)^T\\(e_3-e_4)^T\end{bmatrix}. \tag{1.21}$$

The inequality (1.17) comes from the Jensen inequality lemma (see [5]), and that of (1.19) from Park's [10] lower bounds lemma (see Appendix) as:

$$-\chi^T(t)\left\{\begin{bmatrix}\sqrt{\frac{\beta}{\alpha}}(e_1-e_2)^T\\ -\sqrt{\frac{\alpha}{\beta}}(e_2-e_3)^T\end{bmatrix}^T\begin{bmatrix}R & S_{1,2}\\ * & R\end{bmatrix}\begin{bmatrix}\sqrt{\frac{\beta}{\alpha}}(e_1-e_2)^T\\ -\sqrt{\frac{\alpha}{\beta}}(e_2-e_3)^T\end{bmatrix}\right.$$

$$+\begin{bmatrix}\sqrt{\frac{\gamma}{\alpha}}(e_1-e_2)^T\\ -\sqrt{\frac{\alpha}{\gamma}}(e_3-e_4)^T\end{bmatrix}^T\begin{bmatrix}R & S_{1,3}\\ * & R\end{bmatrix}\begin{bmatrix}\sqrt{\frac{\gamma}{\alpha}}(e_1-e_2)^T\\ -\sqrt{\frac{\alpha}{\gamma}}(e_3-e_4)^T\end{bmatrix}$$

$$\left.+\begin{bmatrix}\sqrt{\frac{\gamma}{\beta}}(e_2-e_3)^T\\ -\sqrt{\frac{\beta}{\gamma}}(e_3-e_4)^T\end{bmatrix}^T\begin{bmatrix}R & S_{2,3}\\ * & R\end{bmatrix}\begin{bmatrix}\sqrt{\frac{\gamma}{\beta}}(e_2-e_3)^T\\ -\sqrt{\frac{\beta}{\gamma}}(e_3-e_4)^T\end{bmatrix}\right\}\chi(t)\leq 0, \tag{1.22}$$

where

$$\alpha = \frac{d_1(t)}{\bar{d}},\ \beta = \frac{d_2(t)}{\bar{d}},\ \gamma = \frac{\bar{d}-d(t)}{\bar{d}}. \tag{1.23}$$

Note that when $\alpha = 0$ or $\beta = 0$ or $\gamma = 0$, we have $\chi^T(t)\ (e_1 - e_2) = 0$ or $\chi^T(t)(e_2 - e_3) = 0$ or $\chi^T(t)(e_3 - e_4) = 0$, respectively. So the relation (1.19) still holds. Finally, we shall remove the uncertainty constraint in (1.4):

$$0 \leq q^T(t)q(t) - \gamma^2 p^T(t)p(t)$$
$$= \chi^T(t)\left\{\left(e_1E^T + e_3E_d^T\right)\left(Ee_1^T + E_d e_3^T\right) - \gamma^2 e_5 e_5^T\right\}\chi(t), \tag{1.24}$$

which can be handled through the so-called $\mathcal{S}$-procedure (see [2]). This leads to the following theorem. Note that the non-negative scalar variable in the $\mathcal{S}$-procedure for the uncertainty constraint can be scaled to 1 without loss of generality.

Theorem 1.1. *For a given γ, the delayed system (1.4) is asymptotically stable if there exist matrices P, Q_1, Q_2, Q_3, R, $S_{1,2}$, $S_{1,3}$ and $S_{2,3}$ such that the following conditions hold:*

$$\Omega_1 + \Omega_2 + \left(e_1 E^T + e_3 E_d^T\right)\left(E e_1^T + E_d e_3^T\right) - \gamma^2 e_5 e_5^T < 0, \tag{1.25}$$

$$\begin{bmatrix} R & S_{1,2} \\ * & R \end{bmatrix} \geq 0, \begin{bmatrix} R & S_{1,3} \\ * & R \end{bmatrix} \geq 0, \begin{bmatrix} R & S_{2,3} \\ * & R \end{bmatrix} \geq 0, \tag{1.26}$$

$$P > 0,\ Q_1 > 0,\ Q_2 > 0,\ Q_3 > 0,\ R > 0, \tag{1.27}$$

where

$$\begin{aligned} \Omega_1 =& e_6 P e_1^T + e_1 P e_6^T + e_1 (Q_1 + Q_2 + Q_3) e_1^T - e_4 Q_3 e_4^T \\ & -(1-\bar{\tau}_1) e_2 Q_1 e_2^T - (1-\bar{\tau}) e_3 Q_2 e_3^T + \bar{d}^2 e_6 R e_6^T, \end{aligned} \tag{1.28}$$

$$\Omega_2 = -\begin{bmatrix} (e_1-e_2)^T \\ (e_2-e_3)^T \\ (e_3-e_4)^T \end{bmatrix}^T \begin{bmatrix} R & S_{1,2} & S_{1,3} \\ * & R & S_{2,3} \\ * & * & R \end{bmatrix} \begin{bmatrix} (e_1-e_2)^T \\ (e_2-e_3)^T \\ (e_3-e_4)^T \end{bmatrix}. \tag{1.29}$$

If we replace (1.25) with $\Omega_1 + \Omega_2 < 0$, it gives the criterion derived in [6] that does not take into account the uncertainty of the system.

3 Examples

Example 1.2. Consider the system (1.4) taken from [7] with

$$A = \begin{bmatrix} -2.0 & 0.0 \\ 0.0 & -0.9 \end{bmatrix}, \quad A_d = \begin{bmatrix} -1.0 & 0.0 \\ -1.0 & -1.0 \end{bmatrix}. \tag{1.30}$$

The maximum upper bounds on the delays (*MUBDs*) under the assumption:

$$\dot{d}_1(t) \leq 0.1, \quad \dot{d}_2(t) \leq 0.8 \tag{1.31}$$

are listed in Tables 1.1–1.2. Comparing with the existing results, we can see that Theorem 1.1 is less conservative with relatively low decision variables.

Example 1.3. Consider the system (1.4) taken from [3] with

$$A = \begin{bmatrix} -1.7073 & 0.6856 \\ 0.2279 & -0.6368 \end{bmatrix}, \quad A_d = \begin{bmatrix} -2.5026 & 1.0540 \\ -0.1856 & -1.5715 \end{bmatrix}. \tag{1.32}$$

Table 1.1 MUBDs of d_2 for given $\bar{d}_1$ in Example 1.2

Method	$\bar{d}_1=1.0$	$\bar{d}_1=1.2$	$\bar{d}_1=1.5$	Number of variables
[9]	0.378	0.178	Infeasible	32
[7]	0.415	0.340	0.248	59
[4]	0.512	0.406	0.283	85
[3]	0.512	0.406	0.283	15
[11]	0.665	0.465	0.165	18
Theorem 1.1	0.873	0.673	0.373	27

Table 1.2 MUBDs of d_1 for given $\bar{d}_2$ in Example 1.2

Method	$\bar{d}_2=1.0$	$\bar{d}_2=1.2$	$\bar{d}_2=1.5$	Number of variables
[7]	0.212	0.090	Infeasible	59
[4]	0.378	0.178	Infeasible	85
[9]	0.378	0.178	Infeasible	32
[3]	0.378	0.178	Infeasible	15
[11]	0.665	0.465	0.165	18
Theorem 1.1	0.873	0.673	0.373	27

Table 1.3 MUBDs of d_2 for given $\bar{d}_1$ in Example 1.3

Method	$\bar{d}_1=0.1$	$\bar{d}_1=0.2$	$\bar{d}_1=0.3$	Number of variables
[7]	0.412	0.290	0.225	59
[9]	0.484	0.384	0.284	32
[4]	0.484	0.385	0.293	85
[3]	0.484	0.385	0.293	15
[11]	0.577	0.477	0.377	18
Theorem 1.1	0.684	0.584	0.484	27

Table 1.4 MUBDs of d_1 for given $\bar{d}_2$ in Example 1.3

Method	$\bar{d}_2=0.1$	$\bar{d}_2=0.2$	$\bar{d}_2=0.3$	Number of variables
[9]	0.484	0.384	0.284	32
[7]	0.547	0.343	0.185	59
[11]	0.577	0.477	0.377	18
[4]	0.585	0.419	0.292	85
[3]	0.585	0.419	0.292	15
Theorem 1.1	0.684	0.584	0.484	27

The derivatives of the delays are assumed to be

$$\dot{d}_1(t) \le 0.3, \quad \dot{d}_2(t) \le 0.8, \tag{1.33}$$

and the improvement of Theorem 1.1 is shown in Tables 1.3–1.4.

Example 1.4. Consider the system (1.4) taken from [11] with

$$A = \begin{bmatrix} -0.5 & -2 \\ 1 & -1 \end{bmatrix},\ A_d = \begin{bmatrix} -0.5 & -1 \\ 0 & 0.6 \end{bmatrix},\ E = E_d = \begin{bmatrix} 0.2 & 0 \\ 0 & 0.2 \end{bmatrix},\ D = I.$$

Assuming $\gamma = 1$, the MUBD of $\bar{d}_1 + \bar{d}_2$ is guaranteed to be 0.476 for any admissible values of $\dot{d}_i(t)$ in [11]. However, based on Theorem 1.1, the system is robustly stable up to 0.902.

4 Conclusions

This chapter proposed an efficient stability criterion for networked control systems via the time-delayed system approach. The network-induced delays were modeled as two additive time-varying delays in the closed-loop system. To check the stability of such particular featured systems, an appropriate Lyapunov–Krasovskii functional was constructed and the Jensen inequality lemma was applied to the integral terms that were derived from the derivative of the Lyapunov–Krasovskii functional. Here, owing to the cascaded structure of delays in the system, the domain of the integral terms could be partitioned into three parts, which produced a linear combination of positive functions weighted by inverses of convex parameters. This was handled efficiently by modifying Park's [10] lower bounds lemma.

Examples showed the resulting criterion outperforms all existing ones with relatively low decision variables.

Acknowledgments This research was supported by Basic Science Research Program through the National Research Foundation of Korea (NRF) funded by the Ministry of Education, Science and Technology (2010-0009743).

Poogyeon Park also gratefully acknowledges the LG YONAM Foundation for its financial support through the Professors' overseas research program for sabbatical research leave at University of Maryland.

Appendix A: Reciprocally Convex Combination and Its Lower Bounds Lemma

Definition A.1. Let $\Phi_1, \Phi_2, \ldots, \Phi_N : \mathbf{R}^m \mapsto \mathbf{R}^n$ be a given finite number of functions such that they have positive values in an open subset $\mathbf{D}$ of $\mathbf{R}^m$. Then, a *reciprocally convex combination* of these functions over $\mathbf{D}$ is a function of the form

$$\frac{1}{\alpha_1}\Phi_1 + \frac{1}{\alpha_2}\Phi_2 + \cdots + \frac{1}{\alpha_N}\Phi_N : \mathbf{D} \mapsto \mathbf{R}^n, \tag{A.34}$$

where the real numbers α_i satisfy $\alpha_i > 0$ and $\sum_i \alpha_i = 1$.

Theorem A.2. *Let $f_1, f_2, \ldots, f_N : \mathbf{R}^m \mapsto \mathbf{R}$ have positive values in an open subset $\mathbf{D}$ of $\mathbf{R}^m$. Then, the reciprocally convex combination of f_i over $\mathbf{D}$ satisfies*

$$\min_{\{\alpha_i | \alpha_i > 0, \sum_i \alpha_i = 1\}} \sum_i \frac{1}{\alpha_i} f_i(t) \tag{A.35}$$

$$= \sum_i f_i(t) + \max_{g_{i,j}(t)} \sum_{i \neq j} g_{i,j}(t) \text{ subject to}$$

$$\left\{ g_{i,j} : \mathbf{R}^m \mapsto \mathbf{R}, g_{j,i}(t) = g_{i,j}(t), \begin{bmatrix} f_i(t) & g_{i,j}(t) \\ g_{i,j}(t) & f_j(t) \end{bmatrix} \geq 0 \right\}. \tag{A.36}$$

References

1. Antsaklis P, Baillieul J (2004) Guest editorial special issue on networked control systems. IEEE Trans Autom Contr 31(9):1421–1423
2. Boyd S, Ghaoui LE, Feron E, Balakrishnan V (1994) Linear matrix inequalities in system and control theory. SIAM, Philadelphia
3. Du B, Lam J, Shu Z, Wang Z (2009) A delay-partitioning projection approach to stability analysis of continuous systems with multiple delay components. IET Contr Theor Appl 3(4):383–390
4. Gao H, Chen T, Lam J (2008) A new delay system approach to network-based control. Automatica 44(1):39–52
5. Gu K, Kharitonov VL, Chen J (2003) Stability of time-delay systems, 1st edn. Birkhäuser Boston
6. Ko JW, Lee WI, Park PG (2011) Delayed system approach to the stability of networked control systems. In: Proceedings of the international multiconference of engineers and computer scientists 2011 (IMECS 2011), Hong Kong. Lecture notes in engineering and computer science, pp 772–774
7. Lam J, Gao H, Wang C (2007) Stability analysis for continuous systems with two additive time-varying delay components. Syst Contr Lett 56(1):16–24
8. Li H, Chow MY, Sun Z (2009a). State feedback stabilisation of networked control systems. IET Contr Theor Appl 3(7):929–940
9. Li T, Guo L, Wu L (2009b). Simplified approach to the asymptotical stability of linear systems with interval time-varying delay. IET Contr Theor Appl 3(2):252–260
10. Park PG, Ko JW, Jeong C (2011) Reciprocally convex approach to stability of systems with time-varying delays. Automatica 47:235–238
11. Peng C, Tian YC (2008) Improved delay-dependent robust stability criteria for uncertain systems with interval time-varying delay. IET Contr Theor Appl 2(9):752–761

Chapter 2
Space Robot Control for Unknown Target Handling

Shinichi Tsuda and Takuro Kobayashi

Abstract Space robot is now playing very important role in space activities. Especially in space station a few robot arms are working for construction and repairing. However these robots are so-called remote manipulators operated by astronauts. For future applications the space robot must be autonomous and is expected to maintain the failed satellites and to capture the space debris. This chapter deals with space robot control for unknown target in robust manner. To cope with unknown characteristics the sliding mode control is applied in this study.

Keywords Robust control • Sliding mode control • Space robot • Target handling

1 Introduction

Space robot technology has been rapidly developed and extensively used in the space station program. Most of these space robots are a kind of remote manipulator systems controlled by astronauts from inside or outside of space station. In the space application more intelligent system is desirable to reduce the workload and hazardous risk of those astronauts. Therefore in the near future this technology will be expected to perform the wider range of operations, such as to maintain failed satellites and to capture space debris in the autonomous manner by the space robot.

S. Tsuda (✉)
Department of Aeronautics and Astronautics, School of Engineering, Tokai University, 4–1–1 Kitakaname, Hiratsuka, Kanagawa, 259–1292, Japan
e-mail: stsuda@keyaki.cc.u-tokai.ac.jp

T. Kobayashi
INEC Engineering Limited, Higashi-shinagawa 4–10–27, Shinagawa, Tokyo 140–0002, Japan
e-mail: 9amjm005@mail.tokai-u.jp

S.I. Ao et al. (eds.), *Intelligent Control and Innovative Computing*, Lecture Notes in Electrical Engineering 110, DOI 10.1007/978-1-4614-1695-1_2,

This capability will tremendously decrease the extravehicular operations of astronauts, which are most time consuming and terribly exhausting. In this respect the autonomy will be mandatory.

In the space robot operation there are a few features like the reactive behavior of attitude motion of the space robot by robot arm operation and the parameter change in attitude dynamic equations of motion by capturing the target and so on. Generally speaking the failed target and debris will not be accurately known a priori and freely rotating, that is, some of physical parameters are unknown. In the above respect some kind of robustness of the space robot control must be incorporated [1].

This study deals with the space robot operation, i.e., controlling the attitude of the space robot and controlling the robot arm under the changed mass property. The sliding mode control [2] is applied to the control of attitude motion and the robot arm in which the absolute supremum value method [3] was used to assure the robustness.

2 Model of Space Robot

The model of a space robot is illustrated in Fig. 2.1. A robot arm is mounted on the body of the spacecraft. The robot arm is articulated with three rotary joints and the motion of the robot arm is assumed to be two dimensional.

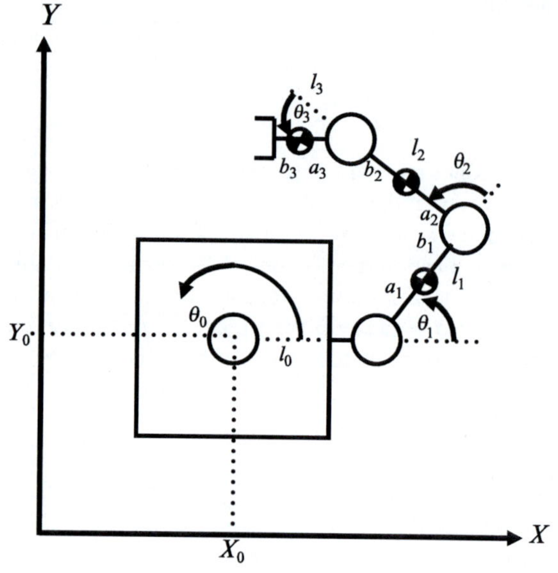

Fig. 2.1 Model of space robot

3 Equations of Motion

Dynamical equations of motion for space robot are derived using Lagrange formula. It will be obtained as follows.

K is the kinetic energy and P is the potential energy, then, Lagrange equations of motion is expressed in the following:

$$Q_{ib} = \frac{d}{dt}\left(\frac{\partial K}{\partial \dot{q}_i}\right) - \frac{\partial K}{\partial q_i} + \frac{\partial P}{\partial q_i}. \tag{2.1}$$

Both energies are given as below:

$$K = \frac{1}{2}mv^T v + \frac{1}{2}\omega^T I\omega \tag{2.2}$$

$$P = mgl. \tag{2.3}$$

where $m, \vec{v}, \vec{\omega}$ and I are mass, velocity vector, angular velocity vector and moment of inertia, respectively.

Detailed Geometry of the space robot is illustrated in the Fig. 2.1.

Center of mass for the space robot and each link is given by $s_{iX}, s_{iY}(i = 0, 1, 2, 3)$. And the velocity is v_i. a_i expresses the length between the joint and link center of mass, and b_i gives the length between the joint and link center of mass.

Then we have the following relationships:

$$\begin{aligned}
s_{0X}(t) &= X_0(t)\\
s_{0Y}(t) &= Y_0(t)\\
v_0(t) &= \dot{s}_{0X}{}^2(t) + \dot{s}_{0Y}{}^2(t)\\
s_{1X}(t) &= X_0(t) + l_0C_0 + a_1C_{01}\\
s_{1Y}(t) &= Y_0(t) + l_0S_0 + a_1S_{01}\\
v_1(t) &= \dot{s}_{1X}{}^2(t) + \dot{s}_{1Y}{}^2(t)\\
s_{2X}(t) &= X_0(t) + l_0C_0 + l_1C_{01} + a_2C_{012}\\
s_{2Y}(t) &= Y_0(t) + l_0S_0 + l_1S_{01} + a_2S_{012}\\
v_2(t) &= \dot{s}_{2X}{}^2(t) + \dot{s}_{2Y}{}^2(t)\\
s_{3X}(t) &= X_0(t) + l_0C_0 + l_1C_{01} + l_2C_{012} + a_3C_{0123}\\
s_{3Y}(t) &= Y_0(t) + l_0S_0 + l_1S_{01} + l_2S_{012} + a_3S_{0123} \quad v_3(t) = \dot{s}_{3X}{}^2(t) + \dot{s}_{3Y}{}^2(t)
\end{aligned} \tag{2.4}$$

The kinetic energies are described as below:

$$K_0 = \frac{1}{2}m_0 v_0^2 + \frac{1}{2}I_0\omega_0^2$$
$$K_1 = \frac{1}{2}m_1 v_1^2 + \frac{1}{2}I_1(\omega_0 + \omega_1)^2$$
$$K_2 = \frac{1}{2}m_2 v_2^2 + \frac{1}{2}I_2(\omega_0 + \omega_1 + \omega_2)^2$$
$$K_3 = \frac{1}{2}m_3 v_3^2 + \frac{1}{2}I_3(\omega_0 + \omega_1 + \omega_2 + \omega_3)^2 \tag{2.5}$$

The potential energies for free floating bodies on the orbit are given by the following:

$$P_0 = P_1 = P_2 = P_3 = 0 \tag{2.6}$$

Those equations are summarized as in (2.7) by substituting the above relations, where $M(\theta)$ is the inertia matrix and $h(\theta, \dot{\theta})$ includes centrifugal and Coriolis terms. $u(t)$ is translational control force, attitude control and joint control torque vector for space robot.

$$M(\theta)\ddot{q}(t) + h(\theta, \dot{\theta}) = u(t) \tag{2.7}$$

where

$$q = \begin{bmatrix} X \; Y \; \theta_0 \; \theta_1 \; \theta_2 \; \theta_3 \end{bmatrix}^T$$
$$\theta = \begin{bmatrix} \theta_0 \; \theta_1 \; \theta_2 \; \theta_3 \end{bmatrix}^T.$$

Further we assume the following relations:

$$M(\theta) = M^0(\theta) + \Delta M(\theta) \tag{2.8}$$
$$h(\theta, \dot{\theta}) = h^0(\theta, \dot{\theta}) + \Delta h(\theta, \dot{\theta}) \tag{2.9}$$

In which $M^0(\theta)$ and $h^0(\theta, \dot{\theta})$ are defined as nominal value matrix and vector, $\Delta M(\theta)$ and $\Delta h(\theta, \dot{\theta})$ are called deference from nominal values and absolute supremum values are defined as bellow;

$$\left|\Delta M_{ij}(\theta)\right| \le \hat{M}_{ij}(\theta) \tag{2.10}$$
$$\left|\Delta h_i(\theta, \dot{\theta})\right| \le \hat{h}_i(\theta, \dot{\theta}). \tag{2.11}$$

And further, absolute supremum values of elements of time derivative $\dot{M}_{ij}(\theta)$ of matrix $M(\theta)$ was also defined in the following manner;

$$\left|\dot{M}_{ij}(\theta)\right| \le \hat{\dot{M}}_{ij}(\theta). \tag{2.12}$$

The absolute supremum value $\hat{v}_i(q, t)$ will be given as follows;

$$|\{M(\theta)\ddot{q}_d(t)\}_i| \le \hat{v}_i(\theta, t). \tag{2.13}$$

4 Sliding Mode Control

The sliding mode control restricts the trajectory of plant states on a hyper plane by the control and slides it to the equilibrium point in an asymptotic manner.

First let us design the switching hyper plane. The target trajectory is given by q_d and controlling errors are defined by the followings;

$$e(t) = q(t) - q_d(t) \tag{2.14}$$

$$\dot{e}(t) = \dot{q}(t) - \dot{q}_d(t). \tag{2.15}$$

And then we give the switching hyper plane as an (2.10).

$$\sigma(t) = \Lambda e(t) + \dot{e}(t) \tag{2.16}$$

where

$$\Lambda = diag(\lambda_1, \cdots, \lambda_n) \quad \lambda_i > 0.$$

If $\sigma(t) = 0$ holds, then, $e(t)$ in (2.16) satisfies the asymptotic stable differential equation and $e(\infty) \to 0$ is assured. In order to secure the state is approaching to the hyper plane, the following Lyapunov function is introduced. And the negative definiteness of its time derivative will be proved.

$$V(\sigma) = \frac{1}{2}\sigma^T M \sigma \tag{2.17}$$

The time derivative of (2.17) is given by

$$\begin{aligned} V &= \frac{1}{2}\sigma^T \dot{M} \sigma + \sigma^T M \dot{\sigma} \\ &= \frac{1}{2}\sigma^T \dot{M} \sigma + \sigma^T (M\Lambda\dot{e} + M\ddot{q} - M\ddot{q}_d) \\ &= \frac{1}{2}\sigma^T \dot{M} \sigma + \sigma^T (M\Lambda\dot{e} - h + u - M\ddot{q}_d). \end{aligned} \tag{2.18}$$

Let us define $u(t)$ as follows;

$$u(t) = -M^0(\theta)\Lambda\dot{e} + h^0(\theta, \dot{\theta}) - P\sigma - Q\mathrm{sgn}(\sigma) \tag{2.19}$$

where

$$P := diag(P_{11}(t), \cdots, P_{nn}(t)) \quad Q := diag(Q_{11}(t), \cdots, Q_{nn}(t)),$$

then, we obtain

$$\begin{aligned} \dot{V} &= \frac{1}{2}\sigma^T \dot{M} \sigma + \sigma^T \left[M\Lambda\dot{e} - h - M\ddot{q}_d\right] \\ &\quad + \sigma^T \left[-M^0\Lambda\dot{e} + h^0 - P\sigma - Q\mathrm{sgn}(\sigma)\right] \end{aligned}$$

$$= -\sigma^T \left[P - \frac{1}{2}\dot{M} \right] \sigma$$
$$+ \sigma^T \left[-Q\mathrm{sgn}(\sigma) + \Delta M \Lambda \dot{e} + \Delta h - M\ddot{q}_d \right]. \tag{2.20}$$

Here we choose P and Q which satisfy $\dot{V}(s) < 0$.

In the first place elements of the diagonal matrix Q are determined as below;

$$Q_{ii}(t) = \sum_{j=1}^{n} \left\{ \hat{M} \Lambda \right\}_{ij} |\dot{e}_j| + \hat{h}_i + \hat{v}_i. \tag{2.21}$$

Then we have

$$\sigma^T Q\mathrm{sgn}(\sigma) \geq \sigma^T \left[\Delta M \Lambda \dot{e} - \Delta h - M\ddot{q}_d \right], \tag{2.22}$$

And the second term of (2.20) becomes negative semi-definite. In the next place if we define elements of diagonal matrix P as follows;

$$P_{ii}(t) = \sum_{j=1}^{n} \hat{\dot{M}}_{ij}/2 + k_i, \quad k_i > 0, \tag{2.23}$$

then, the first term $P - \frac{1}{2}\dot{M}$ of (2.20) is given by the following,

$$\frac{1}{2} \begin{bmatrix} \sum_{j=1}^{n} \hat{\dot{M}}_{1j} - \dot{M}_{11} & -\dot{M}_{12} & \cdots & -\dot{M}_{1n} \\ -\dot{M}_{21} & \sum_{j=1}^{n} \hat{\dot{M}}_{2j} - \dot{M}_{22} & & -\dot{M}_{2n} \\ \vdots & & \ddots & \vdots \\ -\dot{M}_{n1} & -\dot{M}_{n2} & \cdots & \sum_{j=1}^{n} \hat{\dot{M}}_{nj} - \dot{M}_{nn} \end{bmatrix} + K. \tag{2.24}$$

By the Gershgorin's theorem, for an arbitrary matrix $A = [a_{ij}]$, if the following inequality is satisfied;

$$a_{ij} \geq \sum_{k=1, k\neq i}^{n} |a_{ik}| \tag{2.25}$$

then, the matrix A is positive semi-definite. Therefore if we apply $k_i > 0$ to the (2.24), then, we have the negative definiteness of the first term in (2.20). This means $\dot{V} < 0$. The above concludes the proof of the negative definiteness of the Lyapunov function.

And in order to avoid the chattering phenomena, we introduce saturation function in place of sgn function.

$$sat(\sigma/\varepsilon) = \begin{cases} 1 & \sigma > \varepsilon \\ \sigma/\varepsilon & |\sigma| \leq \varepsilon \\ -1 & \sigma < -\varepsilon \end{cases} \tag{2.26}$$

5 Numerical Simulations

We conducted numerical simulations for the space robot model defined in Fig. 2.1. And to perform the mission two phases are introduced.

5.1 *Phase I*

To capture the target the robot arm follows the motion of the target for 10 s. By this operation grasping operation will be completed.

In order to realize to follow the target, a goal trajectory $r_d(t)$ for the position of endeffector of the robot arm is defined and then, the joint trajectory for $q_d(t)$ is calculated. The position of the center of target is X_t and Y_t, and the distance between the center of the target and the grasping point is given by r_t. And the target has the rotational motion. Then we have the following relations;

$$r_d(t) = \begin{bmatrix} Xt + r_t \cdot \cos\left(\frac{\pi t}{360} + \frac{3\pi}{2}\right) \\ Y_t + r_t \cdot \sin\left(\frac{\pi t}{360} + \frac{3\pi}{2}\right) \\ \frac{\pi t}{360} + \frac{\pi}{2} \end{bmatrix} \tag{2.27}$$

$$q_d(t) = \begin{bmatrix} 0 \; 0 \; 0 \; \theta_{1d} \; \theta_{2d} \; \theta_{3d} \end{bmatrix} \tag{2.28}$$

5.2 *Phase II*

After the grasping operation the velocity of the endeffector will be controlled to be 0 [m/s].

To realize the above operation a goal trajectory for the joint velocity is given by linear functions of time which reduce the velocity to 0 [m/s] after the 30 [s]. The joint velocity vector is given by (2.29).

The supremum value is determined by Table 2.1.

$$\dot{q}_d(t) = \begin{bmatrix} 0 \; 0 \; 0 \; \dot{\theta}_{1d}(t) \; \dot{\theta}_{2d}(t) \; \dot{\theta}_{3d}(t) \end{bmatrix} \tag{2.29}$$

In Table 2.2 parameters for the space robot are defined.

Table 2.1 Parameters of the target

	Target	Assumed value for determining the suremum value
Mass [Kg]	500	600
Moment of inertia [Kgm^2]	333.33	400
Rotational velocity [deg/s]	0.5	0.5
Size	2 [m] × 2 [m]	2 [m] × 2 [m]

Table 2.2 Parameters of the space robot

	Body	Link 1	Link 2	Link 3
Mass [Kg]	1,500	40	40	30
Link length [m]	1.5	1.5	1.5	1.0
Moment of inertia [Kgm^2]	1,000	30	30	10
Initial angle [deg]	0	45	90	−45

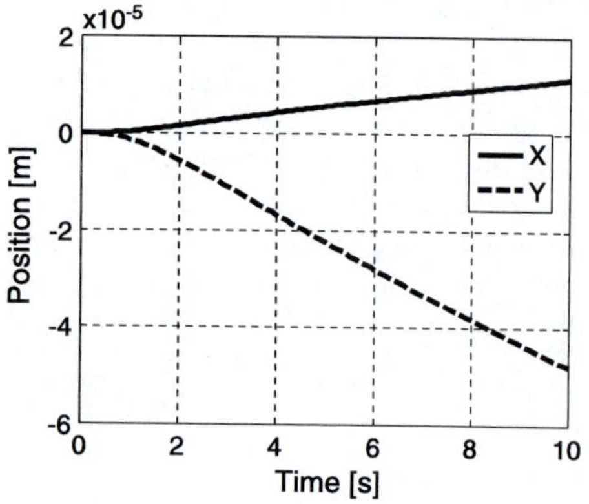

Fig. 2.2 History of space robot position for phase 1 operation

Other parameters are assumed as follows;

$$k_1 = k_2 = k_3 = k_4 = k_5 = k_6 = 100$$
$$\lambda_1 = \lambda_2 = 15, \lambda_3 = 10, \lambda_4 = \lambda_5 = \lambda_6 = 5$$
$$\varepsilon_1 = \varepsilon_2 = \varepsilon_3 = \varepsilon_4 = \varepsilon_5 = \varepsilon_6 = 0.05.$$

Some of the above parameters are determined by iterative manner.

Results of the *Phase I* are shown in Figs. 2.2–2.8. The performance of tracking the target is satisfactory and the error of tracking was below 1 [mm].

Results of the *Phase I* are shown in Figs. 2.2–2.8. The performance of tracking the target is satisfactory and the error of tracking was below 1 [mm].

Results of Phase II control are given by Figs. 2.9–2.18.

The control of position and velocity of the space robot is satisfactory and control input for spacecraft position and joint angles is sufficiently small, for instance, the maximum torque for both the space robot attitude control and joint control is smaller than 1 [Nm]. These values are consistent with the space application.

In general mounted thruster forces are from 1 N to 10 N for thousand kg class spacecrafts and typical arm length for the torque will be 2 m or 3 m. Furthermore, typical toque capability by reaction wheel for the attitude control of spacecraft is 1 Nm. These facts validate the applicability of our approach.

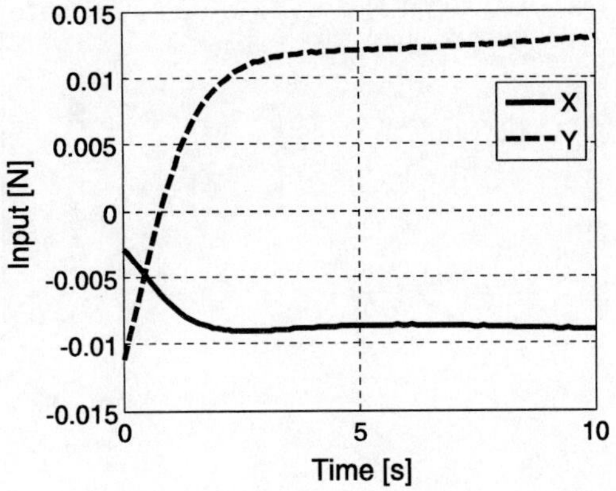

Fig. 2.3 History of control force for space robot

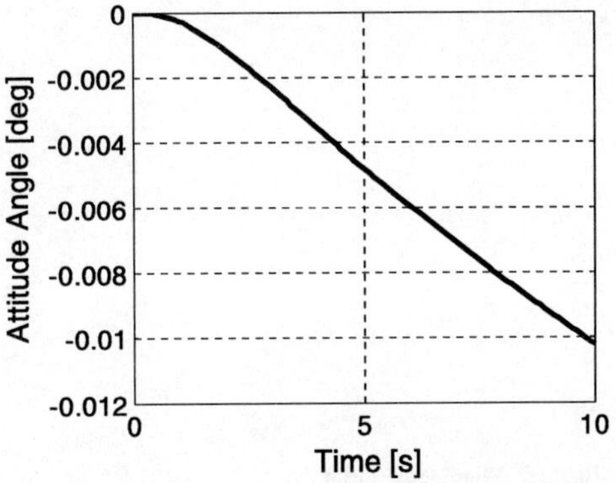

Fig. 2.4 History of space robot attitude angle for phase 1 operation

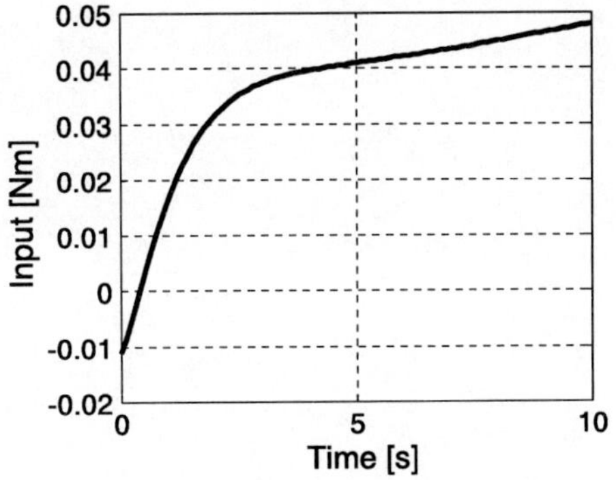

Fig. 2.5 History of control input torque

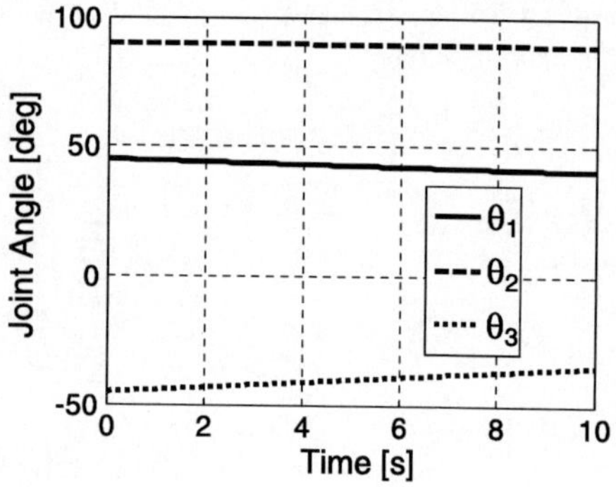

Fig. 2.6 History of joint angles for phase 1 operation

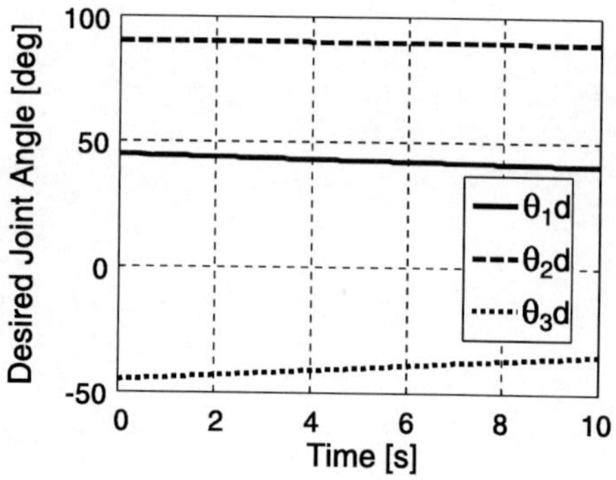

Fig. 2.7 History of desired joint angles

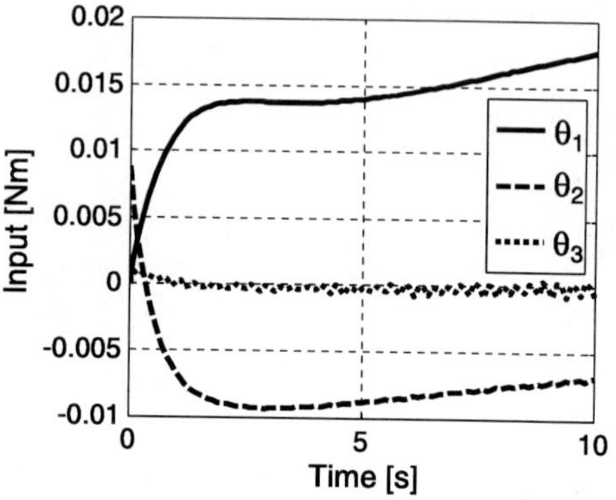

Fig. 2.8 History of joint torque input

Fig. 2.9 History of space robot position

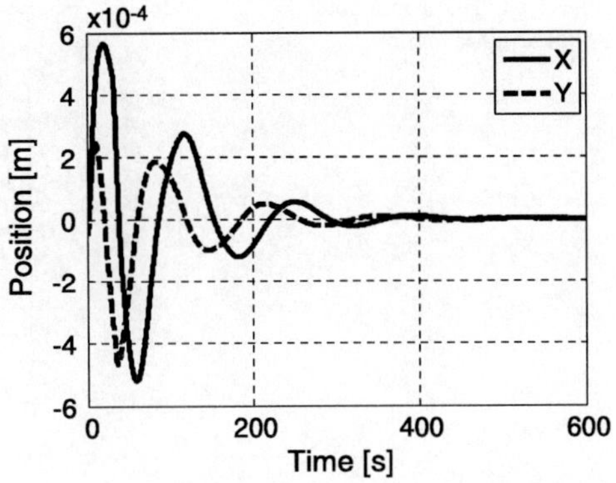

Fig. 2.10 History of space robot velocity

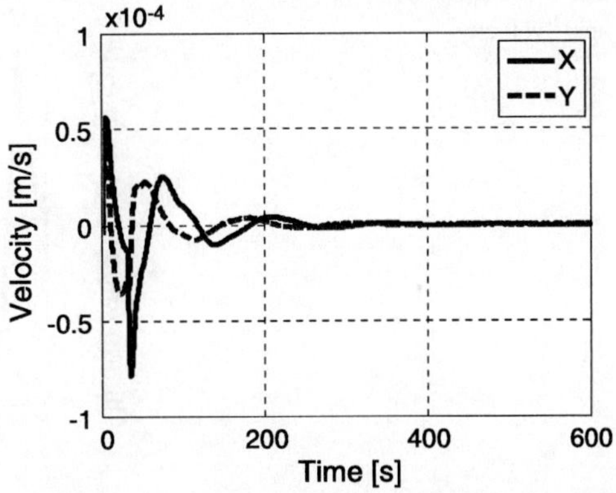

Fig. 2.11 History of translational control input

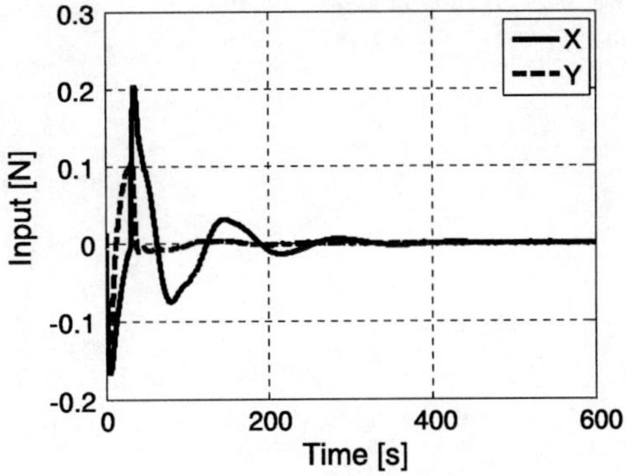

Fig. 2.12 History of space robot attitude angle

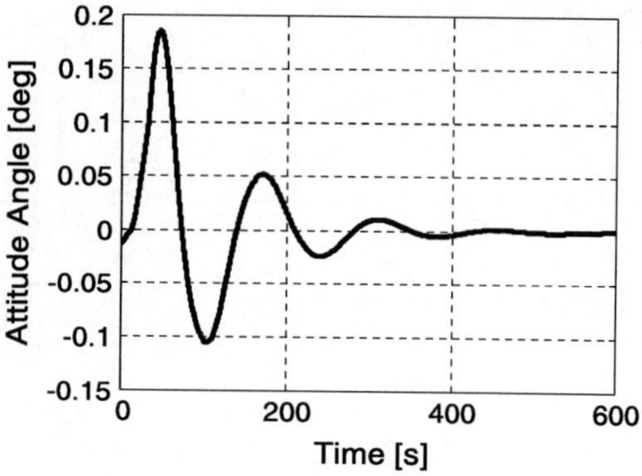

Fig. 2.13 History of attitude angle velocity

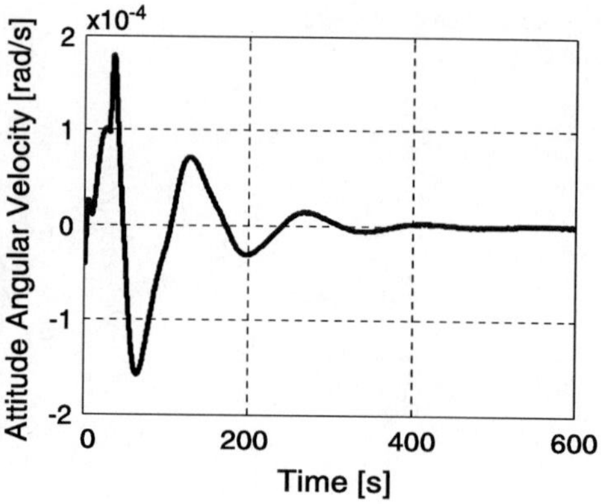

Fig. 2.14 History of torque control input for space robot

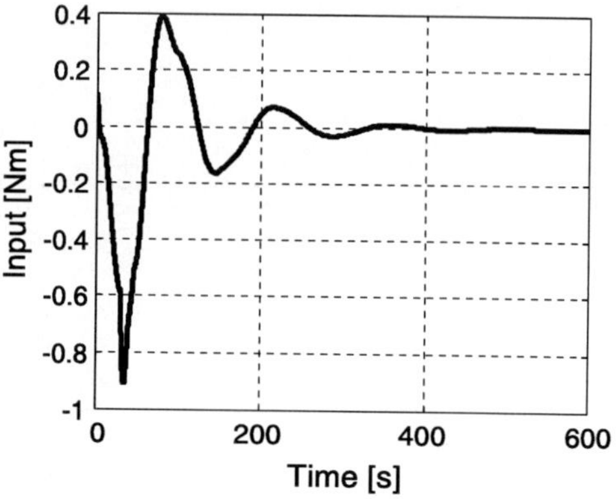

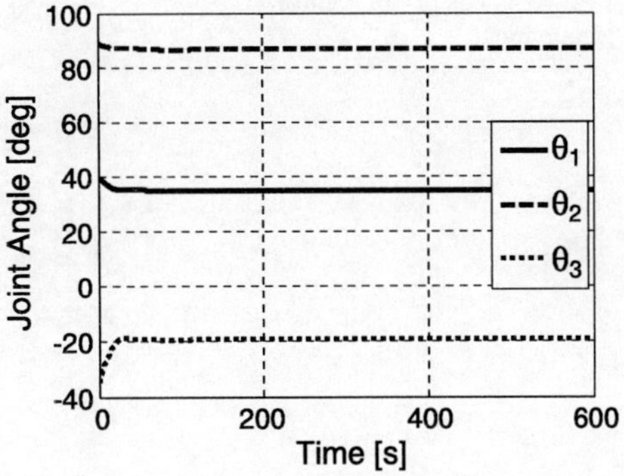

Fig. 2.15 History of joint angles for space robot

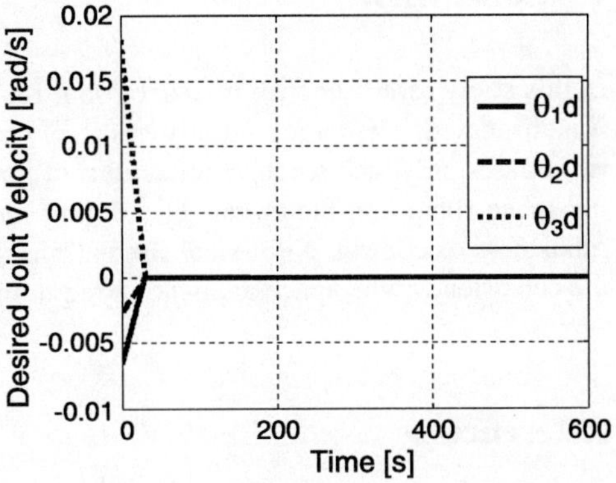

Fig. 2.16 History of desired joint velocities

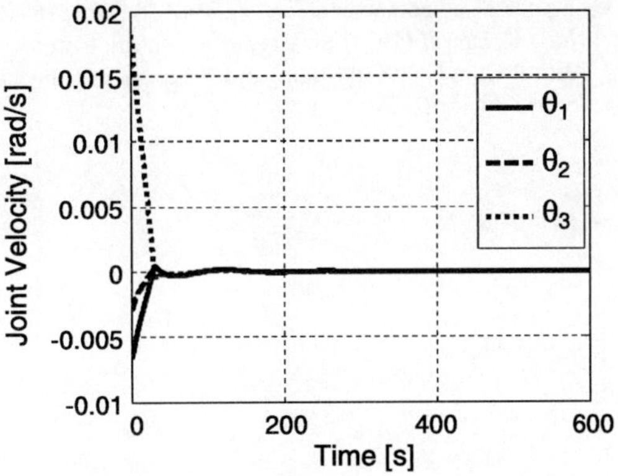

Fig. 2.17 History of joint angles

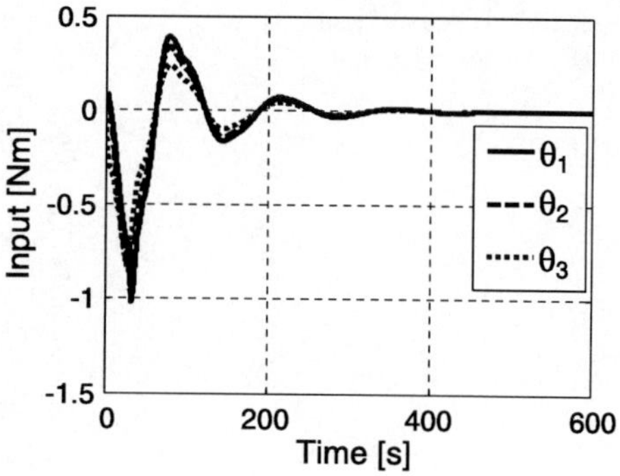

Fig. 2.18 History of control torque input

6 Conclusion

In this study robust control of space robot for unknown target capturing operation was discussed. The target initially has freely rotating motion, therefore we defined two phases, in which we have operations of grasping the target and stabilizing both the space robot and the target. The sliding mode control was applied to have the robustness of control. Numerical simulations were conducted and the results show the consistency with space application requirement. This validates our approach.

References

1. Kobayashi T, Tsuda S (2011) Robust control of space robot for moving target capturing, Lecture notes in engineering and computer science: proceedings of the international multiconference of engineers and computer scientists, IMECS 2011, 16–18, Hong Kong, pp 834–838
2. Noba K, Den H (1994) Sliding mode control. Corona, Tokyo
3. Mita T, Chin G (1990) Sliding mode control and trajectory control of robot arm.System Control Inform 34(1):50–55

Chapter 3
Multi-Sensor Architectures

D.M. Akbar Hussain, Zaki Ahmed, and M.Z. Khan

Abstract The use of multiple sensors typically requires the fusion of data from different type of sensors. The combined use of such a data has the potential to give an efficient, high quality and reliable estimation. Input data from different sensors allows the introduction of target attributes (target type, size) into the association logic. This requires a more general association logic, in which both the physical position parameters and the target attributes can be used simultaneously. Although, the data fusion from a number of sensors could provide better and reliable estimation but abundance of information is to be handled. Therefore, more extensive computer resources are needed for such a system. The parallel processing technique could be an alternative for such a system. The main objective of this research is to provide a real time task allocation strategy for data processing using multiple processing units for same type of multiple sensors, typically radar in our case.

Keywords Target tracking • Data fusion • Sensor level • Parallel processing

1 Introduction

The use of multiple sensors typically requires the fusion of data from different type of sensors. The combined use of such a data has the potential to give an

D.M.A. Hussain (✉)
Department of Energy Technology, Aalborg University Denmark, Niels Bohrs Vej 8, 6700 Esbjerg, Denmark
e-mail: akh@et.aau.dk

Z. Ahmed
Pakistan Institute of Laser and Optics, Rawalpindi, Pakistan
e-mail: zaki786@ieee.org

M.Z. Khan
Consultant
e-mail: emadeyyal@hotmail.com

S.I. Ao et al. (eds.), *Intelligent Control and Innovative Computing*, Lecture Notes in Electrical Engineering 110, DOI 10.1007/978-1-4614-1695-1_3,

efficient, high quality and reliable estimation. Input data from different sensors allows the introduction of target attributes (target type, size) into the association logic. This requires a more general association logic, in which both the physical position parameters and the target attributes can be used simultaneously. Although, the data fusion from a number of sensors could provide better and reliable estimation but abundance of information is to be handled. Therefore, more extensive computer resources are needed for such a system. The parallel processing technique could be an alternative for such a system. The main objective of this research is to provide a real time task allocation strategy for data processing using multiple processing units for same type of multiple sensors, typically radar in our case.

Basically multi-sensor system is a net formed by number of different or similar types of sensors such as radar, sonar, infrared sensor and cameras, etc. Multiple sensors can be deployed as co-located and non-co-located systems. The co-located system for example can be described by a warship which is mounted with different sensors; radar, sonar, infrared and cameras. The netting of radars-sensors is an example of non-co-located sensors which is considered in this research. The first step in developing a multiple sensor system is the choice of the architecture of the system for the data processing. In fact only two architectures [1–3] are available for the data processing firstly, the distributed architecture which consists of a computer at each radar-sensor site performing the tracking functions as an independent single target tracking system. The tracks formed at each radar site are then sent to a command (computer) site where a single multi-sensor track for each target is maintained. Secondly, the centralized architecture consists of single (main) computer which receive the observation-measurement data from the different sensor sites. These observations are then processed to obtain a single multi-sensor track for each target. There is a possibility of a third architecture which could be combination of the two architectures described above. In text the distributed architecture is sometimes called as sensor level tracking and the centralized architecture is known as central level tracking and we are using and investigating the later description for the architecture, similar kind of work has also been investigated in reference [4, 5]. Figure 3.1 gives architecture for general data processing using different kind of input sensor data that is kinematics as well as attributes. Next the two levels for the data fusion problem are discussed.

1.1 Sensor Level Fusion

In this setup each sensor is coupled with an independent tracking system, which is responsible for track initialization, data association, track prediction and track update. The architecture for a sensor level tracking system is shown in Fig. 3.2 [1]. Each tracking system individually performs the above mentioned tracking functions and finally all the updated tracks are sent to a common place (computer) for fusion of tracks to obtain a single global track for each target.

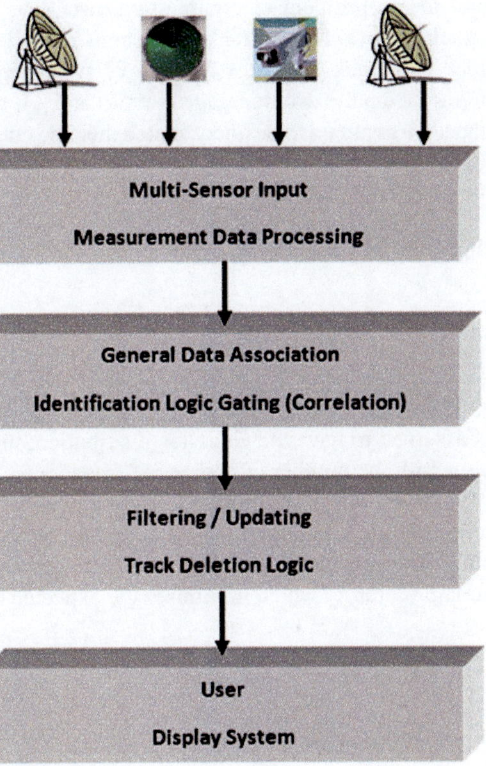

Fig. 3.1 Multiple sensor data processing

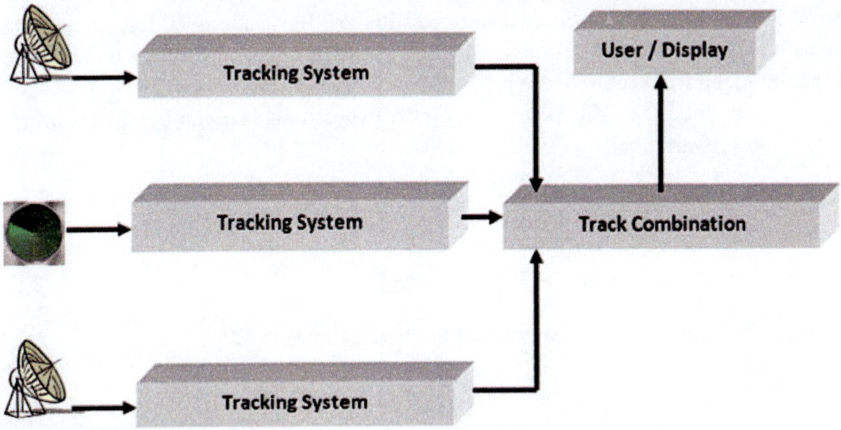

Fig. 3.2 Sensor level tracking

Fundamental to the problem of combining sensor level tracks is determining whether two tracks from different tracking systems potentially represent the same target. Consider two track estimates $\underline{\hat{x}}_i$ and $\underline{\hat{x}}_j$ with corresponding covariance matrices P_i and P_j from the two tracking systems i and j. If the estimation errors on the two estimates are assumed to be uncorrelated then the common test statistics [6],

$$d_{ij}P^{-1}d_{ij}^{T} \leq D_{th} \tag{3.1}$$

can be used to decide whether the two estimates are from the same target.

Where

$$d_{ij} = \left(\underline{\hat{x}}_i - \underline{\hat{x}}_j\right) \tag{3.2}$$

and

$$P_{ij} = P_i + P_j \tag{3.3}$$

Because d_{ij} is assumed to have the Gaussian distribution, therefore (3.1) will have a χ^2 distribution with the number of degrees of freedom equal to the dimension of the estimate vector. Therefore, when (3.1) (using a threshold D_{th} obtained from a χ^2 distribution) is satisfied then the two estimates represent the same target. When the estimates are determined to be from the same target, they are combined. The combined estimate vector $\underline{\hat{x}}_c$ which minimizes the expected error is

$$\underline{\hat{x}}_c = \underline{\hat{x}}_i + C\left(\underline{\hat{x}}_i - \underline{\hat{x}}_j\right) \tag{3.4}$$

Where

$$C = \frac{P_i}{P_i + P_j} \tag{3.5}$$

When the estimation errors of the two estimates are not independent the covariance defined above in (3.3) for the two estimates is not strictly valid because of the error correlation between the two sensor estimates. Bar-shalom [7] has proposed a statistical test, to account for the correlation between the two estimates this method is summarized from references [1,8].

In this method the difference between the two estimates given by d_{ij} is normalized by the covariance;

$$\begin{aligned} E\left[d_{ij}d_{ij}^{T}\right] &= E\left[\left\{(\underline{\hat{x}}_i - \hat{x})(\underline{\hat{x}}_j - \hat{x})\right\}\left\{(\underline{\hat{x}}_i - \hat{x})(\underline{\hat{x}}_j - \hat{x})\right\}^{T}\right] \\ &= P_i + P_j - P_{ij} - P_{ij}^{T} \end{aligned} \tag{3.6}$$

where $\underline{x}$ is the true state (noise free) of the two estimates and

$$P_{ij} \approx E\left[(\underline{\hat{x}}_i - \hat{x})(\underline{\hat{x}}_j - \hat{x})^{T}\right]^{T} \tag{3.7}$$

reflects the correlation between the two estimates. Therefore, the new proposed (by Bar-Shalom) test statistics to check two estimates represent the same target is

$$\left(\underline{\hat{x}}_i - \underline{\hat{x}}_j\right)\left(P_i + P_j - P_{ij} - P_{ij}^T\right)^{-1}\left(\underline{\hat{x}}_i - \underline{\hat{x}}_j\right)^T \leq D_{th} \tag{3.8}$$

When the estimates are determined to be from the same target they are combined using (3.4) and the value of C is now given as

$$C = \frac{(P_i - P_{ij})}{P_i + P_j - P_{ij} - P_{ij}^T} \tag{3.9}$$

Initially the cross covariance matrix P_{ij} is assumed zero that is

$$P_{ij}(0/0) = 0.0 \tag{3.10}$$

after that for n > 0 the values of $P_{ij}(n/n)$ are computed recursively by the following relationship

$$P_{ij}(n/n) = A_i(n)F(n-1)A_j^T(n) \tag{3.11}$$

$$A_i(n) = I - K_i(n)H \tag{3.12}$$

$$A_j(n) = I - K_j(n)H \tag{3.13}$$

$$F(n-1) = \Phi P_{ij}(n-1/n-1)\Phi + \Gamma Q \Gamma^T \tag{3.14}$$

Matrices Φ, K, H, Q and Γ are the same state transition, Kalman gain, measurement, covariance of measurement noise and excitation respectively used for standard Kalman filter [9]. It is evident that if the number of sites are more than two or if more than two tracks represent the same target the above procedure will be quite complex due to the number of calculations involved. The above described procedure has one big advantage that is apart from the straight line motion target path, it can also be used for maneuvering target as well. The target estimates strictly for targets moving in a straight line path from two tracking systems i and j can also be combined using the following relationships [2].

$$\underline{\hat{x}}_c = P_c\left(\frac{\underline{\hat{x}}_i}{P_i} + \frac{\underline{\hat{x}}_j}{P_j}\right) \tag{3.15}$$

Where

$$P_c = \left(\frac{1}{P_i} + \frac{1}{P_j}\right)^{-1} \tag{3.16}$$

1.2 Central Level Fusion

In this setup the measurements from different sensor sites are received at a central system as shown in Fig. 3.3. In contrast to a sensor level tracking system initialization of tracks, data association, track prediction and track update is processed only at a central place (computer). The data processing is done by considering all the information (measurements) available from all the sensors. The track loss

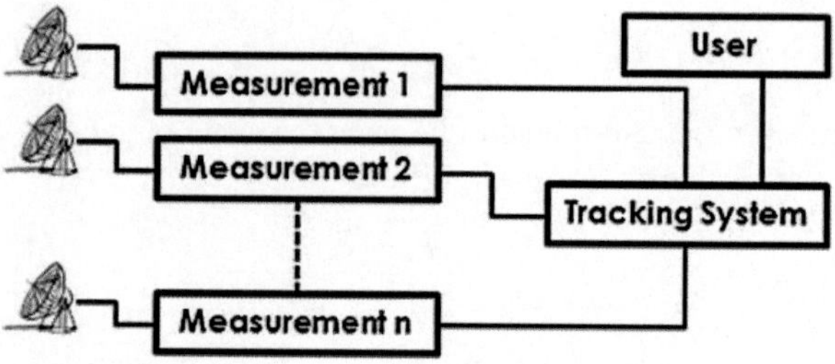

Fig. 3.3 Central level architecture

rate and mis-correlations in central level tracking are fewer than in the case with sensor level tracking. Also more accurate tracking should be expected if all the data (measurements) are processed at the same place. A target track that consists of measurements from more than one sensor should be more accurate than the track which could be established on the partial data received by the individual tracking system. Finally, the approach whereby all data are sent directly to the central processor should, in principle, lead to faster and efficient computation. The overall time required in developing sensor level tracks and then to combine these tracks is generally greater than the time required for central level processing of all data at once [1]. On the other hand branching of tracks may occur in a central level tracking if similar tracks are not merged due to some reasons and multiple tracks may also be initiated at the time of track initialization. Also, if the data from one of the sensor is degraded it will affect the central level tracks. In this study a straight line model for the target dynamics have been assumed, therefore using the same technique given in (3.15) and (3.16), the combined measurement vector and its covariance are obtained instead of the combined estimate vector and the covariance of the estimate.

Consider for example the two measurement vectors $\underline{z}_i$ and $\underline{z}_j$ with the corresponding measurement error co variances R_i and R_j from the two sensor sites, then the combined measurement vector $\underline{z}_c$ and the combined measurement error covariance matrix are

$$\underline{z}_c = R_c \left(\frac{\underline{z}_i}{R_i} + \frac{\underline{z}_j}{R_j} \right) \tag{3.17}$$

Where

$$R_c = \left(\frac{1}{R_i} + \frac{1}{R_j} \right)^{-1} \tag{3.18}$$

The approximate statistics given by (3.1) is now modified as to test the nearness of two measurements instead of the estimates that is

$$(\underline{z}_i - \underline{z}_j) R_a^{-1} (\underline{z}_i - \underline{z}_j)^T \leq D_{th} \tag{3.19}$$

where

$$R_c = R_i + R_j \tag{3.20}$$

When the two measurements using the above statistical test (3.19) are determined to be from the same target they are merged using (3.15) and (3.16).

2 Implementation

The trade-off between the sensor level and central level systems and the availability of the resources can normally dictate the best choice of the architecture for a particular application. In this study central level architecture is simulated on a network of computers for the measurement to measurement fusion. For simplicity only two sensors are considered for the system typically sensors provide data in polar coordinates that is the range and bearing of the targets. The tracking however, is performed in Cartesian coordinates so the data is transformed from polar to Cartesian coordinates. The dynamics of the target considered in this study are modeled as straight line motion target. The data from the host computer is sent in the form of array called the measurement vector to the other processing units. The first processor which is directly connected with the host computer has seven modules/processes namely merge measurements, initialization of tracks, distribution of data (measurements and tracks), state estimation update, local similarity, storage of tracks and unused measurements and finally sending of track estimates to host. Merging of measurements and data distribution modules are explained in the following paragraphs.

2.1 Merging of Measurement

After receiving the two measurement vectors corresponding to the two radars the merge measurement task first of all sets the used/unused flag of each measurement in the two measurement vectors to 0.0, then it starts performing the statistical test described in (3.19) for each measurement pair in the two measurement vectors. For a successful test the two measurements are merged using (3.17) and (3.18) provided the used/unused flags of each measurement is 0.0. After that the used/unused flags of the two merged measurements are set to 1.0 which indicates that these measurements have been used in merging. Therefore, a measurement from a sensor (1) can be merged only once with another measurement of sensor (2), this option was used to avoid merging of a single measurement from sensor (1) with a number of measurements in the second sensor (2) when multiple measurements from other targets occurs in the same neighborhood. For example, consider a case of four crossing targets and assume that the used/unused flag for the two measurement vectors are not set to 1.0, when these measurements are merged. As long as the targets are distinct (separate from each other) there will be no problem, because only the two corresponding measurements from a common target will be merged. However, consider the situation at the time of crossing, the merge measurement task will take the first measurement from sensor (1) and starts performing the statistical test given in (3.19) with every measurement in the second sensor (2) and because the measurements are very close to each other therefore possibly every measurement of sensor (2) will be likely candidate for merging with the first measurement of sensor (1). But if the used/unused flags are set to 1.0 after each successful comparison it will prevent merging of the same measurement with another measurement. However, the setting of used/unused (1.0) flag do not guarantee for the correct merging of two measurements when ambiguity occurs, but probably provides equal chance to other measurements belonging to the two sensors for merging. For situations when the two measurements from the same target do not satisfy the statistical test (3.19) both measurements from the two sensors are kept, this means track splitting will occur but the similarity criterion should take care of such situations. Finally all the merged measurements are stored in a new measurement vector.

2.2 *Distribution of Tracks and Observations*

For the distribution of observations and tracks among the different processors an intelligent procedure described in reference [4] is used. In which the new measurement vector is checked for data ambiguity and if there is no ambiguity data distribution task divides the number of tracks among the available processors as equally as possible and sends all measurements of the new measurement vector to all the processors. In the case of ambiguity, measurements of the new measurement vector are divided as equally as possible to the available processors and all the tracks are sent to every processor. After the distribution of data, the state estimation update is done using the standard Kalman filter and after updating, tracks are compared to eliminate similar tracks. Finally, last module store the tracks from all processors and also sends the position estimates to the host computer for real time display. The performance of the multi-sensor algorithms implemented is investigated by considering tracking accuracy & the speed up achieved using multiple processing units as parallel processors. Therefore, simulations were performed by considering typical scenarios and the processing time with a single as well as with four processing units were obtained for comparison.

3 Simulations

First part of the simulations was concerned with the comparison of the tracking accuracy for the two levels of data fusion that is sensor level (estimate estimate fusion) and the central level (measurement measurement fusion). For the estimate–estimate fusion (3.15) and for the measurement–measurement fusion

Table 3.1 Parameters for the two radars

Parameters	Values
Scan sector w.r.t y-axis 1–2	$\pm$90 degrees
Bearing resolution 1–2	5.0 degrees
Probability of detection 1–2	1.0 (100%)
Range resolution 1–2	0.02 Km
Clutter density 1–2	0.0
Maximum range 1–2	50.0 Km
Minimum range 1–2	5.0 Km
Scan interval 1–2	1.0 s
Range variance σ_r^2 1–2	0.001 Km^2
Bearing variance σ_θ^2 1–2	0.004 Radians^2

(3.17) are used as a straight line motion target path is being considered in this study. Simulations are performed by considering typical scenarios and the processing time with a single as well as with four processing units was monitored for comparison. Initial parameters for the two radars to simulate the data for targets are given in Table 3.1.

3.1 *Scenario 1*

The first scenario for the simulation was a single target moving from the origin (0.0, 0.0) Km with a speed of 0.015 Km/s making an angle of 90 degrees with x-axis, the position of each sensor which is fixed are also shown in Fig. 3.4. In this scenario the target is equidistant from the two sensors there is no correlation between the measurements obtained from the two sensors as different random numbers were used for each sensor.

Both algorithms sensor level and central level were used to obtained the track estimates and at each time instant (scan interval) the error between the actual target position (noise free) and the estimated position was monitored using the following relation,

$$Error = \sqrt{(\hat{x}_n - x_a)^2 + (\hat{y}_n - y_a)^2} \quad \mathrm{n = i, j} \tag{3.21}$$

where $\hat{x}$ and $\hat{y}$ are estimated x, y positions and x_a and y_a are actual noise free positions respectively at each scan interval. It is observed that the error obtained by the central level tracking system is less than the sensor level case especially at the start of the run there is big difference between the two errors obtained. In the case of sensor level tracking each sensor measurement is initialized/updated independently and then the initialized/updated estimates from the two tracking systems are merged, therefore the error which occurs at the time of initialization and depend on the measurement error covariance R is seen at the start of the run and afterwards because of the filtering process the error is reduced at each time step. In the case of the central level tracking the measurements from the two sensors are first merged

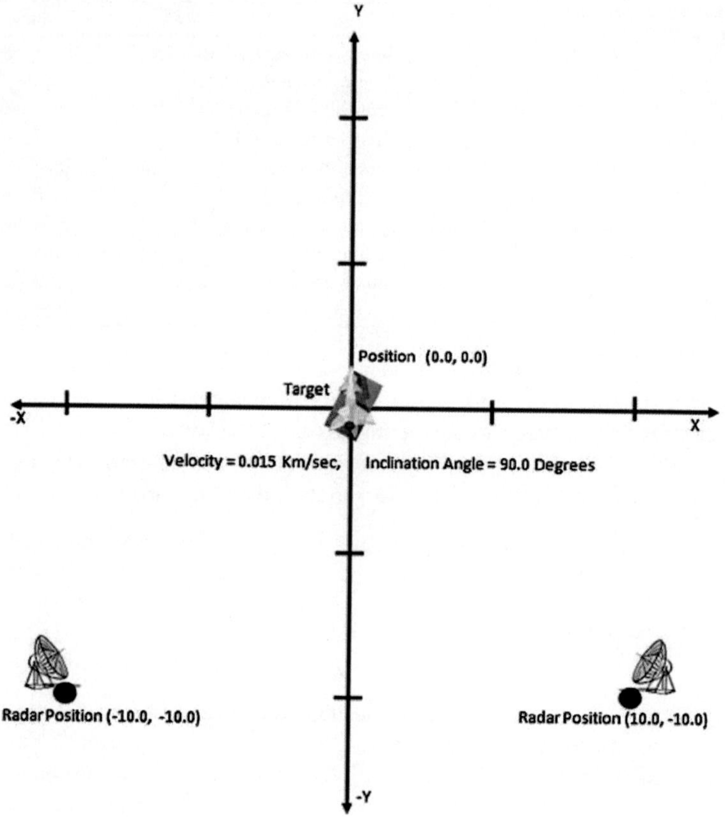

Fig. 3.4 Scenario 1

before the initialization process and the merged measurement has more confidence (small *R*) as compared with the two independent measurements, therefore a smaller error occurs at the start of the run.

3.2 Scenario 2

In this scenario again a single target is moving from an initial position (0.0, 0.0) Km with a velocity of 0.015 Km, making an angle of 90 degrees with the x-axis, in this scenario one sensor is relatively more close to the target as compared to the second as shown in Fig. 3.5. Again it was observed that the error obtained by the central level tracking system is less compared with the sensor level case especially at the start of the run there is big difference for the errors obtained with sensor level tracking, even though one sensor measurements have less noise. This is because

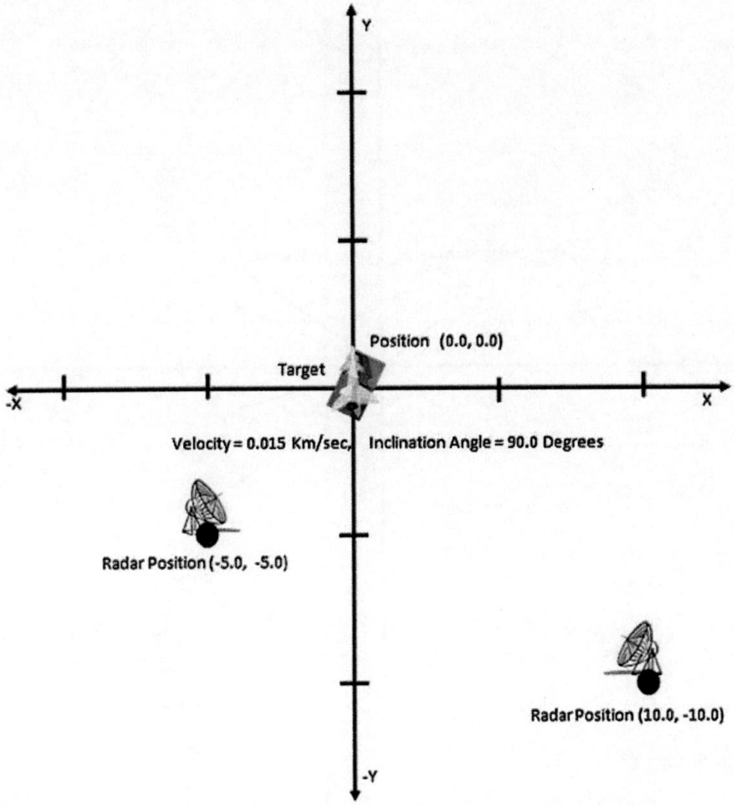

Fig. 3.5 Scenario 2

of the independent initialization of tracks in the sensor level tracking as explained in the previous scenario. Although in this case the merged estimate because of weighting matrix P_c (3.15) is biased towards the more accurate data (Sensor 1) but the individual initialization still makes it difficult to reduce the error up to the extent achieved with the central level tracking.

3.3 *Scenario 3*

In this scenario simulation are carried out with two stationary sensors placed at (20.0, −30.0) Km and (−20.0, −30.0) Km respectively. There are four targets which cross each other after 30 scan at the origin (0.0, 0.0) as shown in Fig. 3.6. The simulations were performed 5 times by generating data with 5 random seeds. In every run four targets were readily initialized in the second scan and tracking

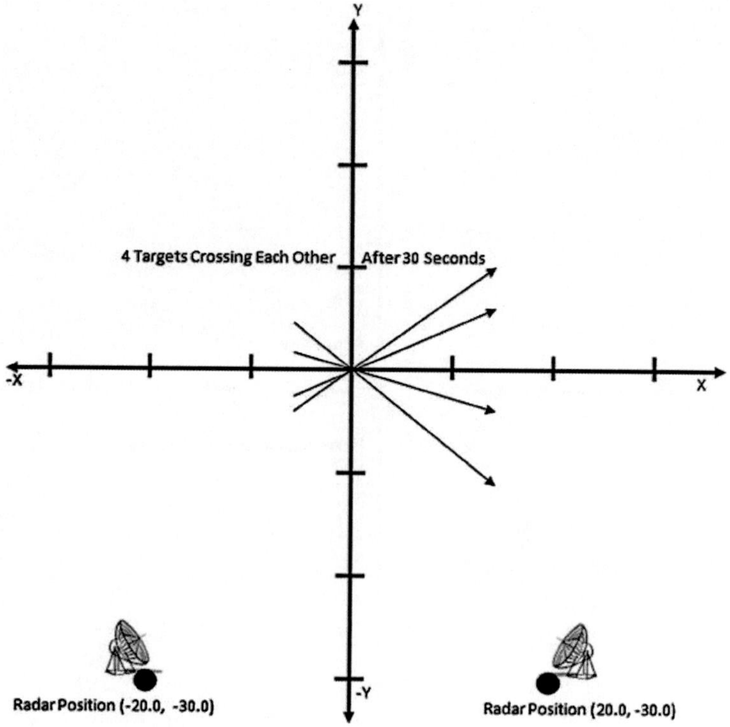

Fig. 3.6 Scenario 3

was accomplished successfully in the remaining scans. The processing time with a single and with a network of four processing units is shown in Fig. 3.7 and it can be seen that a speed up of 70% can be achieved.

4 Summary

In this investigation a multi-sensor system using two similar kinds of radars was implemented on a network. The main objective was to make comparison of tracking accuracy with different sensor/central level architectures and also to demonstrate that parallel architecture can be used for a real time development of a multi-sensor system. Although a very simple approach was used for the implementation but in a similar way a more general data association procedure can be developed for the fusion of data coming from different type of sensors. The merging of measurements instead of tracks is better approach when crossing target scenarios are under consideration, because if the later approach is used the merging process

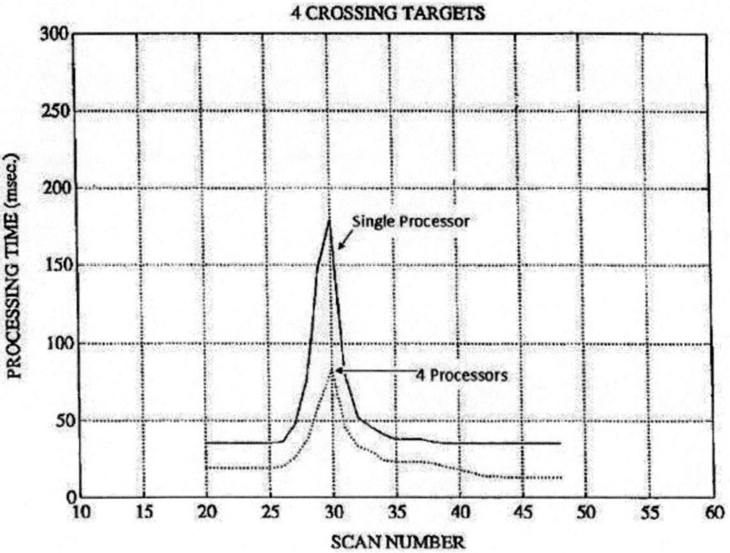

Fig. 3.7 Processing time

at the crossing point can merge all the target tracks into one global track which can lead to track instability (loss). The simulations have shown that a reasonable speed up of up to 50% is achievable.

References

1. Blackman SS (1986) Multiple target tracking with radar applications. Dedham, MA, Artech House, Inc.
2. Farina A, Studer FA (1986) Radar data processing. Research Studies. Letchworth, England/ New York, Research Studies Press, Ltd./John Wiley and Sons, Inc.
3. Hall DL, McMullen SAH (2004) Mathematical techniques in multi-sensor data fusion. Artech house inc, USA
4. Akbar Hussain DM (1991) Some implementations of multiple target tracking algorithms on transputers. PhD thesis
5. Akbar Hussain DM, Ahmed Z, Khan MZ, Valente A (2011) Proceedings of the international multi-conference of engineers and computer scientists 2011 Vol. II, IMECS 2011, Hong Kong, 16–18 March 2011
6. Singer RA, Stein JJ (1971) An optimal tracking filter for processing sensor data of imprecisely determined origin in surveillance systems Prods. of the 1971 IEEE conference on decision and control, Dec. 1971, Miami beach
7. Bar-Shalom Y (1981) On the track to track correlation problem, IEEE Trans Automatic Control, 26:571–572
8. Bar-Shalom Y, Fortmann TE (1988) Tracking and data association. Academic Press
9. Bridgewater AW (1978) Analysis of second and third order steady state tracking filter. AGARD conference proceedings no. 252

Chapter 4
Five-Axis Curve Machining for Computer Control Machine Tools

Rong-Shine Lin and Shyh-Leh Chen

Abstract Five-axis machining has been widely used in aerospace, automobile and mold/die industries. It has many advantages compared to conventional three-axis machining, including higher metal removal rates, better surface finish, and more precise part surface in one setup. This research presents a new method for curve machining on five-axis Computer Numerical Control (CNC) machine tools. It uses the CNC interpolator approach, or called curve interpolator, which can produce accurate tool position as well as tool orientation. The interpolator calculates a new command in real time, the same time period needed for sampling the control-loop feedback devices. It performs trajectory planning and inverse kinematics transformation in each sampling period. To implement this curve interpolator, a 3-D parametric curve g-code must be defined for five-axis CNC machining. The comparisons for this real-time method and the conventional off-line method in terms of trajectory accuracy and feedrate variation are demonstrated in the end of this chapter.

Keywords Computer control • Five-axis machining • Machine tools • CNC

R.-S. Lin (✉)
Associate Professor in the Department of Mechanical Engineering and Advanced Institute of Manufacturing for High-tech Innovations, National Chung Cheng University, Chia-Yi, Taiwan 62102, R.O.C.
e-mail: imelin@ccu.edu.tw

S.-L. Chen
Professor in the Department of Mechanical Engineering and Advanced Institute of Manufacturing for High-tech Innovations, National Chung Cheng University, Chia-Yi, Taiwan 62102, R.O.C.
e-mail: imeslc@ccu.edu.tw

S.I. Ao et al. (eds.), *Intelligent Control and Innovative Computing*, Lecture Notes in Electrical Engineering 110, DOI 10.1007/978-1-4614-1695-1_4,

1 Introduction

Multi-axis machine tools have been widely used in aerospace, automobile and mold/die industries. Usually, additional two rotary axes are set up on top of a traditional three-axis Computer Numerical Control (CNC) machine tool, which is called five-axis machine. Five-axis machining has many advantages compared to three-axis one, including higher metal removal rates, better surface finish, and more precise part surface in one setup [1–4]. However, the current methods for five-axis machining utilize off-line part programming approaches by which the CAD system divides the surface into a set of line segments that approximates the surface at the desired tolerance [5–8]. These line segments are further processed by post processors to produce straight-line g-codes which constitute the commands needed to control the machine. In the CNC system, these linear g-codes are fed into the interpolator that makes a linear motion for the curve. Chou and Yang [9] provided a mathematical formulation for an Euler-angle-type five-axis machine to track a parametric curve.

This off-line approach for five-axis machining either assumes a constant tool orientation along each segment, or assumes a linear change in the tool orientation between successive end-points. The constant orientation algorithm causes severe roughness around the end-points along the surface since the orientation changes are abrupt at these points. The linear orientation algorithm produces a better surface, but still cannot interpolate the orientations accurately between end points. This is because that the changes of the orientation along a curve are not necessarily linear and substantially causes machining errors. An additional drawback of the off-line methods is that the cutter accelerates and decelerates at each segment, which increases the surface non-uniformity and subsequently increases the cutting time [10, 11].

To overcome these drawbacks, this research presents an algorithm for real-time five-axis interpolator that generates both tool orientation and position precisely. This interpolator can produce desired trajectories for five -axis curve machining. The interpolator calculates new commands for five-axis controllers in the same time period, which is needed for sampling the control-loop feedback devices. The interpolator performs two steps in each sampling period: (1) trajectory planning based on a constant feedrate and (2) inverse kinematics transformation based on the structure of the machine.

The input to the interpolator is a new defined g-code (i.e., a CNC instruction), which contains the geometric information of the part surface as well as the cutting conditions, such as the feedrate, the spindle speed, the specific tool, etc. The interpolator begins with trajectory planning portion, which generates the position (x, y, z) and the orientation $(\mathrm{O}_x, \mathrm{O}_y, \mathrm{O}_z)$ of the cutter based on the curve geometry and the specified constant feedrate in each sampling period. This is a generic, machine independent algorithm. Based on the calculated cutter's position and orientation, the reference values of the five axes can be obtained by the inverse kinematics transformation, an algorithm which depends upon the structure of each particular machine.

To use this interpolator for five-axis curve machining, new g-codes must be defined for programming the part program [12]. In this research, the 3-D parametric curve g-code for five-axis machining will be given and illustrated by examples. The accuracy and feedrate analyses for the interpolating results will be discussed in the end of the chapter.

2 Five-Axis CNC Systems

The motion control kernel of the CNC system consists of a real-time interpolator and a servo-controller that dominates the axis coordination and motion accuracy. Figure 4.1 shows the structure of the CNC motion control kernel. As shown in the figure, the input to the CNC is the g-code type of part program. The major function of the real-time interpolator is to generate the reference commands for the servo-controller, which coordinates the multi-axis motion simultaneously.

The real-time interpolator for five-axis CNCs consists of the trajectory/orientation planning module and the inverse kinematics transformation module. The trajectory/orientation planning generates the tool position (x, y, z) and orientation (O_x, O_y, O_z) based on the machining trajectory, or called tool path. The six (x, y, z, O_x, O_y, O_z) variables have to be transformed into five (X, Y, Z, A, B), which are the reference inputs for five servo-controllers. This transformation is called inverse kinematics transformation where the solution depends on the structure of a particular machine.

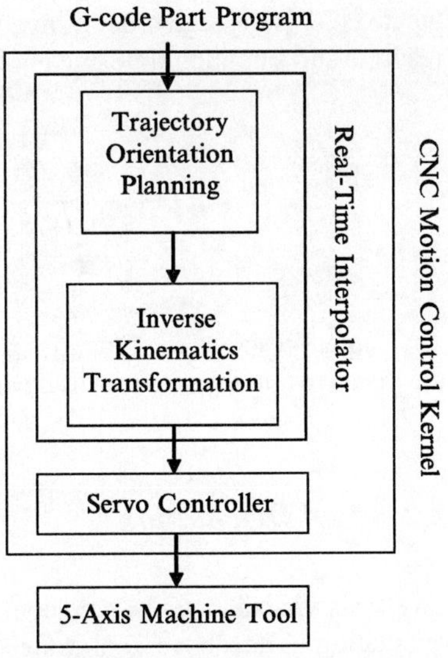

Fig. 4.1 The interpolator scheme for five-axis curve machining

2.1 Tool Trajectory

In order to produce smooth curves for a feed-drive CNC system, the machining feedrate (V) must be constant when the cutter tracks the desired trajectory. To keep the constant feedrate, the cutting tool has to move a constant distance relative to the workpiece in each sampling period of the interpolator. The desired tool trajectory can be expressed as a parametric curve $C\ (u)$ and denoted as,

$$C(u) = x(u)\hat{i} + y(u)\hat{j} + z(u)\hat{k} \tag{4.1}$$

$$\text{where} \begin{cases} x(u) = a_n u^n + a_{n-1} u^{n-1} + \cdots + a_1 u + a_0 \\ y(u) = b_n u^n + b_{n-1} u^{n-1} + \cdots + b_1 u + b_0 \\ z(u) = c_n u^n + c_{n-1} u^{n-1} + \cdots + c_1 u + c_0 \end{cases}$$

Based on the parametric curve, the interpolator calculates the tool position at each sampling period (T). However, the constant change in the parameter u in the parametric domain does not guarantee the constant length in the Cartesian domain, and consequently does not guarantee a constant feedrate. To obtain the constant feedrate, the conversion between the parameter u and the constant distance that the tool moves at each sampling period has to be derived, A solution based on Taylor's expansion is used to obtain the value of u that corresponds to equal trajectory length of the parametric curve.

$$u_{i+1} = u_i + \left.\frac{du}{dt}\right|_{t=t_i} (t_{i+1} - t_i) + \left.\frac{1}{2}\frac{d^2u}{dt^2}\right|_{t=t_i} (t_{i+1} - t_i)^2 + \text{HOT} \tag{4.2}$$

where HOT is the high order term and usually can be neglected. By substituting into feedrate and sampling time, the above equation becomes,

$$u_{i+1} = u_i + \frac{V \cdot T}{\left\| \frac{dC(u)}{du} \right\|_{u=u_i}} - \frac{V^2 \cdot T^2 \left(\frac{dC(u)}{du} \cdot \frac{d^2C(u)}{du^2} \right)}{2 \left\| \frac{dC(u)}{du} \right\|^4_{u=u_i}} \tag{4.3}$$

This equation shows the function of u in terms of a constant length (VT), which is the distance that the tool moves during one sampling period T.

2.2 Tool Orientation

Five-Axis machine tools can control not only the tool position, but the tool orientation. The above tool trajectory planning calculates tool positions. The tool orientation can be obtained by calculating the main normal for the curve.

The normal direction is perpendicular to the curve tangent. The unit tangent vector can be determined by,

$$\vec{T} = \frac{C'(u)}{\|C'(u)\|} \tag{4.4}$$

From the differential geometry, the rate change of tangent along the curve is the curve normal and denoted as,

$$\frac{d\vec{T}}{dS} = \kappa\vec{N}, \tag{4.5}$$

where κ is the radius of curvature and $\vec{N}$ is the unit normal vector. By rearranging the above equation, the unit normal can be obtained,

$$\vec{N} = \frac{1}{\kappa}\frac{d\vec{T}}{dS} \quad \text{with } \kappa = \left\|\frac{d\vec{T}}{dS}\right\| \tag{4.6}$$

This normal vector is the tool orientation for a desired curve and can be used for five-axis end-milling machining.

2.3 *Inverse Kinematics*

The above curve trajectory and normal represents the tool position (x, y, z) and orientation (O_x, O_y, O_z), respectively. However, the five-axis machine tool has only five axes to be controlled. The transformation from the tool position and orientation into five-axis reference positions uses inverse kinematics techniques.

The inverse kinematics transformation depends on the structure of the machine. The machine structure used for our derivations is shown in Fig. 4.2, which is a horizontal five-axis machine tool. As shown in Fig. 4.2, two rotation axes are place on top of the Z-axis table. The tilting axis is rotating along the X axis, and denoted as A axis. The rotation axis is installed on top of the tilting axis and rotating along the Y axis, which denoted as B axis. In this case, the tool axis is fixed on the orientation [0, 0, 1]. To produce the part surfaces with tool orientation control on this five-axis machine, the part surface normal direction needs to be rotated into the fixed tool orientation [0, 0, 1].

Since the rotation axes are placed on top of the translation axes, the rotation and translation motions can be de-coupled. Therefore, the first step of the inverse kinematics transformation is to calculate the required rotation angles for tool orientations. We assume that the required tilting and rotating angles for this motion are A and B, respectively. A transformation equation with two rotations can be obtained in terms of the homogeneous transformation matrices,

$$R(x, A) \bullet R(y, B) \bullet [O_x, O_y, O_z, 1]^T = [0, 0, 1, 1]^T \tag{4.7}$$

where $R(x, A)$ is the homogeneous transformation matrix that rotates along the x axis with an angle A, shown in (4.8) in detail. Likewise, $R(y, B)$ represents the

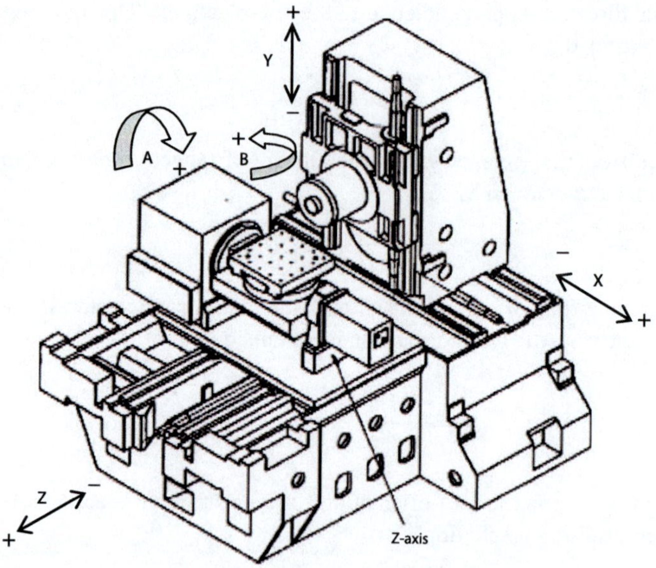

Fig. 4.2 The structure of a horizontal five-axis machine tool

notation of rotating along Y axis with an angle B. $[0,0,1,1]^T$ is the transpose matrix of [0,0,1,1], likewise $[O_x, O_y, O_z, 1]^T$. The detailed expression for (4.7) is shown below,

$$\begin{bmatrix} 1 & 0 & 0 & 0 \\ 0 & \cos A & -\sin A & 0 \\ 0 & \sin A & \cos A & 0 \\ 0 & 0 & 0 & 1 \end{bmatrix} \bullet \begin{bmatrix} \cos B & 0 & \sin B & 0 \\ 0 & 1 & 0 & 0 \\ -\sin B & 0 & \cos B & 0 \\ 0 & 0 & 0 & 1 \end{bmatrix} \bullet \begin{bmatrix} O_x \\ O_y \\ O_z \\ 1 \end{bmatrix} = \begin{bmatrix} 0 \\ 0 \\ 1 \\ 1 \end{bmatrix} \tag{4.8}$$

Two unknowns, A and B, in (4.8) can be solved for

$$A = \tan^{-1}\left(\frac{O_y}{O_z}\right) \quad \text{and} \quad B = -\tan^{-1}\left(\frac{O_x}{\sqrt{O_y{}^2 + O_z{}^2}}\right) \tag{4.9}$$

After two simultaneously rotating motions, the tool position has been moved to a new location. Therefore, the next step is to find the distance between this tool location and the desired trajectory location. The determined distance in terms of XYZ coordinate system is the reference position for three translational axes. Using homogeneous transformation notation, this distance can be calculated as

$$\begin{aligned}[X, Y, Z, 1]^T = {} & T(OTx, OTy, OTz) \bullet R(x, A) \bullet T(Tx, Ty, Tz) \bullet \\ & R(y, B) \bullet T(Qx, Qy, Qz) \bullet [x, y, z, 1]^T \end{aligned} \tag{4.10}$$

where $T(a,b,c)$ represents a linear translation along a vector of $[a,b,c]$ in the XYZ coordinate system. In (4.10) (x,y,z) is the tool location on the curve in terms of the part coordinate system. (Q_x, Q_y, Q_z) is a relative distance between the part coordinate system and the rotating table (B-axis) coordinate system. (T_x, T_y, T_z) is a relative distance between the rotating coordinate system and the tilting table (A-axis) coordinate system. (OT_x, OT_y, OT_z) is a relative distance between the tilting coordinate system to the tool coordinate system. Therefore, (4.10) can be expressed as

$$\begin{bmatrix} X \\ Y \\ Z \\ 1 \end{bmatrix} = \begin{bmatrix} 1 & 0 & 0 & OTx \\ 0 & 1 & 0 & OTy \\ 0 & 0 & 1 & OTz \\ 0 & 0 & 0 & 1 \end{bmatrix} \bullet \begin{bmatrix} 1 & 0 & 0 & 0 \\ 0 & \cos A & -\sin A & 0 \\ 0 & \sin A & \cos A & 0 \\ 0 & 0 & 0 & 1 \end{bmatrix} \bullet \begin{bmatrix} 1 & 0 & 0 & Tx \\ 0 & 1 & 0 & Ty \\ 0 & 0 & 1 & Tz \\ 0 & 0 & 0 & 1 \end{bmatrix} \bullet$$

$$\begin{bmatrix} \cos B & 0 & \sin B & 0 \\ 0 & 1 & 0 & 0 \\ -\sin B & 0 & \cos B & 0 \\ 0 & 0 & 0 & 1 \end{bmatrix} \bullet \begin{bmatrix} 1 & 0 & 0 & Qx \\ 0 & 1 & 0 & Qy \\ 0 & 0 & 1 & Qz \\ 0 & 0 & 0 & 1 \end{bmatrix} \bullet \begin{bmatrix} x \\ y \\ z \\ 1 \end{bmatrix} \tag{4.11}$$

The (X,Y,Z) in (4.11) and the (A,B) in (4.9) are the solutions of inverse kinematics transformation, which are the reference commands for five-axis servo-controllers. The controllers usually perform a specified control algorithm, e.g., PID control, to reduce the axial tracking errors.

3 Motion Code

3.1 Rational Bezier Curve

Free-form curves or surfaces have been widely used for product design in recent years. One of the convenient models for free-form curve is the parametric form. This research uses rational Bezier curve, a parametric curve, for the CNC input format [13]. A rational Bezier curve can be denoted as,

$$C(u) = \frac{\sum_{i=0}^{n} w_i p_i Br_i^n(u)}{\sum_{i=0}^{n} w_i Br_i^n(u)}, w_i \geq 0 \tag{4.12}$$

where u is the parameter; p is the control point that controls the shape of the curve. Br is the blending function, which is a Bernstein polynomial with order $n-1$. w is a weight that adjusts the curve's weighting.

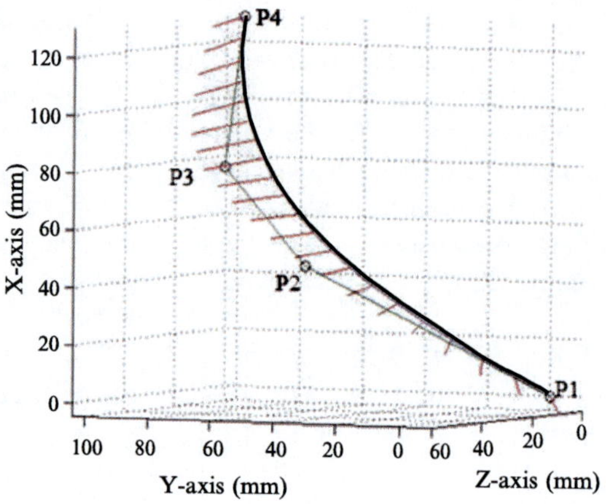

Fig. 4.3 The interpolating result of a rational Bezier curve

3.2 G-Code Format

In order to describe a rational Bezier curve, the motion code, or called g-code, includes the control points and weights. A newly defined g-code format for the curve is shown below,

N01	G701	p1x	p1y	p1z	w1	m
		p2x	p2y	p2z	w2	
		p3x	p3y	p3z	w3	
		p4x	p4y	p4z	w4	

G701 is used for the curve interpolator motion word and *N* is the block number. The control point in the Cartisent coordinated system is followed by the motion word, as well as the associated weight. m is the number of control points to be used in this motion code. For this example, *m* is 4 and therefore, 3 more control points will be needed for this g-code.

An example is demonstrated for the above definition, as shown below.

N01	G701	X0	Y0	Z0	W.8	m4
		X50	Y30	Z60	W.5	
		X80	Y80	Z30	W.2	
		X130	Y90	Z10	W.7	

The result of the curve interpolation for this g-code is shown in Fig. 4.3.

4 Results and Discussions

4.1 Feedrate Analyses

The CNC system uses feed drive control technologies that control the motion by the specified program feedrate. Using the same sample above, the programmed feedrate is 3.6 m/min (0.6 m/s) and the system sampling time is 0.01 s. By making the time derivative of the axial position equation (X, Y, Z axes), the axial velocity can be obtained and shown in Fig. 4.4. Likewise, Fig. 4.5 shows the angular velocities for the rotating axes (A and B).

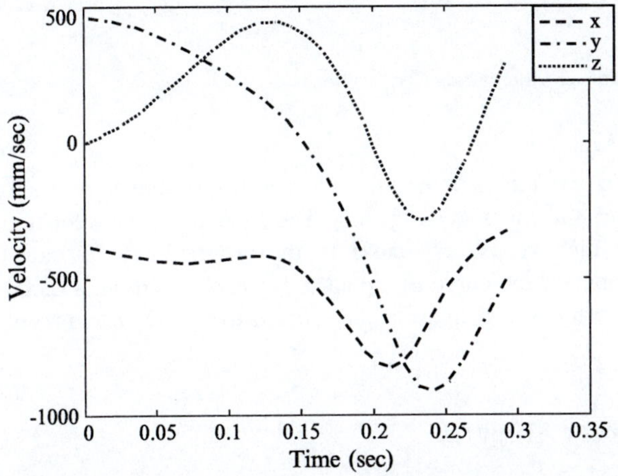

Fig. 4.4 The velocities for three translational axes (x, y, and z)

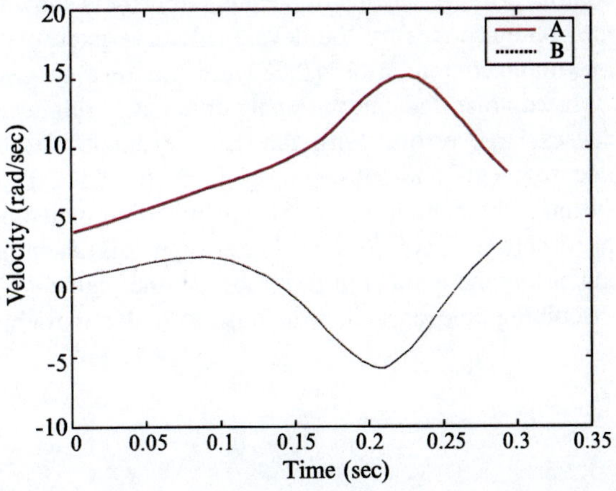

Fig. 4.5 The angular velocities for two rotating axes (A and B)

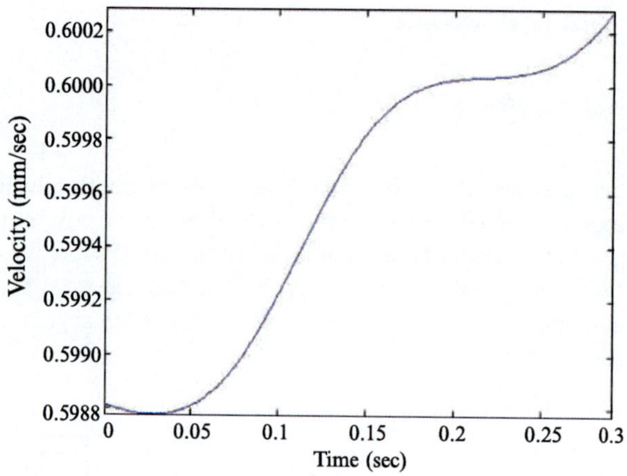

Fig. 4.6 The tool velocities during CNC motion

In order to see the tool motion velocity, the distance for each sampling is calculated and shown in the Fig. 4.6. The velocity is between 0.5988 m/s and 0.6002 m/s, which keeps very close to the programmed feedrate 0.6 m/s. The calculation time for the curve interpolator for each iteration is 0.0044 s, which is less than the system sampling time and satisfies the CNC system regulation.

4.2 *Accuracy Analyses*

The interpolating accuracy depends on the precision of the tool motion relative to the desired curve. The error between these two trajectories is defined as the contour error. A scheme of the contour error is shown in the Fig. 4.7. By analyzing the contour errors for the developed curve interpolator, as shown in Fig. 4.8, the maximum contour error is 0.683 mm and the average contour error is 0.118 mm.

In contrast, the conventional curve machining uses off-line decomposition that causes large errors. With the same example above, it is decomposed into 21 line segments and subsequently feeds the 21 G01 motion codes into the CNC system. The contour errors for this machining method are shown in Fig. 4.9. The maximum error is 4.369 mm. The error, in this machining method, can be reduced by increasing the number of decomposed line segments; however, it causes the longer machining time, as well as the mismatched tool orientations [14].

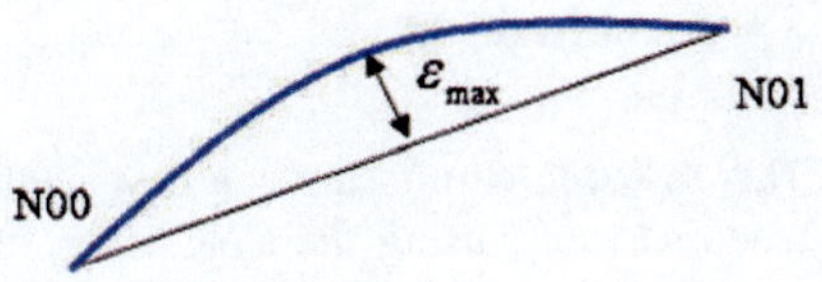

Fig. 4.7 The contour error between the desired curve and the interpolated trajectory

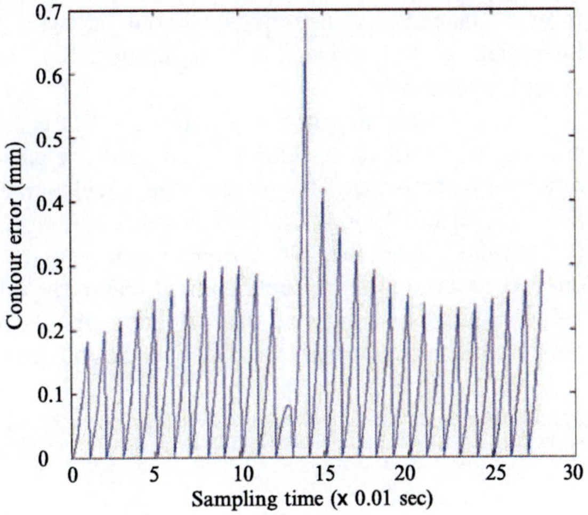

Fig. 4.8 The contour errors for the interpolated results

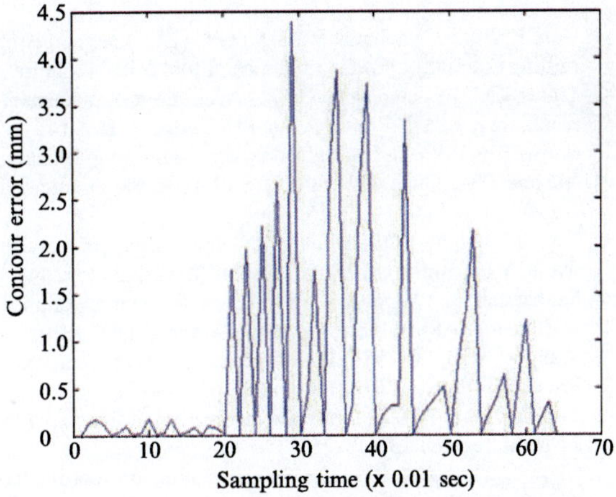

Fig. 4.9 The contour errors for the conventional G01 method

5 Conclusions

This research demonstrates a new method for curve machining on five-axis CNC machine tools, using the CNC interpolator approach. The interpolator calculates in real time a new command in the same time period needed for sampling the control-loop feedback devices. It performs two steps in each sampling period: (1) tool trajectory and orientation planning based on a constant feedrate and (2) inverse kinematics transformation based on the structure of the machine. The interpolating result shows that it can produce high accurate tool positions as well as tool orientations.

To implement this curve interpolator, a 3-D parametric curve g-code must be defined for five-axis machining. This paper defines the parametric curve in the rational Bezier format. The interpolating calculation time for a curve with 4 control point for each iteration is 0.0044 s, which is less than the system sampling time (0.01 s) and satisfies the CNC system regulation. If a curve with more control points, the computation load will be increased due to the complexity of the higher order polynomials. It is suggested that the curve should be described under 6 control points, which uses the blending function with 5th order polynomial.

Acknowledgment The authors would like to acknowledge the financial support of the National Science Council, Taiwan, R. O. C. under the grant: NSC 98–2221-E-194 -045 -MY2.

References

1. Gani E, Jruth J, Vanherck P, Lauwers B (1997) A geometrical model of the cut in the five-axis milling accounting for the influence of tool orientation. Int J Manuf Technol 13:677–684
2. Jensen CG, Anderson DC (1992) Accurate tool placement and orientation for finish surface machining, ASME Winter Annual Meeting, pp 127–145
3. Sprow EE (1993) Set up to five-axis programming, Manufacturing Engineering, pp 55–60
4. Vickers GW, Quan KW (1989) Ball-mills versus end-mills for curved surface machining, ASME J Eng Ind 111:22–26
5. Huang Y, Oliver J (1995) Integrated simulation, error assessment, and tool-path correction for five-axis NC milling. J Manuf Syst. 14(5):331–344
6. Sanbandan K, Wang K (1989) Five-axis swept volumes for graphic NC simulation and verification, ASME the 15th Design Automation Conference, DE, vol. 19–1, pp 143–150
7. Wang W, Wang K (1986) Real-time verification of multiaxis NC programs with raster graphics. Proc IEEE Int Conf Robot Automation 166–171
8. You C, Chu C (1997) Tool-path verification in five-axis machining of sculptured surfaces. Int J Manuf Technol 13:248–255
9. Chou J, Yang D (1992) On the generation of coordinated motion of five-axis CNC/CMM machines, J Eng Ind 114:15–22
10. Shpitalni M, Koren Y, Lo CC (1994) Real-time curve interpolator. Comput Aid Des 26(11):832–838
11. Lin R, Koren Y (1996) Real-time interpolators for multi-axis CNC machine tools, CIRP J Manuf Syst 25(2):145–149

12. Lin R-S, Chen S-L, Liao J-H (2011) Advanced curve machining method for 5-axis CNC machine tools. In: Lecture notes in engineering and computer science: proceedings of the international multiconference of engineers and computer scientists 2011, IMECS 2011, Hong Kong, 16–18 March 2011
13. Liu XB, Fahad A, Kazuo Y, Masahiko M (2005) Adaptive interpolation scheme for NURBS curves with the integration of machining dynamics, Int J Machine Tools Manuf 45(2005):433–444
14. So BS, Jung YH, Park JW, Lee DW (2007) Five-axis machining time estimation algorithm based on machine characteristics. J Mater Process Technol 187–188:37–40

Chapter 5
Approximate Linear Quadratic Regulator Problem and Its Application to Optimal Control in Discrete-Time LTI Systems

Kazuyoshi Mori

Abstract It is considered to approximately solve an linear quadratic (LQ) regulator problem in the case of discrete-time LTI systems. It leaves parameters as symbols in the evaluation function. The notion of the approximate LQ regulator problem is introduced. Also a computation method to solve the problem is proposed. A numerical example of the approximate LQ regulator problem is also presented, which is applied to an inverted pendulum on a cart.

Keywords Linear system theory • Optimal control • Discrete-time LTI system • Symbolic computation

1 Introduction

Optimal control problem is one of the most essential problems in the control theory [2, 5, 7]. However, the solution of an optimal control problem for a linear system is generally based on the maximization or minimization of an evaluation function. However, it is not straightforwardly clear what sort of influence the evaluation function has to the input–output relation [4]. Hence, it is necessary to do a control design with a trial evaluation function in order to obtain the appropriate evaluation function we need.

In this chapter, a method of solving an optimal control problem approximately in the case of discrete-time linear systems [5] is given. This method leaves parameters as symbols in the evaluation function, that is, postpones the determination of the evaluation function until obtaining the optimal input. Thereby, the relation between the evaluation function and the input–output relation becomes explicitly clear.

K. Mori (✉)
School of Computer Science and Engineering, The University of Aizu,
Aizu-Wakamatsu 965-8580, Japan
e-mail: Kazuyoshi.MORI@IEEE.ORG

S.I. Ao et al. (eds.), *Intelligent Control and Innovative Computing*, Lecture Notes in Electrical Engineering 110, DOI 10.1007/978-1-4614-1695-1_5,

We consider the Linear Quadratic (LQ) method and propose the notion of approximate LQ method. In this chapter, we especially consider the LQ regulator problem. In the classical case of this problem, we need to solve the Riccati equation of matrices. To include parameters, we will introduce the notion of the approximate Riccati equation.

This work is an extended and revised work of Mori [3].

2 Approximate LQ Regulator Problem

First, we review the classical LQ regulator problem [5]. Consider the state space matrix equation and the output matrix equation as follows:

$$\mathbf{x}(k+1) = A_d\mathbf{x}(k) + B_d\mathbf{u}(k), \tag{5.1}$$

$$\mathbf{y}(k) = C_d\mathbf{x}(k) + D_d\mathbf{u}(k), \tag{5.2}$$

respectively, where $\mathbf{x} \in \mathbf{R}^n$, $\mathbf{u} \in \mathbf{R}^m$, and $\mathbf{y}_d \in \mathbf{R}^l$ ($\mathbf{R}$ denotes the set of real numbers). Let $Q_d \in \mathbf{R}^{n\times n}$ and $R_d \in \mathbf{R}^{m\times m}$ are positive semidefinite and positive definite matrices, respectively, which are weighting matrices. The evaluation function J_d is

$$J_d = \sum_{i=0}^{\infty}(\mathbf{x}(i)^t Q_d\mathbf{x}(i) + \mathbf{u}(i)^t R_d\mathbf{u}(i)).$$

Then, it is known that the feedback matrix F_d in which J_d is minimized can be obtained by solving the Riccati equation

$$P_d = Q_d + A_d^t P_d A_d - A_d^t P_d B_d (R_d + B_d^t P_d B_d)^{-1} B_d^t P_d A_d \tag{5.3}$$

and as follows:

$$F_d = \left(R_d + B_d^t P_d B_d\right)^{-1} B_d^t P_d A_d.$$

See, for example, [5]. The Riccati equation of (5.3) can be solved by numerical iteration.

Based on this classical LQ regulator problem, we propose the notion of *approximate* LQ regulator problem. Consider again the state space matrix equation and the output matrix equation of (5.1) and (5.2), respectively. Suppose here that we have v parameters $a_1, a_2, \ldots, a_v$. Let $\widehat{Q}_d \in \mathbf{R}[a_1,\ldots,a_v]^{n\times n}$ and $\widehat{R}_d \in \mathbf{R}[a_1,\ldots,a_v]^{m\times m}$. We assume, without loss of generality, that when all parameters a_i's are set to 0, then the matrices $\widehat{Q}_d$ and $\widehat{R}_d$ are positive semidefinite and positive definite, respectively.

Now we introduce *rth approximate LQ regulator problem*, where r is a non-negative integer. To state it, we use the expression $(\mathrm{mod}\ a_1^{r+1},\ldots,a_v^{r+1})$ as an abbreviation of $(\mathrm{mod}\ a_1^{r+1})\ (\mathrm{mod}\ a_2^{r+1})\ \cdots\ (\mathrm{mod}\ a_v^{r+1})$. For example,

$$\alpha = \beta \quad (\mathrm{mod}\ a_1^{r+1},\ldots,a_v^{r+1})$$

is equivalent to

$$\alpha = \beta \quad (\text{mod } a_1^{r+1}) \ (\text{mod } a_2^{r+1}) \ \cdots \ (\text{mod } a_v^{r+1}).$$

The rth approximate LQ regulator problem is to find the matrix $\widehat{F}_d$ such that the following matrix equations hold:

$$\widehat{P}_d = Q_d + A_d^t \widehat{P}_d A_d - A_d^t \widehat{P}_d B_d \left(\widehat{R}_d + B_d^t \widehat{P}_d B_d\right)^{-1} B_d^t \widehat{P}_d A_d \quad (\text{mod } a_1^{r+1}, \ldots, a_v^{r+1}), \tag{5.4}$$

$$\widehat{F}_d = \left(\widehat{R}_d + B_d^t \widehat{P}_d B_d\right)^{-1} B_d^t \widehat{P}_d A_d \ (\text{mod } a_1^{r+1}, \ldots, a_v^{r+1}). \tag{5.5}$$

When r is obvious, we may omit "rth" and use "approximate LQ regulator problem."

The definition of the rth approximate LQ regulator problem requires the equality modulo $a_1^{r+1}, \ldots, a_v^{r+1}$. This and the sense of "approximate" follow the definitions of the approximate eigenvalue and eigenvector proposed by Kitamoto [1] as well as the approximate factorization proposed by Sasaki et al. [6].

We note that the rth approximate LQ regulator problem defined above is possible to compute naively. For example, we can apply Taylor's expansion. However, this is too ineffective to employ it because this uses derivatives many times in the rational functions, which is much ineffective.

3 Computing the Approximate LQ Method

In this section, we present a computing method of the rth approximate LQ regulator problem. First, we consider $\widehat{P}_d$ of (5.4). Here, observe that in (5.4), matrices A_d and B_d are matrices over $\mathbf{R}$. Weighting matrices Q_d and R_d are matrices over $\mathbf{R}[a_1, \ldots, a_v]$. Further P_d is a matrix over $\mathbf{R}[a_1, \ldots, a_v]$ such that degree of every entry of P_d is less than or equal to r with respect to each a_i $(1 \leq i \leq r)$.

Observe further that, in (5.4), we have a matrix inverse $(\widehat{R}_d + B_d^t \widehat{P}_d B_d)^{-1}$. If we do not use "(mod $a_1^{r+1}, \ldots, a_v^{r+1}$)," this will be a matrix over the formal power series $\mathbf{R}[[a_1, \ldots, a_v]]$. This fact suggests us the possible computation of the matrix inverse $(\widehat{R}_d + B_d^t \widehat{P}_d B_d)^{-1}$.

If all a_i's are set to zero, (5.4) is equal to (5.3). This means that the matrix $(\widehat{R}_d + B_d^t \widehat{P}_d B_d)$ with all a_i's being zero has its inverse. We denote by Φ_0 the inverse of the matrix $(\widehat{R}_d + B_d^t \widehat{P}_d B_d)$ with all a_i's being zero. This Φ_0 is a matrix over $\mathbf{R}$.

Consider the matrix $(\widehat{R}_d + B_d^t \widehat{P}_d B_d)\Phi_0$. If all a_i's are zero, this matrix is an identity matrix. Thus, denote by $I - \Psi$ the matrix $(\widehat{R}_d + B_d^t \widehat{P}_d B_d)\Phi_0$. Then, its inverse can be expressed as:

$$I + \Psi + \Psi^2 + \Psi^3 + \cdots$$

in a formal power series. More precisely, by letting $\Psi = I - (\widehat{R}_d + B_d^t \widehat{P}_d B_d)\Phi_0$, we have

$$\left(\widehat{R}_d + B_d^t \widehat{P}_d B_d\right)^{-1} = \Phi_0(I + \Psi + \Psi^2 + \Psi^3 + \cdots)$$

and further

$$\left(\widehat{R}_d + B_d^t \widehat{P}_d B_d\right)^{-1} = \Phi_0(I + \Psi + \Psi^2 + \Psi^3 + \cdots) \\ (\bmod\ a_1^{r+1}, \ldots, a_v^{r+1}). \tag{5.6}$$

Because Ψ does not have nonzero constant entry, (5.6) can be rewritten as:

$$\left(\widehat{R}_d + B_d^t \widehat{P}_d B_d\right)^{-1} = \Phi_0(I + \Psi + \Psi^2 + \Psi^3 + \cdots + \Psi^r) \\ (\bmod\ a_1^{r+1}, \ldots, a_v^{r+1}). \tag{5.7}$$

Now, we can use the right hand side of (5.7) instead of $(\widehat{R}_d + B_d^t \widehat{P}_d B_d)^{-1}$ in (5.4). By this relation, (5.4) can be rewritten as:

$$\begin{aligned} \widehat{P}_d = {} & Q_d + A_d^t \widehat{P}_d A_d \\ & - A_d^t \widehat{P}_d B_d \Phi_0 (I + \Psi + \Psi^2 + \Psi^3 + \cdots + \Psi^r) B_d^t \widehat{P}_d A_d \\ & (\bmod\ a_1^{r+1}, \ldots, a_v^{r+1}). \end{aligned} \tag{5.8}$$

Also (5.5) can be rewritten as:

$$\widehat{F}_d = \Phi_0(I + \Psi + \Psi^2 + \Psi^3 + \cdots + \Psi^r) B_d^t \widehat{P}_d A_d \\ (\bmod\ a_1^{r+1}, \ldots, a_v^{r+1}).$$

Let $\widehat{P}_{d0}$, $\widehat{Q}_{d0}$, and $\widehat{R}_{d0}$ be the $\widehat{P}_d$, $\widehat{Q}_d$, and $\widehat{R}_d$, respectively, with all a_i's being zero. Further let $\widehat{P}_{d+}$, $\widehat{Q}_{d+}$, and $\widehat{R}_{d+}$ be the $\widehat{P}_d - \widehat{P}_{d0}$, $\widehat{Q}_d - \widehat{Q}_{d0}$, and $\widehat{R}_d - \widehat{R}_{d0}$, respectively. Then $\widehat{P}_{d0}$ can be obtained by solving the following matrix equation:

$$\begin{aligned} \widehat{P}_{d0} = {} & Q_{d0} + A_d^t \widehat{P}_{d0} A_d \\ & - A_d^t \widehat{P}_{d0} B_d \left(\widehat{R}_{d0} + B_d^t \widehat{P}_{d0} B_d\right)^{-1} B_d^t \widehat{P}_{d0} A_d. \end{aligned} \tag{5.9}$$

This is done by the computation same as in (5.3).

Then the matrix $\left(\widehat{R}_d + B_d^t \widehat{P}_d B_d\right)$ can be rewritten as:

$$\begin{aligned}\widehat{R}_d + B_d^t \widehat{P}_d B_d &= \widehat{R}_{d0} + \widehat{R}_{d+} + B_d^t \left(\widehat{P}_{d0} + \widehat{P}_{d+}\right) B_d \\ &= \widehat{R}_{d0} + \widehat{R}_{d+} + B_d^t \widehat{P}_{d0} B_d + B_d^t \widehat{P}_{d+} B_d.\end{aligned}$$

Thus, we have

$$\Phi_0 = \widehat{R}_{d0} + B_d^t \widehat{P}_{d0} B_d. \tag{5.10}$$

Now we have

$$\Psi = I - \left(\widehat{R}_d + B_d^t \widehat{P}_d B_d\right) \Phi_0 \;(\text{mod } a_1^{r+1}, \ldots, a_v^{r+1}), \tag{5.11}$$

$$\widehat{P}_d = Q_d + A_d^t \widehat{P}_d A_d - A_d^t \widehat{P}_d B_d \Phi_0 \left(I + \sum_{i=1}^{r} \Psi^r\right) B_d^t \widehat{P}_d A_d \quad (\text{mod } a_1^{r+1}, \ldots, a_v^{r+1}). \tag{5.12}$$

Now we can write the procedure to obtain $\widehat{P}_d$ of (5.4) and $\widehat{F}_d$ of (5.5):

1: Obtain P_{d0} from (5.9).
2: Let Φ_0 be $\widehat{R}_{d0} + B_d^t \widehat{P}_{d0} B_d$.
3: Let $\widehat{P}_d = \widehat{P}_{d0}$.
4: Let $\Psi = I - \left(\widehat{R}_d + B_d^t \widehat{P}_d B_d\right) \Phi_0$ (mod $a_1^{r+1}, \ldots, a_v^{r+1}$).
5: Let $\widehat{P}_{d\text{new}} = Q_d + A_d^t \widehat{P}_d A_d - A_d^t \widehat{P}_d B_d \Phi_0 \left(I + \sum_{i=1}^r \Psi^i\right) B_d^t \widehat{P}_d A_d$ (mod $a_1^{r+1}, \ldots, a_v^{r+1}$).
6: If $\widehat{P}_{d\text{new}}$ is sufficiently equal to $\widehat{P}_d$, then let $\widehat{P}_d = \widehat{P}_{d\text{new}}$ and go to 9.
7: Let $\widehat{P}_d = \widehat{P}_{d\text{new}}$.
8: Go to 4.
9: Let $\widehat{F}_d = \Phi_0 \left(I + \sum_{i=1}^r \Psi^i\right) B_d^t \widehat{P}_d A_d$ (mod $a_1^{r+1}, \ldots, a_v^{r+1}$).
10: Return $\widehat{P}_d$ and $\widehat{F}_d$.

In the following section, we will present an example of the computation of the rth approximate LQ regulator problem.

4 Example

In this section, we present an example of our result.

We employ, as the example, an inverted pendulum on a cart, shown in Fig. 5.1. The following symbols are used:

M: Mass of the cart
r: Location of the cart
u: Voltage to the motor of the cart
a: Gain of the voltage to the force
R: Coefficient of viscosity for the motor
H: Horizontal motion of center of gravity of the pendulum rod
J: Moment of inertia of the rod about its center of gravity
θ: Angle of the rod from the vertical line
V: Vertical motion of center of gravity of the pendulum rod
l: Length of the pendulum rod
c: Coefficient of viscosity for the rod
m: Mass of the rod of the pendulum
g: Gravitational acceleration.

This physical model can be described as follows:

$$M\ddot{r} + R\dot{r} = au - H, \tag{5.13}$$

$$J\ddot{\theta} = Vl\sin(\theta) - Hl\cos(\theta) - c\dot{\theta}, \tag{5.14}$$

$$m\frac{\mathrm{d}^2}{\mathrm{d}t^2}(r + l\sin(\theta)) = H, \tag{5.15}$$

$$m\frac{\mathrm{d}^2}{\mathrm{d}t^2}(l\cos(\theta)) = V - mg. \tag{5.16}$$

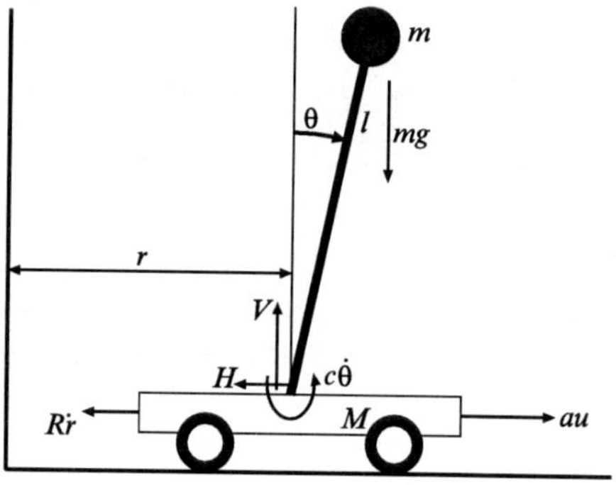

Fig. 5.1 An inverted pendulum on a cart

Deleting V and H from the equations above, we have

$$(M+m)\ddot{r} + ml\cos(\theta)\ddot{\theta} + R\dot{r} - ml\sin(\theta)\dot{\theta}^2 = au, \tag{5.17}$$

$$ml\cos(\theta)\ddot{r} + (J+ml^2)\ddot{\theta} + c\dot{\theta} - mlg\sin(\theta) = 0. \tag{5.18}$$

Now we assume that both θ and $\dot{\theta}$ are nearly equal to 0. Then we have $\sin(\theta) = \theta$, $\cos(\theta) = 1$, $\sin(\theta)\dot{\theta} = 0$. Now the equations can be modified as:

$$(M+m)\ddot{r} + ml\ddot{\theta} + R\dot{r} = au, \tag{5.19}$$

$$ml\ddot{r} + (J+ml^2)\ddot{\theta} + c\dot{\theta} - ml\theta g = 0, \tag{5.20}$$

which can be further modified as:

$$\begin{bmatrix} \ddot{r} \\ \ddot{\theta} \end{bmatrix} = \frac{1}{\alpha_0} \begin{bmatrix} 0 & 0 \\ -m^2l^2g & mlg(M+m) \\ -R(J+ml^2) & mlR \\ mlc & -c(M+m) \\ a(J+ml^2) & -mla \end{bmatrix}^t \begin{bmatrix} r \\ \theta \\ \dot{r} \\ \dot{\theta} \\ u \end{bmatrix}, \tag{5.21}$$

where $\alpha_0 = J(M+m) + mMl^2$.

This can be described as in the state space matrix equation and the output matrix equation as follows:

$$\dot{\mathbf{x}} = A\mathbf{x} + B\mathbf{u},$$

$$\mathbf{y} = C\mathbf{x} + D\mathbf{u},$$

where

$$\mathbf{x} = \begin{bmatrix} r \\ \theta \\ \dot{r} \\ \dot{\theta} \end{bmatrix}, \quad \mathbf{u} = \begin{bmatrix} u \end{bmatrix}, \quad \mathbf{y} = \begin{bmatrix} r \\ \theta \end{bmatrix},$$

$$A = \frac{1}{\alpha_0} \begin{bmatrix} 0 & 0 & \alpha_0 & 0 \\ 0 & 0 & 0 & \alpha_0 \\ 0 & -m^2l^2g & -R(J+ml^2) & mlc \\ 0 & mlg(M+m) & mlR & -c(M+m) \end{bmatrix},$$

$$B = \frac{1}{\alpha_0} \begin{bmatrix} 0 \\ 0 \\ a(J+ml^2) \\ -mla \end{bmatrix},$$

$$C = \begin{bmatrix} 1 & 0 & 0 & 0 \\ 0 & 1 & 0 & 0 \end{bmatrix},$$

$$D = O.$$

We here use the following values:

$$\begin{aligned} M &= 2.52\ [\text{kg}] \\ a &= 88.0\ [\text{N/V}] \\ R &= 88.89\ [\text{kg/s}] \\ l &= 0.115\ [\text{m}] \\ c &= 185.95 \times 10^{-6}\ [\text{kg} \cdot \text{m}^2/\text{s}] \\ m &= 0.1\ [\text{kg}] \\ g &= 9.8\ [\text{m/s}^2]. \end{aligned}$$

Now we modify this continuous-time system to a discrete-time system with the sampling period $T_s = 6.0 \times 10^{-3}$ [s]. This is described as in the state space matrix equation and the output matrix equation as follows:

$$\mathbf{x}(k+1) = A_d \mathbf{x}(k) + B_d \mathbf{u}(k),$$
$$\mathbf{y}(k) = C_d \mathbf{x}(k) + D_d \mathbf{u}(k),$$

where

$$A_d = \begin{bmatrix} 1 & -0.000012155 & 0.0053894 & -4.6894 \times 10^{-9} \\ 0 & 1.0030 & 0.0095869 & 0.0060010 \\ 0 & -0.0039102 & 0.80361 & -5.7031 \times 10^{-6} \\ 0 & 0.98480 & 3.0841 & 1.0013 \end{bmatrix},$$

$$B_d = \begin{bmatrix} 0.00060449 \\ -0.0094909 \\ 0.19442 \\ -3.0532 \end{bmatrix},$$

$$C_d = \begin{bmatrix} 1 & 0 & 0 & 0 \\ 0 & 1 & 0 & 0 \end{bmatrix},$$

$$D_d = \begin{bmatrix} 0 & 0 \end{bmatrix}^t.$$

We consider the following weight matrices:

$$Q_d = \begin{bmatrix} 1+a & 0 & 0 & 0 \\ 0 & 1 & 0 & 0 \\ 0 & 0 & 1 & 0 \\ 0 & 0 & 0 & 1 \end{bmatrix},$$

$$R_d = \begin{bmatrix} 0.01 \end{bmatrix}.$$

Then, from (5.9), P_{d0} can be computed as:

$$P_{d0} = \begin{bmatrix} 246.35 & 187.01 & 97.138 & 6.6726 \\ 187.01 & 976.62 & 268.88 & 19.886 \\ 97.138 & 268.88 & 124.67 & 8.6264 \\ 6.6726 & 19.886 & 8.6264 & 1.6057 \end{bmatrix}.$$

In the following, we consider the case $r = 5$ and present the result of the 5th approximate LQ regulator problem.

Based on Sect. 3, we can compute $\widehat{P}_d$ and $\widehat{F}_d$. Let $\widehat{P}_d = (\hat{p}_{dij})$ $\widehat{F}_d = (\hat{f}_{dij})$. Then we have:

$$\hat{p}_{d11} = 246.35 + 161.42a - 21.877a^2 + 10.612a^3 - 6.5893a^4,$$

$$\hat{p}_{d12} = 187.01 + 96.242a - 22.751a^2 + 11.349a^3 - 7.0914a^4,$$

$$\hat{p}_{d13} = 97.138 + 56.044a - 10.632a^2 + 5.1617a^3 - 3.1967a^4,$$

$$\hat{p}_{d14} = 6.6726 + 3.8126a - 0.73782a^2 + 0.35910a^3 - 0.22257a^4,$$

$$\hat{p}_{d21} = 187.01 + 96.242a - 22.751a^2 + 11.349a^3 - 7.0914a^4,$$

$$\hat{p}_{d22} = 976.62 + 81.458a - 23.816a^2 + 12.146a^3 - 7.6329a^4,$$

$$\hat{p}_{d23} = 268.88 + 42.694a - 11.075a^2 + 5.5198a^3 - 3.4401a^4,$$

$$\hat{p}_{d24} = 19.886 + 2.9327a - 0.76895a^2 + 0.38404a^3 - 0.23953a^4,$$

$$\hat{p}_{d31} = 97.138 + 56.044a - 10.632a^2 + 5.1617a^3 - 3.1967a^4,$$

$$\hat{p}_{d32} = 268.88 + 42.694a - 11.075a^2 + 5.5198a^3 - 3.4401a^4,$$

$$\hat{p}_{d33} = 124.67 + 23.349a - 5.1708a^2 + 2.5108a^3 - 1.5510a^4,$$

$$\hat{p}_{d34} = 8.6264 + 1.5979a - 0.35888a^2 + 0.17467a^3 - 0.10798a^4,$$

$$\hat{p}_{d41} = 6.6726 + 3.8126a - 0.73782a^2 + 0.35910a^3 - 0.22257a^4,$$

$$\hat{p}_{d42} = 19.886 + 2.9327a - 0.76895a^2 + 0.38404a^3 - 0.23953a^4,$$

$$\hat{p}_{d43} = 8.6264 + 1.5979a - 0.35888a^2 + 0.17467a^3 - 0.10798a^4,$$

$$\hat{p}_{d44} = 1.6057 + 0.10939a - 0.024909a^2 + 0.012152a^3 - 0.0075185a^4$$

Fig. 5.2 Error of $\hat{f}_{d11}$

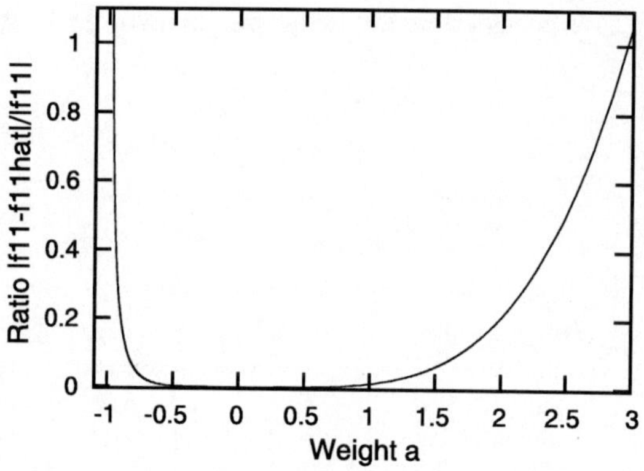

Fig. 5.3 Error of $\hat{f}_{d21}$

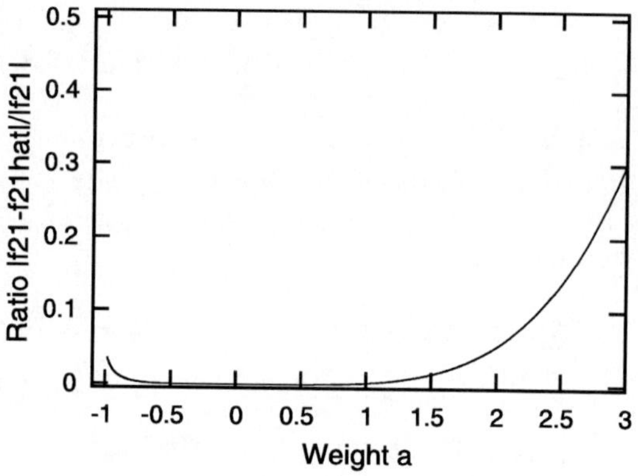

and

$$\hat{f}_{d11} = -0.32118 - 0.160208a + 0.0402236a^2 - 0.0201227a^3 + 0.0125831a^4,$$
$$\hat{f}_{d21} = -2.16539 - 0.141132a + 0.0421333a^2 - 0.0215371a^3 + 0.0135442a^4,$$
$$\hat{f}_{d31} = -1.48484 - 0.0731351a + 0.0195841a^2 - 0.00978713a^3 + 0.00610421a^4,$$
$$\hat{f}_{d41} = -0.363138 - 0.00502926a + 0.00135979a^2 - 0.000680941a^3 + 0.000425018a^4.$$

We compare this result and the classical numerical result. The error e_i of $\hat{f}_{di1}$ for $(i = 1\text{–}4)$ is given as:

$$e_i(a) = \frac{|f_{di1} - \hat{f}_{di1}|}{|f_{di1}|}.$$

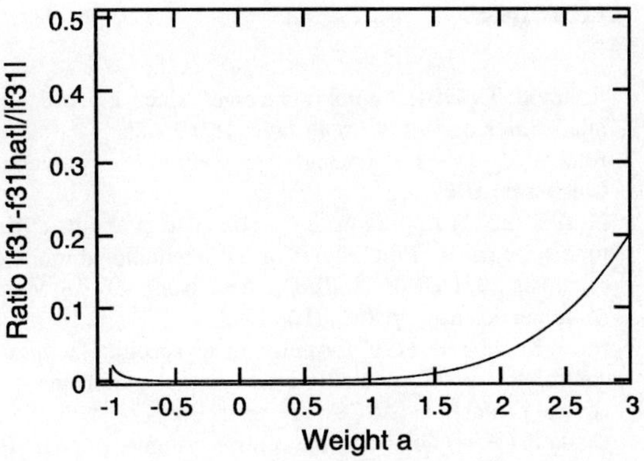

Fig. 5.4 Error of $\hat{f}_{d31}$

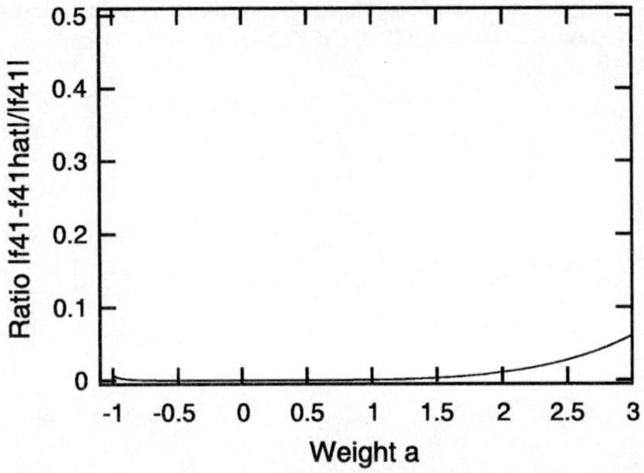

Fig. 5.5 Error of $\hat{f}_{d41}$

Figures 5.2–5.5 show these errors. From these figures, we have obtained that when a is between -0.94 and 2.48, all errors are less than 0.5. This shows that the result of the rth approximate LQ regulator problem gives almost exact result when the parameter is nearly equal to zero.

5 Conclusion

The notion of the rth approximate LQ regulator problem has been introduced, in which the evaluation function can include parameters as symbols. The computation method to solve the rth approximate LQ regulator problem has also been given, in which the evaluation function can include parameters as symbols. Further, a numerical example has been presented, which has shown the effectiveness of the notion of the rth approximate LQ regulator problem.

References

1. Kitamoto T (1994) Approximate eigenvalues, eigenvectors and iverse of a matrix with polynomial entries. Jpn J Ind Appl Math 11(1):73–85
2. Kučera V (1979) Discrete linear control: the polynomial approach. John Wiley & Sons, Chichester, UK
3. Mori K (2011) Approximate riccati equation and its application to optimal control in discrete-time systems. In: Proceedings of the international multiconference of engineers and computer scientists 2011 (IMECS 2011), Hong Kong, 16–18 March. Lecture notes in engineering and computer science, pp 808–812
4. Mori K, Abe K (1997) Approximate spectral factorization and its application to optimal control—case of discrete-time siso with 1 parameter—. In: Proceedings of the second asian control conference (ASCC '97), pp 421–424
5. Ogata K (1994) Discrete-time control systems, 2nd edn. Prentice Hall, Upper Saddle River, NJ
6. Sasaki T, Suzuki M, Kolář M, Sasaki M (1991) Approximate factorization of multivariate polynomials and absolute irreducibility testing. Jpn J Ind Appl Math 8(3):357–375
7. Zhou K, Doyle J, Glover K (1996) Robust and optimal control. Prentice Hall, Upper Saddle River, NJ

Chapter 6
Structured Robust Control for MIMO Systems Using Artificial Intelligent Techniques

Somyot Kaitwanidvilai, Piyapong Olranthichachat, and Natthaphob Nimpitiwan

Abstract In this chapter, a new technique called fixed-structure robust control using artificial intelligent (AI) technique is proposed to enhance both performance and stability of controlled system. In this approach, the structure of controller is specified and the robust stabilization problem is solved by AI techniques. The advantages of simple structure, robustness and high performance can be achieved. In this paper, two multiple inputs multiple outputs (MIMO) controller design problems, i.e., robust control design for electro-hydraulic servo system and fixed-structure robust mixed-sensitivity approach for three phase inverter are illustrated. Simulation results show that the proposed controller has simpler structure than that of conventional robust controller, and the stability margin obtained indicates the robustness of the proposed controller.

Keywords Artificial intelligent techniques • H_∞ optimal control • Three phase inverter • Electro-hydraulic servo system

1 Introduction

Recently, many artificial intelligent (AI) techniques have been adopted to design robust controllers. In [1], the robust H_∞ optimal control problem with a structure specified controller was solved by genetic algorithm (GA). Mixed-sensitivity approach was adopted for indicating the performance of the designed controller. As

S. Kaitwanidvilai (✉) • P. Olranthichachat
Faculty of Engineering, King Mongkut's Institute of Technology Ladkrabang, Bangkok 10520, Thailand
e-mail: drsomyotk@gmail.com; olranthichachat.piyapong@gmail.com

N. Nimpitiwan
Faculty of Engineering, Bangkok University, Thailand
e-mail: natthaphob.n@bu.ac.th

S.I. Ao et al. (eds.), *Intelligent Control and Innovative Computing*, Lecture Notes in Electrical Engineering 110, DOI 10.1007/978-1-4614-1695-1_6,

results indicated, GA is a feasible method to design a structure specified H_∞ optimal controller. Chen et al. [2] proposed a PID design algorithm using mixed H_2/H_∞ control. In their chapter, H_2 is mixed with H_∞ to indicate both the performance and robustness of entire system. In addition, the controller parameters were tuned in the stability domain evaluated by Routh–Hurwitz stability criterion and sampling technique. Similar method was proposed by applying the intelligent GA to solve the robust H_2/H_∞ control problem [3]. Clearly, the results in their papers [1–3] clarified the robustness of the designed systems.

The above mentioned methods [1–3] were adopted to design the structure specified robust controllers. Their techniques are based on the concept of H infinity optimal control. Alternatively, McFarlane and Glover [4] proposed an alternative technique called H_∞ loop shaping control to design a robust controller. This technique is based on the concept of loop shaping which only two compensator weights need to be selected. In this control scheme, several researchers [5–7] proposed several fixed-structure methods to design robust controllers.

In this chapter, the fixed-structure robust control techniques are adopted to design the robust controllers for multiple inputs multiple outputs (MIMO) systems, i.e., a MIMO electro-hydraulic servo system and a three phase inverter system. The remainder of this chapter is organized as follows. Section 2 illustrates the modeling and control design of electro-hydraulic servo system. Particle swarm optimization (PSO) and robust loop shaping techniques are adopted to design the robust controller. Section 3 describes the robust control design for a three phase inverter system. GA and robust mixed-sensitivity method are adopted to design the robust controller. Finally, Sect. 4 summarizes the paper.

2 Fixed-structure Robust Loop Shaping Control for Electro-hydraulic Servo System

Electro-hydraulic servo system is an attractive choice for being used in both industrial and non-industrial applications. The advantages of this actuator are fast dynamic response, high power to inertia ratio, etc. Many approaches have been proposed to control this system to achieve good performance and robustness. One among them is robust control which the controlled system can perform well even under the conditions of disturbance and uncertainties. Normally, in the robust control design problem, several linear mathematic equations need to be solved to find the optimal robust controller. Unfortunately, resulting controllers determined by the conventional techniques are normally complicated with high order.

Typical electro-hydraulic servo system is shown in Fig. 6.1 which consists of a position control system and a force control system [8]. The position control system is adopted to control the actuator movement and the force control system is applied to supply a required force to the system load. The main objective of the servo system is to satisfy the specified requirements; for example, zero steady state errors

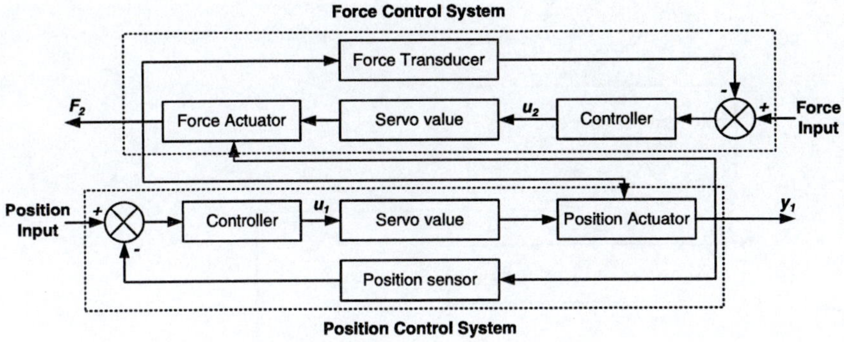

Fig. 6.1 Electro-hydraulic servo system

in motion of the actuator and force output. The state-space model of the electro-hydraulic servo system can be shown in the following equation [8]:

$$\begin{aligned} \dot{x} &= Ax + Bu \\ y &= Cx \end{aligned} \tag{6.1}$$

As seen in the above mentioned dynamic, the system is MIMO system which has two outputs, F_2—force of the system and y_1—position of the actuator, and two inputs (u), u_1—input servo valve of the position control system, and u_2—input servo valve of the force control system.

2.1 PSO Based Robust Loop Shaping Control

Based on the concept of PSO technique, in this problem, a set of controller parameters p is formulated as a particle and the fitness can be written as:

$$\varepsilon = \|T_{zw}\|_\infty^{-1} = \left\| \begin{bmatrix} I \\ W_1^{-1}K(p) \end{bmatrix} \left(I + G_s W_1^{-1} K(p)\right)^{-1} M_s^{-1} \right\|_\infty^{-1} \tag{6.2}$$

Fitness value is specified as a very small value if the controlled system is unstable. The flow chart diagram of the proposed technique is shown in Fig. 6.2.

For the design, the controller $K(p)$ will be designed to minimize the infinity norm from disturbance to state or maximize the stability margin (ε) by using the PSO. The PSO is based on the concept of swarm's movement. When applying the PSO, PSO parameters, i.e. the population of swam(n), lower and upper boundaries ($p_{\min}, p_{\max}$) of problem space, minimum and maximum velocities of particles ($v_{\min}, v_{\max}$), maximum iteration($i_{\max}$), minimum and maximum inertia weights, need to be specified. In an iteration of the PSO, the value of fitness or objective

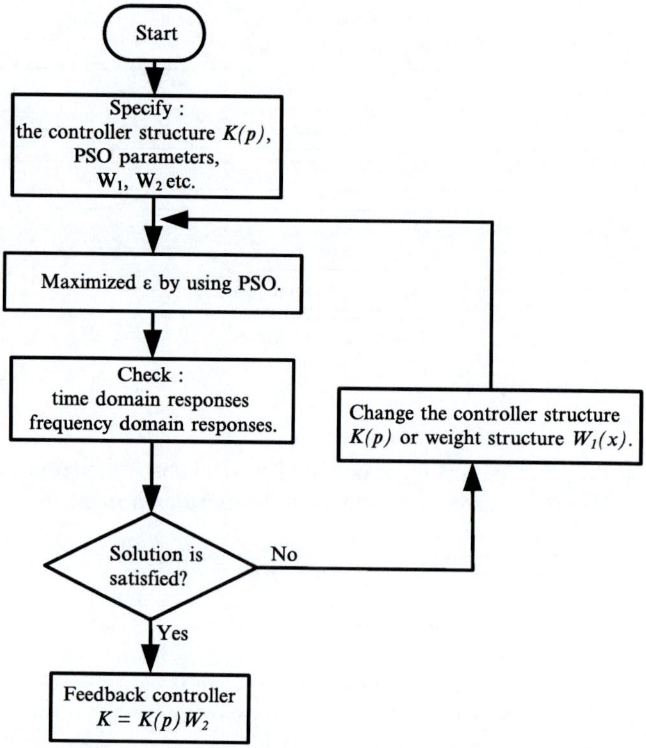

Fig. 6.2 The design of uncertainty weight of DC motor speed control plant

function (*fs*) of each particle is evaluated. The particle which gives the highest fitness value is kept as the answer of current iteration. The inertia weight (Q), value of velocity (v) and position (p) of each particle in the current iteration (i) are updated by using equations (6.3)–(6.5).

$$Q = Q_{\max} - \left(\frac{Q_{\max} - Q_{\min}}{i_{\max}}\right) i \tag{6.3}$$

$$v_{i+1} = Qv_i + \alpha_1[\gamma_{1i}(P_b - p_i)] + \alpha_2[\gamma_{2i}(U_b - p_i)] \tag{6.4}$$

$$p_{i+1} = p_i + v_{i+1} \tag{6.5}$$

where α_1, α_2 are acceleration coefficients.

The state-space model of nominal plant of electro-hydraulic servo system can be seen in the reference [8]. The state vector of this plant consists of four variables which are the supply pressure in force control system, the supply pressure in position control system, the position and the velocity of the actuator. Based on the reference [4], the pre- and post-compensator weights can be chosen as

Table 6.1 PSO parameters and controller parameter ranges

Parameters	Values
Minimum velocity	0
Maximum velocity	0.2
Acceleration coefficient	2.1
Minimum inertia weight	0.6
Maximum inertia weight	0.9
Maximum iteration	100
Population size	500
p_{1-8}	[−60, 60]

$$W_1 = \begin{bmatrix} \dfrac{0.53s + 62}{s + 0.001} & 0 \\ 0 & \dfrac{0.53s + 62}{s + 0.001} \end{bmatrix}, \quad W_2 = \begin{bmatrix} 1 & 0 \\ 0 & 1 \end{bmatrix} \tag{6.6}$$

In this chapter, the structure of controller is selected as a MIMO PI control. By using the conventional robust loop shaping technique, the optimal stability margin of the shaped plant is found to be 0.5522. To design the conventional H_∞ loop shaping controller, stability margin 0.5217 is selected. The final controller (full order H_∞ loop shaping controller) is eighth order.

In this chapter, the structure of controller is selected as:

$$K(p) = \begin{bmatrix} \dfrac{p_1 s + p_2}{s + 0.001} & \dfrac{p_3 s + p_4}{s + 0.001} \\ \dfrac{p_5 s + p_6}{s + 0.001} & \dfrac{p_7 s + p_8}{s + 0.001} \end{bmatrix} \tag{6.7}$$

Next, the proposed technique is adopted to design the robust PI controller. In the optimization problem, the upper and lower bounds of control parameters and the PSO parameters are given in Table 6.1. After running the PSO for 56 iterations, the optimal solution is obtained as:

$$K(p) = \begin{bmatrix} \dfrac{-0.5284s - 36.683}{s + 0.001} & \dfrac{-0.0717s - 8.0878}{s + 0.001} \\ \dfrac{-0.0322s - 3.2293}{s + 0.001} & \dfrac{-0.8966s - 38.441}{s + 0.001} \end{bmatrix} \tag{6.8}$$

Figure 6.3 shows the fitness or stability margin (ε) of the best controller obtained in each iteration. As seen in this figure, the best controller evolved by the PSO has a stability margin of 0.3905.

Figure 6.4 shows the responses of the outputs of the system in two channels (position and force outputs) when the unit step command input is fed into the position command. As seen in this figure, the response from the proposed controller is similar to that of the conventional full order robust controller.

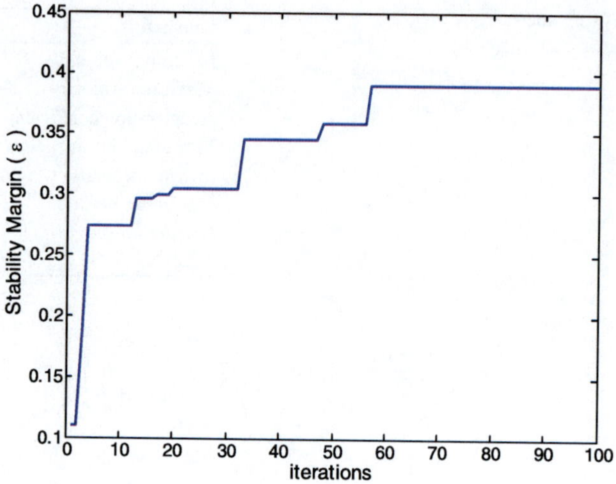

Fig. 6.3 Stability margin (ε) versus iterations

3 Fixed-Structure Robust Control for Three-phase Power Inverter

In this example, a design of grid connected three-phase current control inverters is illustrated. Fixed structured H_∞ optimal control design is applied to design the controller for the inverter. Controllers with robust stability (RS) and nominal performance (NP) are the main concerns of the design. GA is employed to solve for the optimal parameters of controllers. Structure of the controller is a proportional integral (PI) controller with MIMO. A fixed-structure current controller is designed for a grid connected 300 kW three-phase inverter as depicted in Fig. 6.5. Note that the parameters in Fig. 6.5 is represented in *abc* reference frame. The specifications of the inverter are shown in Table 6.2.

3.1 Model of Grid Connected Three-phase Power Inverter

By applying the *0dq-abc* transformation matrix defined by

$$\begin{bmatrix} i_a \\ i_b \\ i_c \end{bmatrix} = \sqrt{\frac{2}{3}} \begin{bmatrix} \frac{1}{\sqrt{2}} & cos\omega t & sin\omega t \\ \frac{1}{\sqrt{2}} & cos\left(\omega t - \frac{2\pi}{3}\right) & sin\left(\omega t - \frac{2\pi}{3}\right) \\ \frac{1}{\sqrt{2}} & cos\left(\omega t + \frac{2\pi}{3}\right) & sin\left(\omega t + \frac{2\pi}{3}\right) \end{bmatrix} \begin{bmatrix} i_0 \\ i_d \\ i_q \end{bmatrix}$$

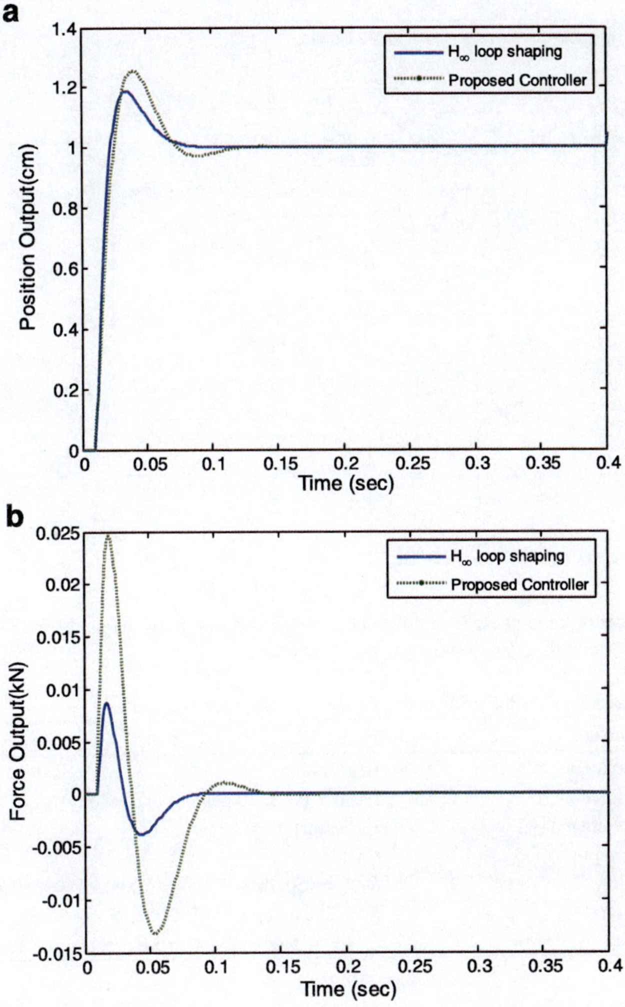

Fig. 6.4 Output responses of the system when the unit step is entered to position command

A three-phase inverter as shown in Fig. 6.5 can be represented by a linearized state space model around "*a normal operating point*" in rotational reference frame as [9–12]

$$\Delta \dot{x}_{Lr} = A_{Lr}\Delta x_{Lr} + B_{Lr}\Delta u_{Lr} \tag{6.9}$$

where

$$\Delta x_{Lr} = \left[\, \Delta I_d \ \Delta I_q \ \Delta I_s \ \Delta V_d \ \Delta v_{cfd} \ \Delta v_{cfq} \ \Delta I_{LD} \ \Delta I_{LQ} \,\right]^T$$

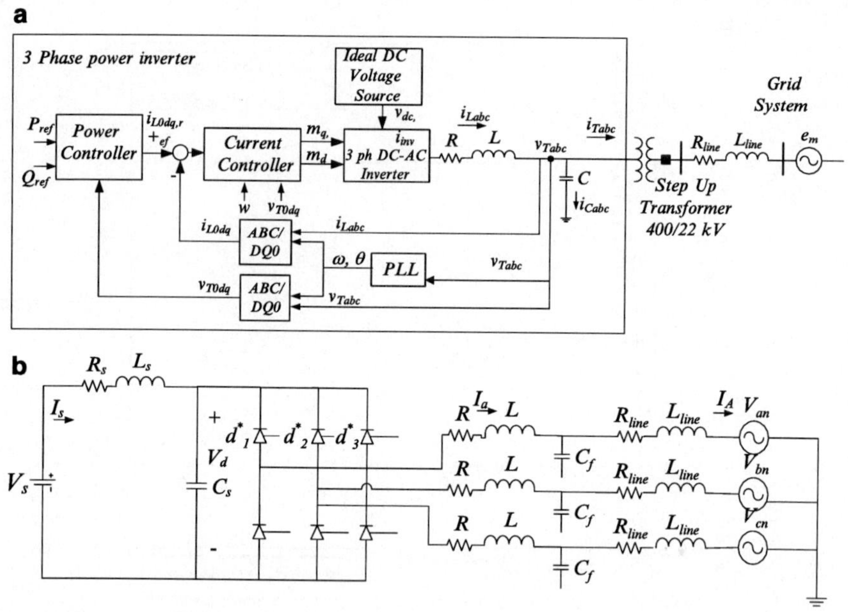

Fig. 6.5 Model of three-phase inverter grid connected inverter in *abc* reference frame: (**a**) System diagram. (**b**) Circuit diagram of the three-phase inverter [9]

Table 6.2 Parameters of the grid connected three-phase power converter

Control parameters	Values
Switching frequency, f	10 kHz
DC voltage source (V_d)	1,100 V (with 20 percent uncertainty)
Rated output voltage (V_T)	380 Vrms, $f = 50$ Hz
Nominal output power (P_L)	300 kW
Grid connection	$R_{Line} = 50\,\text{m}\Omega,\ L_{Line} = 4\,\mu\text{H},\ C_f = 850\,\mu\text{F}$
DC–AC inverter	$R = 200\,\text{m}\Omega,\ L = 300\,\mu\text{H},\ R_s = 3\,\text{m}\Omega,\ L_s = 3\,\mu\text{H},$ $C_s = 5{,}000\,\mu\text{F}$ (with 5.0 percent uncertainty)

and

$$\Delta u_{Lr} = \left[\Delta m_d\ \Delta m_q\right]^T$$

$$A_{Lr} = \begin{bmatrix} -\frac{R}{L} & -\omega & 0 & \frac{\sqrt{6}}{4L}m_d & -\frac{1}{L} & 0 & 0 & 0 \\ \omega & -\frac{R}{L} & 0 & \frac{\sqrt{6}}{4L}m_q & 0 & -\frac{1}{L} & 0 & 0 \\ 0 & 0 & -\frac{R_s}{L_s} & -\frac{1}{L_s} & 0 & 0 & 0 & 0 \\ -\frac{\sqrt{6}}{4C_s}m_d & \frac{\sqrt{6}}{4C_s}m_q & \frac{1}{C_s} & 0 & 0 & 0 & 0 & 0 \\ \frac{1}{C} & 0 & 0 & 0 & 0 & -\omega & -\frac{1}{C} & 0 \\ 0 & \frac{1}{C} & 0 & 0 & \omega & 0 & 0 & -\frac{1}{C} \\ 0 & 0 & 0 & 0 & \frac{1}{L_{line}} & 0 & -\frac{R_{line}}{L_{line}} & -\omega \\ 0 & 0 & 0 & 0 & 0 & \frac{1}{L_{line}} & \omega & -\frac{R_{line}}{L_{line}} \end{bmatrix}$$

$$B_{Lr} = \begin{bmatrix} \frac{\sqrt{6}}{4L}V_d & 0 \\ 0 & \frac{\sqrt{6}}{4L}V_d \\ 0 & 0 \\ -\frac{\sqrt{6}}{4C_s}I_d & \frac{\sqrt{6}}{4C_s}I_q \\ 0 & 0 \\ 0 & 0 \\ 0 & 0 \\ 0 & 0 \end{bmatrix}, \quad \Delta x_r = \begin{bmatrix} \Delta I_d \\ \Delta I_q \\ \Delta I_s \\ \Delta V_d \\ \Delta v_{cfd} \\ \Delta v_{cfq} \\ \Delta I_{LD} \\ \Delta I_{LQ} \end{bmatrix}, \quad \Delta u_r = \begin{bmatrix} \Delta m_d \\ \Delta m_q \end{bmatrix}$$

From the definition of rotational reference frame, real power and reactive power outputs (P^{ref} and Q^{ref}) of inverter based DGs with balanced three-phase loads can be calculated as

$$I_{LD} = \frac{V_{Td}P^{\mathrm{ref}} - V_{Tq}Q^{\mathrm{ref}}}{V_{Td}{}^2 + V_{Tq}{}^2} \quad \text{and} \quad I_{LQ} = \frac{v_q P^{\mathrm{ref}} + v_d Q^{\mathrm{ref}}}{v_d{}^2 + v_q{}^2}$$

In this application, time domain performance (e.g., speed, overshoot) of the controller is evaluated by applying integral squared error (ISE) or the "cheap" linear quadratic regulator (LQR) cost function as [13]

$$J = \sqrt{\int_0^{\infty} \|e(\tau)\|^2 \, d\tau}$$

Where $e(t)$ is a vector of error signals.

The objective function of the robust control optimization problem is

$$\underset{\Theta}{\text{minimize}}\; J(\Theta)$$

subject to

$$\left\| \begin{matrix} W_m T \\ W_s S \end{matrix} \right\|_{\infty} = \max_{\omega} \sqrt{|W_m T|^2 + |W_s S|^2} < 1$$

Where γ is a pre-specified upper bound which defines the rigorousness of control specifications, W_{m} is a multiplicative weight function of a plant, W_{s} is weighted sensitivity function. For the illustrative plant, the multiplicative weight function (W_{m}) requires a steep bound in the form [14]

$$W_m = \left(\frac{1}{\sqrt[k_m]{\varepsilon}} \left[\frac{s + \dfrac{\omega_{bc}}{\sqrt[k_m]{M_u}}}{s + \dfrac{\omega_{bc}}{\sqrt[k_m]{\varepsilon}}} \right] \right)^{k_m} I_{2\times 2}; \tag{6.10}$$

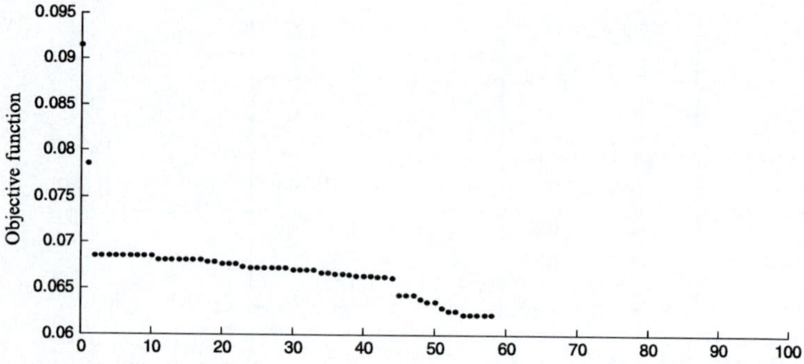

Fig. 6.6 Objective function of the illustrative case

parameters in (6.10) are chosen as follow: $\varepsilon = 0.01$, $k_m = 3$, $M_y = 4.5$, $\omega_{bc} = 2.5\,\text{kHz}$. The weighted sensitivity function (W_S) is chosen as

$$W_S = \left(\frac{1}{\sqrt[k_S]{M_S}}\left[\frac{s + \sqrt[k_S]{M_S}\omega_{bc}}{s + \sqrt[k_S]{\varepsilon}\omega_{bc}}\right]\right)^{k_S} I_{2\times 2}; \qquad (6.11)$$

parameters in (6.11) are chosen as follow: $\varepsilon = 0.01$, $k_s = 1$, $M_S = 10$, $\omega_b = 10\,\text{Hz}$. To avoid difficulty in implementation (i.e., controller with higher order), a PI controller is pre-specified as the configuration of $K(s)$ [1].

$$K(s) = \begin{bmatrix} K_{PD} + \dfrac{K_{ID}}{s} & 0 \\ 0 & K_{PQ} + \dfrac{K_{IQ}}{s} \end{bmatrix}$$

To determine the optimal controller, the proposed technique employs GA to search for the optimal control parameters. The best objective function and the optimal solution from GA optimization are 0.073 with $K_{PD} = 2.22 \times 10^{-4}$, $K_{PQ} = 1.91 \times 10^{-4}$, $K_{ID} = 0.191$ and $K_{ID} = 0.199$. The objective functions from each generation are shown in Fig. 6.6.

The effectiveness of the proposed controller is verified by the time domain responses under step changes of the ideal DC source voltage. The fluctuation of DC voltage imitates operating conditions of renewable resources (e.g., fuel cells, solar cells, wind turbines), which may results from variations of environmental conditions.

In this example, results of a simulation with a reference value of average real power output (P^{ref}) at 300 kW are shown in Fig. 6.7. To demonstrate the stability of the plant, by applying the model of a grid connected three-phase inverter as shown in Fig. 6.5, the ideal DC voltage source (V_{DC}) is perturbed: from 1,100 V (normal value) to 800 V at $t = 0.10$ ms and back to 1,100 V at $t = 0.15$ ms. Note from

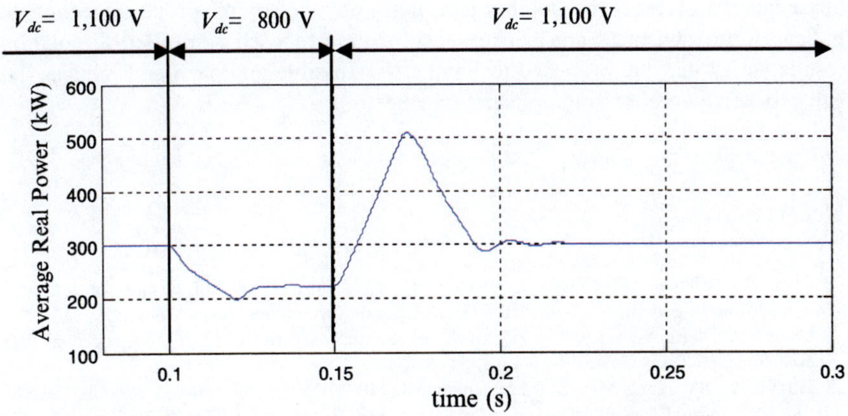

Fig. 6.7 A simulation result of a three phase inverter with $P_{ref} = 300\,\text{kW}$ and V_{DC} perturbations

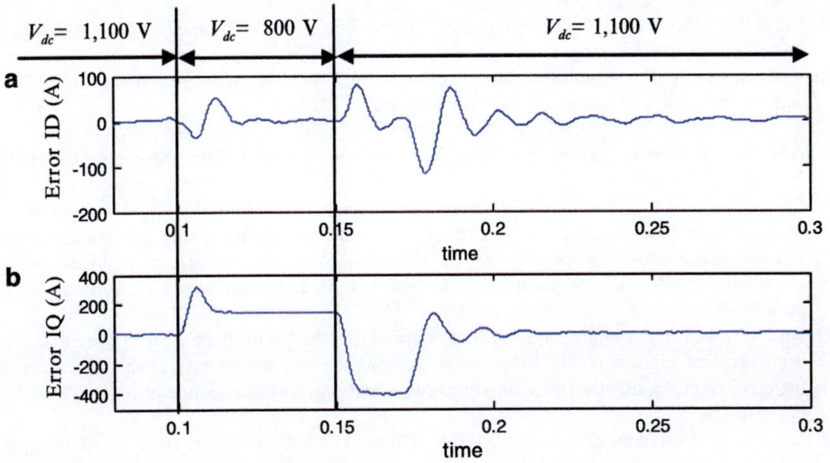

Fig. 6.8 Error of current (**a**) in direct axis (I_{LD}) and (**b**) in quadrature axis (I_{LQ})

Fig. 6.7 that the average output power has a capability to recover from DC voltage drop up to 27 percent within 2–3 cycles. Figure 6.8 shows the error of output current in direct and quadrature axis measure at the terminal of the inverter.

4 Conclusions

This paper proposes several new techniques for designing the fixed structure robust controllers. As shown in the results, the proposed technique can be applied to control the MIMO plants, i.e., the electro-hydraulic servo system and the three phase inverter system. The stability margin (ε) obtained by the proposed technique indicates the robustness and performance of the proposed controller. Simulation results verify that the proposed technique is a feasible approach and reduces the gap between theoretical and practical schemes.

References

1. Chen BS, Cheng YM (1998) A structure-specified optimal control design for practical applications: a genetic approach. IEEE Trans Control Syst Technol 6(6):707–718
2. Chen BS, Cheng Y-M, Lee CH (1995) A genetic approach to mixed H_2/H_∞ optimal PID control. IEEE Trans Control Syst 15(5):51–60
3. Ho SJ, Ho SY, Hung MH, Shu LS, Huang HL (2005) Designing structure-specified mixed H_2/H_∞ optimal controllers using an intelligent genetic algorithm IGA. IEEE Trans Control Syst 13(6):119–124
4. McFarlane D, Glover K (1992) A loop shaping design procedure using H_∞ synthesis. IEEE Trans Automatic Control 37(6):759–769
5. Genc AU (2000) A state-space algorithm for designing H_α loop shaping PID controllers, technical report. Cambridge University Press, Cambridge
6. Patra S, Sen S, Ray G (2008) Design of static H_∞ loop shaping controller in four-block framework using LMI approach. Automatica 44:2214–2220
7. Kaitwanidvilai S, Parnichkun M (2004) Genetic algorithm based fixed-structure robust H infinity loop shaping control of a pneumatic servo system. Int J Robot Mechatronics 16(4): 362–373
8. Olranthichachat P, Kaitwanidvilai S Structure specified robust H infinity loop shaping control of a MIMO electro-hydraulic servo system using particle swarm optimization, lecture notes in engineering and computer science. In: Proceedings of the international multiconference of engineers and computer scientists 2011, IMECS 2011, Hong Kong, 16–18 March 2011, pp. 879–883
9. Nimpitiwan N, Kaitwanidvilai S (2011) Robust control design for three-phase power inverters using genetic algorithm. In: Proceedings of international conference electrical engineering/electronics, computer, telecommunications and information technology (ECTI-CON), pp. 756–759
10. Kroutikova N, Hernandez-Aramburo CA, Green TC (2007) State-space model of grid-connected inverters under current control mode. IET Electr Power Appl 1(3):329–338
11. Abdel-Rahim N, Quaicoe JE (1996) Small-signal model and analysis of a multiple feedback control scheme for three-phase voltage-source UPS inverters. In: Proceedings power electronics specialist conference (PESC), vol. 1.Baveno, Italy, pp. 188–194
12. Wu R, Dewan SB, Slemon GR (1991) Analysis of an AC-to-DC voltage source converter using PWM with phase and amplitude control. IEEE Ind Appl 27(2):355–364
13. Jamshidi M, Krohling RA, Coelho LS, Fleming PJ (2002) Robust control systems with genetic algorithms. Taylor & Francis, New York
14. Rodriguez AA (2005) Analysis and design of multivariable feedback control systems. Control3D, Tempe, AZ: Control3D.

Chapter 7
Fuzzy Robust Trajectory Tracking Control of WMRs

Jafar Keighobadi and Yaser Mohamadi

Abstract Using intelligent robust controllers, perfect trajectory tracking of non-holonomic wheeled mobile robots (WMRs) is developed. A feedback linearizable computed torque controller (FLCTC) may be used to make the convergence of a WMR on preplanned trajectories. Owing to the weak performance against exogenous signals, the FLCTC is replaced by a sliding mode controller (SMC) based on a proportional integration derivative (PID) sliding surfaces. The proposed SMC though results in a robust tracking performance against exogenous inputs as well as model uncertainties; the chattering phenomenon originated from the characteristic of the sign function is unavoidable. As a well-known technique, the sign term could be replaced by a saturation kind which could not remove the high range chattering of tracking errors. Therefore, using the expert knowledge and experiences in the field of WMRs, a Mamdani type fuzzy SMC (FSMC) is designed to perfect trajectory tracking without considerable chattering effects. The asymptotic stability of the proposed control systems is investigated based on the Lyapunov's direct method. Furthermore, the superiority of the proposed pure and Fuzzy SMCs to the recent FLCTC is revealed through software simulation results.

Keywords Sliding mode control • Intelligent control • Robust • Wheeled mobile robot

J. Keighobadi (✉) • Y. Mohamadi
Faculty of Mechanical Engineering, University of Tabriz, 29 Bahman, Tabriz, Iran
e-mail: keighobadi@tabrizu.ac.ir; ymn.urmia@gmail.com

S.I. Ao et al. (eds.), *Intelligent Control and Innovative Computing*, Lecture Notes in Electrical Engineering 110, DOI 10.1007/978-1-4614-1695-1_7,

1 Introduction

Perfect tracking of pre-planned position and orientation trajectories is an important task of nonholonomic WMRs in industrial areas. Hence, a wide-range of trajectory tracking controllers has been proposed in documented literature [1, 2]. However, most of the control systems have been designed for a special kind of WMR whose center of mass coincides to its rotation center between the back driving wheels. In this chapter, to extend the application range of the proposed controllers to the industrial WMRs, the center of mass of the robot is considered to be out of the center of rotation.

Because of the nonlinear dynamics and the nonholonomic constraints on the WMR's kinematics, the controller design is a difficult problem. The kinematic models of non-holonomic systems may be used as the base of feedback controllers. However, the controllers based on dynamic models are preferred owing to including the inertial effects in the control system and using the driving torques as control inputs.

As a well-known nonlinear controller, the FLCTC is developed for perfect tracking control of the WMR's posture variables. However, this method is practical while there are no uncertainties in the WMR dynamics. Therefore, the performance of the controller is highly limited due to persistent excitation condition, uncertainties, and exogenous inputs [2]. Furthermore, even if a nonlinear controller can be designed accurately based on the mathematical model, the controller may be too complicated to be implemented with software [2].

SMCs have been considered recently to improve the performance of nonlinear systems [3–5]. As an enhanced robust method, the SMC containing proportional, integration, and derivative (PID) sliding surfaces provides the system dynamics with an invariance property to the uncertainties [6]. The SMC is robust with respect to the uncertainties; however, the control method produces some drawbacks associated with large control chattering [4, 6]. Most of these drawbacks have been tackled by many researchers through integration with different techniques in recent years [7–10]. To deal with high range chattering, for example, a saturation term is used as an alternative to the sign term of the standard SMC. Although, the chattering may decrease, it cannot be removed due to increasing the tracking errors, on the other hand. Therefore, adaptive methods are considered to increase the tracking performance simultaneous with removing the chattering effects.

Fuzzy logic is used as a strong tool in controlling, estimation, fault detection and classification of nonlinear systems, see [10–12]. In this chapter, using knowledge and experiences of expert engineers, a Mamdani type fuzzy SMC (FSMC) is proposed to make perfect tracking of position and orientation trajectories. Due to using Gaussian membership functions, a simple and robust controller against exogenous inputs and modeling uncertainties is developed. Therefore, the chattering phenomenon decreases and the WMR path tracking performance is improved significantly. Due to the simple structure of the FSMC, it could be implemented in real WMRs.

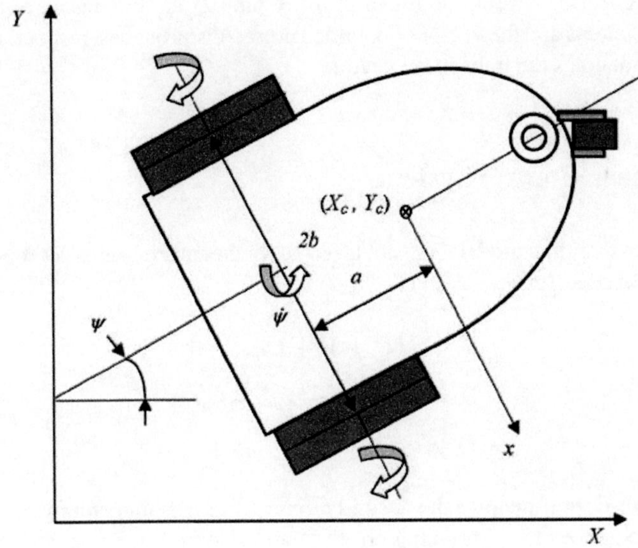

Fig. 7.1 The schematics of WMR [2]

2 WMR Model

The kinematic model of the WMR in Fig. 7.1 shows that its rotation center on the axle of length $2b$ is placed out of the center of mass (X_c, Y_c). In the dynamic modeling, see [6], the orientation angle, ψ the relative coordinates along the body axes, x, y and the fixed coordinates X, Y are used as generalized coordinates. Owing to the nonholonomic constraint, the back wheels of the WMR are fixed parallel to the body and allowed to roll or spin but not slip. The kinematic equations of the WMR are given as follows [6].

$$\begin{pmatrix} \dot{X} \\ \dot{Y} \\ \dot{\psi} \end{pmatrix} = \begin{pmatrix} \sin\psi & -a\cos\psi \\ \cos\psi & -a\sin\psi \\ 0 & 1 \end{pmatrix} \begin{pmatrix} \dot{x} \\ \omega \end{pmatrix} \tag{7.1}$$

where, ω is the rotation rate of the WMR around the normal axis on XY plane. The dynamic model of the WMR is considered as:

$$\begin{aligned} T &= M\ddot{q} + V + C\dot{q} + D \\ q &= [X\ Y]^{\mathrm{t}} \end{aligned} \tag{7.2}$$

where, T is the two-dimensional input torques vector to the independent driving wheels. The complete details of the inertia matrix, M and the vector of centrifugal and coriolis torques, V could be found in [6]. C and D are the matrix of dynamic friction coefficients and the vector of bounded noises/disturbances, respectively. The superscript t denotes the transposed vector.

3 Nonlinear Control Systems

Regarding the dynamic model (7.2) not affected by the uncertainties, D the CTFLC law is considered as follows, see [1].

$$T = MU + V + C\dot{q} \tag{7.3}$$

$$U = [\ddot{X}_d, \ddot{Y}_d]^{\mathrm{t}} - k_d\dot{e} - k_p e$$

$$e(t) = q_d(t) - q(t) \tag{7.4}$$

where, the subscript d denotes the desired reference values of posture variables q.

The convergence rate of tracking errors, e and $\dot{e}$ to zero may be adjusted with respect to the control gains, k_d and k_p.

The FLCTC systems are sensitive to the disturbances and the modeling uncertainties which are bounded and can be simply considered in (7.2) as:

$$T = (M + \Delta M)\ddot{q} + V + \Delta V + (C + \Delta C)\dot{q} + D \tag{7.5}$$

Therefore, if the uncertainty terms, ΔM, ΔV and ΔC together with the exogenous inputs, D could be assumed negligible values the tracking error dynamics is linearized by applying the control torques (7.3) to (7.5) as:

$$\ddot{e} + k_d\dot{e} + k_p e = 0 \tag{7.6}$$

Therefore, the tracking position error vector, e will exponentially converge to zero considering the control gains as:

$$k_d \geq 2k_p > 0 \tag{7.7}$$

However, the uncertainties of (7.5) affect the tracking error dynamics as:

$$M(\ddot{e} + k_d\dot{e} + k_p e) = \Delta M\ddot{q}_d + \Delta V + \Delta C\dot{q} + D \tag{7.8}$$

Therefore, by applying the CTFLC (7.3) to the WMR, the convergence of tracking errors to zero could not be guaranteed. Hence, the robust SMCs are proposed to make perfect trajectory tracking in the presence of uncertainties and noises. The SMC is designed based on the following PID sliding surface, $s(t)$.

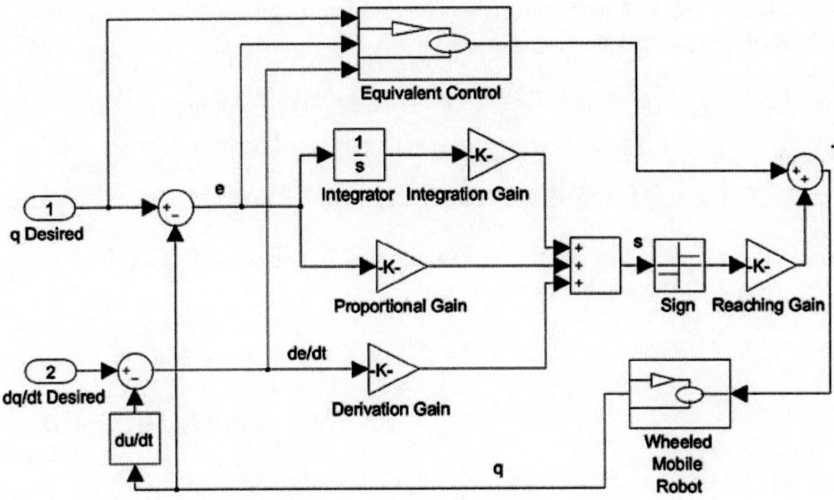

Fig. 7.2 Block diagram of SMC

$$s(t) = K_p e(t) + K_i \int e(t)dt + K_d \dot{e}(t) \tag{7.9}$$

where, K_p, K_d and K_i are the PID positive gains. In this chapter, the SMC is considered in the following form, see [6].

$$T = T_{eq} + K_r sign(s(t)) \tag{7.10}$$

where, the equivalent control term, T_{eq} is obtained using the nominal dynamic model of the WMR which is free of uncertainties. However, the second term as a reaching control torque leads to an acceptable tracking performance of the SMC in the presence of bounded modeling uncertainties and exogenous inputs. Using the SMC, see Fig. 7.2, the convergence of the WMR tracking error, $e(t)$ to zero is obtained through holding it on the stable sliding surface of (7.9). In this way, the tracking error can be moved along the sliding surface to the origin. To guarantee this, the time derivative of the following positive function should be a negative value [6].

$$\mathrm{v}(t) = \frac{1}{2} s^{\mathrm{t}}(t) s(t), \quad \mathrm{v}(0) = 0 \tag{7.11}$$

Replacing $\dot{s}$ from (7.9) in the time derivative of (7.11) gives:

$$\dot{\mathrm{v}}(t) = s^{\mathrm{t}} \left(K_p \dot{e} + K_i e + K_d \ddot{e} \right) \tag{7.12}$$

Regarding $M + \Delta M$ as a nonsingular inertial matrix affected by bounded uncertainties, $\ddot{q}$ is represented using (7.5) as:

$$\begin{aligned}\ddot{q} &= (M + \Delta M)^{-1}T - (M + \Delta M)^{-1}(V + \Delta V)\dot{q} \\ &\quad -(M + \Delta M)^{-1}[(V + \Delta V) + (C + \Delta C)^{-1}\dot{q} + D] \\ &= (M + \Delta M_1)T - (V + \Delta V_1) - (C + \Delta C_1)\dot{q} - D_1\end{aligned} \tag{7.13}$$

where, the subscript 1 stands for the new versions of corresponding uncertainties in (7.5). Now, (7.12) could be rewritten as:

$$\begin{aligned}\dot{v} = s^t[K_p\dot{e} + K_i e + K_d(\ddot{q}_d - \ddot{q})] = s^t[K_p\dot{e} &+ K_i e + K_d\ddot{q}_d + K_d V \\ &+ K_d\Delta V_1 + K_d C\dot{q} + K_d\Delta C_1)\dot{q}] \\ &- s^t K_d(M + \Delta M_1)(T_{eq} + T_{slid} + D_1)\end{aligned} \tag{7.14}$$

Replacing $\ddot{q}$ (7.13) in the time derivative of (7.9) results as:

$$\dot{s}(t) = K_p\dot{e} + K_i e + K_d[\ddot{q}_d + (C + \Delta C_1)\dot{q} + V + \Delta V_1 + D_1 - (M + \Delta M_1)T] \tag{7.15}$$

The equivalent control torque is obtained through the converging solution of $\dot{s}$ for the nominal model in which D_1 and all Δ terms are zero, see [6]:

$$T_{eq} = M^{-1}\left[K_d^{-1}K_p\dot{e} + K_d^{-1}K_i e + \ddot{q}_d + C\dot{q} + V\right] \tag{7.16}$$

Replacing (7.16) in (7.14) gives:

$$\begin{aligned}\dot{v} = s^t[K_p\dot{e} + K_i e + K_d(\ddot{q}_d + V + \Delta V_1 + C\dot{q} + \Delta C_1\dot{q})] - s^t K_d(M + \Delta M_1)(T_{slid} \\ + D_1) - s^t K_d(M + \Delta M_1)(K_d M)^{-1}[K_p\dot{e} + K_i e + K_d\ddot{q}_d + K_d C\dot{q} + K_d V]\end{aligned} \tag{7.17}$$

Through simplifying, see [6], (7.17) changes to:

$$\begin{aligned}s^t\dot{s} = s^t\{K_d[\Delta C_1 - \Delta M_1 M^{-1}C]\dot{q} + K_d(\Delta V_1 - \Delta M_1 M^{-1}V) - K_d(M \\ + \Delta M_1)D_1 - K_d\ddot{q}_d - \Delta M_1 M^{-1}(K_p\dot{e} + K_i e)\} - s^t K_d(M + \Delta M_1)T_{slid}\end{aligned} \tag{7.18}$$

Through using the two-norm operator, (7.18) results as follows.

$$\begin{aligned}s^t\dot{s} \le \|s^t\| \, [K_d(\|\Delta C_1\| + \|\Delta M_1 M^{-1}C\|)\,\|\dot{q}\|] + K_d(\|\Delta V_1\| \\ + \|(M + \Delta M_1)D_1\| + \|\ddot{q}_d\| + \|\Delta M_1 M^{-1}\|\,(K_p\,\|\dot{e}\| + K_i\,\|e\|) \\ + \|\Delta M_1\, M^{-1}\, V\|) - s^t K_d(M + \Delta M_1)T_{slid}\end{aligned} \tag{7.19}$$

Considering the global stability condition of the SMC (7.10), the following switching control term is obtained.

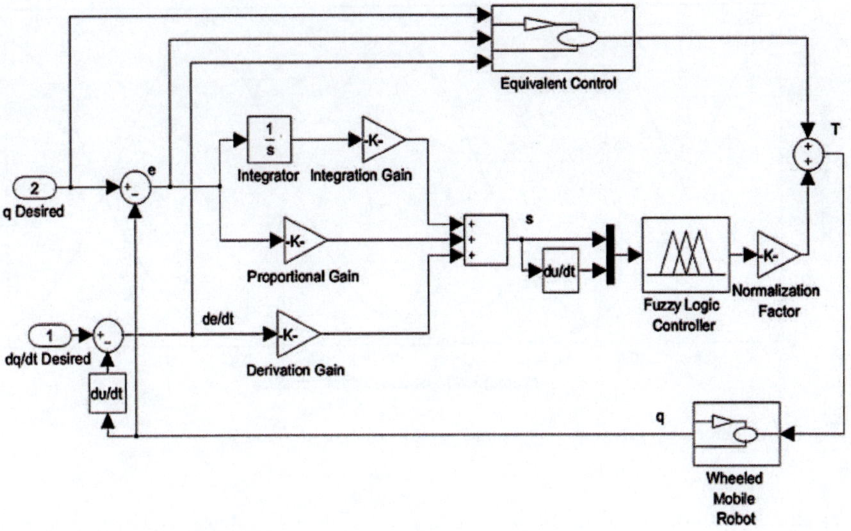

Fig. 7.3 Block diagram of FSMC

$$T_{sl} = \frac{\rho_1}{K_d(M + \Delta M_1)} \text{sign}(s) = \rho \text{sign}(s) \tag{7.20}$$

The performance of the SMC (7.10) strictly depends on the reaching control gain, ρ which should be designed with respect to the upper bound of the uncertainties. The higher the upper bound of the uncertainties, the higher switching gain should be considered. However, a large switching gain increases the chattering range while a small one cannot satisfy the stability condition. Therefore, to accelerate the reaching phase and to reduce the chattering during maintaining the sliding behavior, the switching term of the SMC is replaced by an intelligent fuzzy reaching term.

4 Fuzzy Control System

In fuzzy systems, a number of *if then* rules are used to describe the quantitative relationship between the system's input and output variables in linguistic terms. Owing to better computational efficiency, a Mamdani type knowledge-base is constructed to determine the fuzzy reaching term, T_r as:

$$T_r(t) = -K_f T_f(t) \tag{7.21}$$

where, K_f is the normalizing factor to the fuzzy system output, T_f. As shown in Fig. 7.3, with respect to the inputs, s and $\dot{s}$, the output variable of the FSMC,

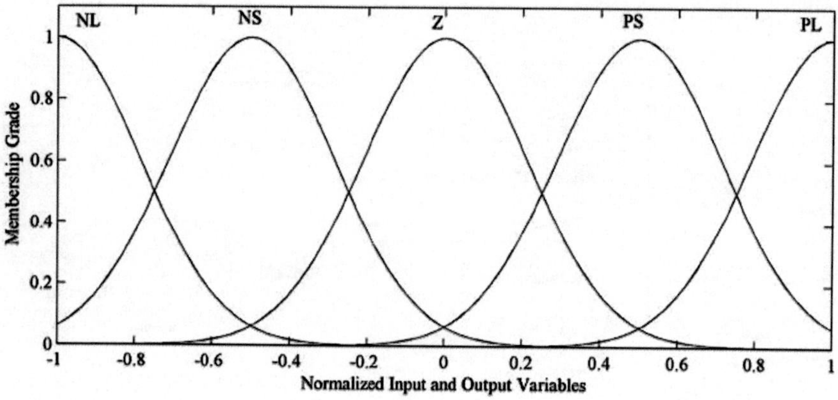

Fig. 7.4 Membership function of input and output variables s, $\dot{s}$ and T_f

T_f should be computed through the designed *if then* rule-base. According to the expertise in design of SMCs, a large switching gain will force the state trajectories to approach the sliding surface rapidly; but at the same time, the chattering is excited. By increasing the tracking errors the value of s will increase; therefore, the switching gain should be correspondingly increased and vice versa. Furthermore, when the sign of $s^t\dot{s}$ changes to positive, the state trajectories deviate from the sliding surface; and therefore, by increasing the magnitude of, $\dot{s}$ the switching gain should be increased to force the trajectories back, and vice versa. On the other hand, when the state trajectories are approaching the sliding surface; if the magnitude of $\dot{s}$ is a large value, the switching gain should be decreased in order to reduce the chattering, and vice versa. The FSMC is completed as:

$$T_t(t) = T_{eq}(t) + T_r(t) \tag{7.22}$$

The membership functions corresponding to the normalized input and output fuzzy sets s, $\dot{s}$ and T_f are presented in Fig. 7.4. In this figure, P, N, Z, S and L stand for *positive*, *negative*, *zero*, *small* and *large*, respectively and thereby for example, NL denotes the *negative-small* fuzzy set.

Therefore, the following 25 *if then* rules are generated using the expert engineering knowledge and experiences in the field of WMRs.

If s is NL and $\dot{s}$ is NL, then T_f is PL.
If s is NL and $\dot{s}$ is NS, then T_f is PL.
If s is NL and $\dot{s}$ is Z, then T_f is PS.
If s is NL and $\dot{s}$ is PS, then T_f is PS.
If s is NL and $\dot{s}$ is PL, then T_f is Z.
If s is NS and $\dot{s}$ is NL, then T_f is PL.
If s is NS and $\dot{s}$ is NS, then T_f is Z.
If s is NS and $\dot{s}$ is Z, then T_f is PS.

If s is NS and $\dot{s}$ is PS, then T_f is Z.
If s is NS and $\dot{s}$ is PL, then T_f is NS.
If s is Z and $\dot{s}$ is NL, then T_f is PS.
If s is Z and $\dot{s}$ is NS, then T_f is PS.
If s is Z and $\dot{s}$ is Z, then T_f is Z.
If s is Z and $\dot{s}$ is PS, then T_f is NS.
If s is Z and $\dot{s}$ is PL, then T_f is NS.
If s is PS and $\dot{s}$ is NL, then T_f is PS.
If s is PS and $\dot{s}$ is NS, then T_f is Z.
If s is PS and $\dot{s}$ is Z, then T_f is NS.
If s is PS and $\dot{s}$ is PS, then T_f is NS.
If s is PS and $\dot{s}$ is PL, then T_f is NL.
If s is PS and $\dot{s}$ is NL, then T_f is Z.
If s is PL and $\dot{s}$ is NS, then T_f is NS.
If s is PL and $\dot{s}$ is Z, then T_f is NS.
If s is PL and $\dot{s}$ is PS, then T_f is NL.
If s is PL and $\dot{s}$ is PL, then T_f is NL.

The fuzzy reaching control torque is computed using the centroid defuzzifier, the well-known Mamdani's minimum inference engine, and the min and max for T- and S-norm operators, respectively, see [2].

The global asymptotic stability of the FSMC is examined using the Lyapunov function given by (7.11). To obtain the conditions of a negative time-derivative of (7.11), first the FSMC (7.22) is replaced in (7.15) as:

$$\dot{\mathrm{v}} = s^{\mathrm{t}}\{K_p\dot{e}+K_i e+K_d[\ddot{q}_d+C\dot{q}+\Delta C_1\dot{q}+V+\Delta V_1+D_1-(M+\Delta M_1)(T_{eq}+T_r)]\} \tag{7.23}$$

Using (7.16), (7.23) is simplified to:

$$\dot{\mathrm{v}} = s^{\mathrm{t}}K_d[-MT_r - \Delta M_1(T_{eq}+T_r) + \Delta C_1\dot{q} + \Delta V_1 + D_1] \tag{7.24}$$

Considering the normalized outputs of the FSMC to $[-1,\ 1]$, (7.24) results in:

$$\begin{aligned} &\dot{\mathrm{v}} \le K_d[-(M+\Delta M)K_f\,\|s\| + \|d(t)\|\,\|s\|] \\ &d(t) = -\Delta M_1 T_{eq} + \Delta C_1\dot{q} + \Delta V_1 + D_1 \end{aligned} \tag{7.25}$$

Now, the fuzzy normalizing gain could be determined as follows.

$$K_f \ge (M+\Delta M)^{-1}\,\|d(t)\| \tag{7.26}$$

Therefore, the following reaching condition is satisfied and the asymptotic stability of the FSMC is guaranteed

$$\dot{\mathrm{v}} = s^{\mathrm{t}}\dot{s}(t) \le 0 \tag{7.27}$$

5 Simulation Results

The effects of the proposed controllers on the convergence of WMR to reference position and orientation trajectories are evaluated through software simulations. A circular sample trajectory is used to produce the desired reference position and heading angle of the WMR in Cartesian coordinates, see [6]. Figures 7.5–7.7 show the performance of the designed controllers in the absence of uncertainties and exogenous inputs. As shown in Fig. 7.5, using all of the controllers, the WMR has tracked the reference circular trajectory in a significant accuracy. It can be seen from Figs. 7.6 and 7.7 that the FLCTC shows a better tracking performance for both position and orientation trajectories compared to those of the SMCs. According to the Fig. 7.7(b), compared with the SMCs, the FLCTC requests larger torques. Due to the natural conservativeness and intelligence of fuzzy systems, the FSMC results small tracking errors compared with that of the SMC.

The importance of the SMCs is understood during affecting the WMR by white Gaussian noises as exogenous inputs. Under these noises, Figs. 7.8–7.11 show the

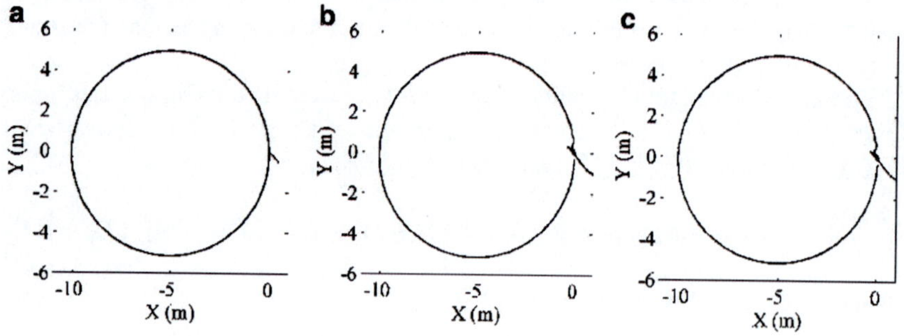

Fig. 7.5 Tracked paths by WMR using: (**a**) FLCTC; (**b**) SMC; (**c**) FSMC

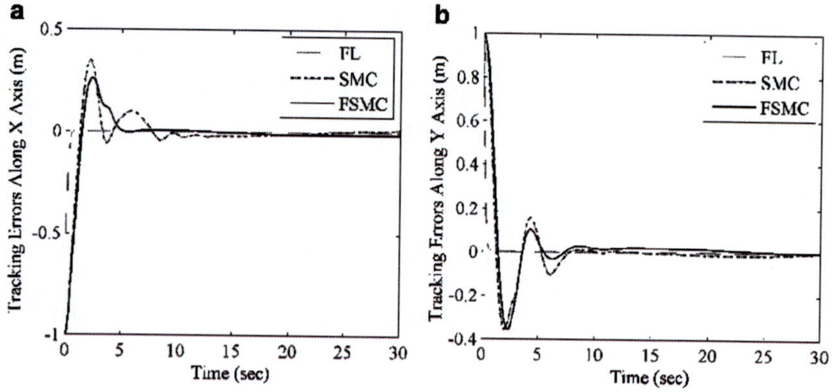

Fig. 7.6 Position trajectory tracking errors of: (**a**) X; (**b**) Y

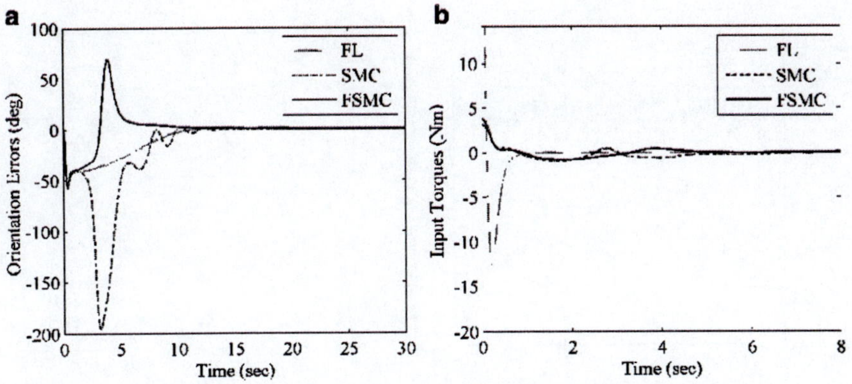

Fig. 7.7 (**a**) Input torques to the right driving wheel; (**b**) orientation errors

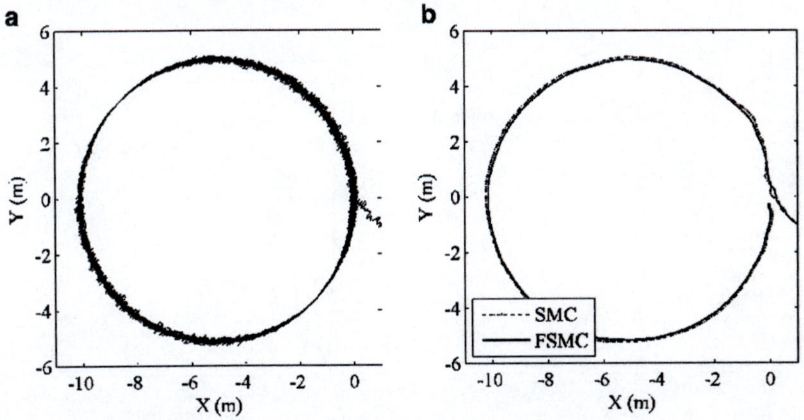

Fig. 7.8 Tracked paths in the presence of uncertainties: (**a**) FLCTC; (**b**) SMCs

simulated tracking performance of the WMR using the proposed controllers. As shown in Fig. 7.8, unlike the FLCTC which is highly sensitive to the measurement noises, the SMCs show a relatively high robustness against the uncertainties and the exogenous inputs. As shown in Figs. 7.9 and 7.10, a higher tracking accuracy along, X and ψ trajectories is resulted using both the FSMC and the SMC compared with that of the FLCTC. Furthermore, due to removing the chattering phenomenon, the FSMC results in a more reliable and accurate trajectory tracking performance compared to the SMC. Figure 7.11 reveals that unlike the FLCTC, the SMCs request small and smooth input torques to the WMR's driving wheels.

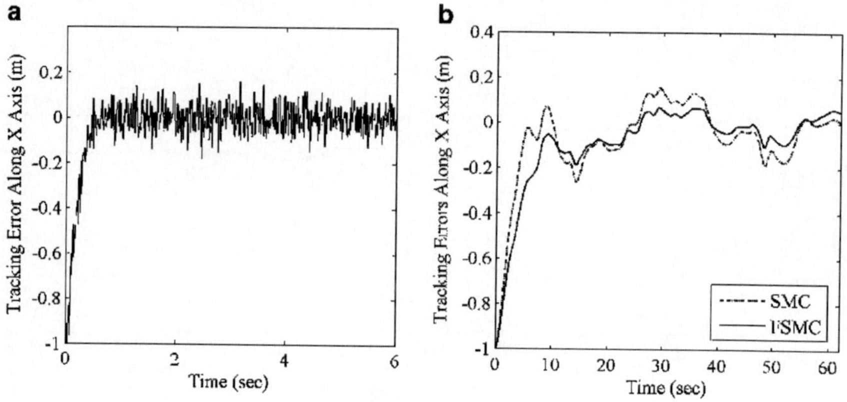

Fig. 7.9 Tracking errors of X with respect to desired values: (**a**) FLCTC; (**b**) SMCs

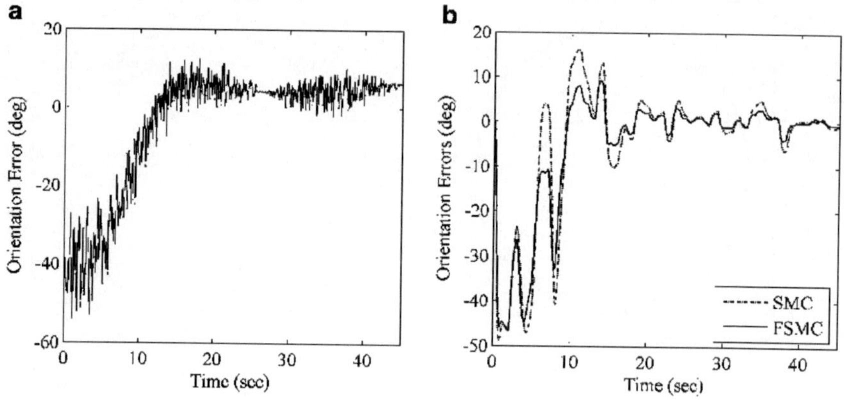

Fig. 7.10 Orientation tracking errors with respect to desired values: (**a**) FLCTC; (**b**) SMCs

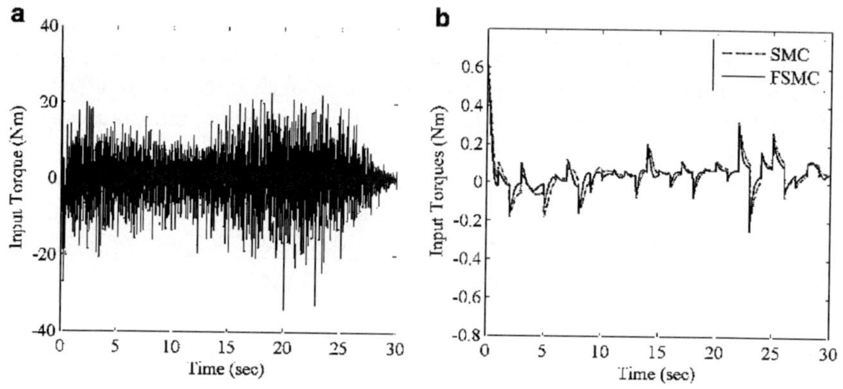

Fig. 7.11 Input torques to the right driving wheel: (**a**) FLCTC; (**b**) SMCs

6 Conclusion and Future Work

In this chapter, three classic and intelligent nonlinear control systems have been designed to perform perfect trajectory tracking of a general kind of nonholonomic WMRs in which the center of mass is out of the center of rotation. Applying the computed input torques by the FLCTC to the driving wheels resulted in a superior tracking performance of the WMR on position and orientation trajectories. However, during affecting exogenous inputs and modeling uncertainties, the FLCTC showed a weak tracking performance associated with high frequency chattering on the tracking errors. By use of the proposed SMC based on the PID sliding surface, the tracking performance of the WMR has been improved compared with that obtained using the FLCTC. The Mamdani type FSMC has been proposed to remove the chattering effects of the sign function and also to further improving the tracking performance against the uncertainties and the exogenous inputs. The asymptotic global stability of the SMC and the FSMC has been examined using direct Lyapunov's method. Unlike a pure fuzzy logic controller which encounters rule expanding problem, see for example [1], the FSMC requires 25 *if-then* rules in the fuzzy knowledge-base. Therefore, it is more suitable for implementations in real WMR systems. For further improvement, a new robust fuzzy control system that directly generates the input torques signals as the outputs of the fuzzy rule-base will be developed in future.

References

1. Keighobadi J, Menhaj MB, Kabganian M (2010) Feedback linearization and fuzzy controllers for trajectory tracking of wheeled mobile robots. Kybernetes 39:83–106
2. Keighobadi J, Menhaj MB (2011) From nonlinear to fuzzy approaches in trajectory tracking control of wheeled mobile robots. Asian J Control DOI: 10.1002/asjc.480
3. Kobayashi T, Tsuda S (2011) Sliding mode control of space robot for unknown target capturing. Eng Lett 19(2):105–112
4. Slotine JJE, Li W (1991) Applied nonlinear control. Prentice-Hall, Englewood Cliffs
5. Yang JM, Kim JH (1999) Sliding mode control for trajectory tracking of nonholonomic wheeled mobile robots. IEEE Trans Robot Automat 15(3):578–587
6. Keighobadi J, Mohamadi Y (2011) Fuzzy sliding mode control of a non-holonomic wheeled mobile robot. In: Lecture notes in engineers and computer science: proceedings of the international multiconference of engineers and computer scientists 2011, IMECS 2011, Hong Kong, 16–18 March 2011
7. Tsai CC, Cheng MB, Lin SC (2007) Robust tracking control for a wheeled mobile robot with dual arms using hybrid sliding-mode neural network. Asian J Control 9(4):377–389
8. Yau HT, Kuo CL, Yan JJ (2006) Fuzzy sliding mode control for a class of chaos synchronization with uncertainties. Int J Nonlinear Sci Numerical Simul 7(3):333–338
9. Li TS, Chang S, Tong W (2004) Fuzzy target tracking control of autonomous mobile robots by using infrared sensors. IEEE Trans Fuzzy Syst 12(4):491–501

10. Li THS, Chang SJ (2003) Autonomous fuzzy parking control of a car-like mobile robot. IEEE Trans Syst Man Cybernet Part A: Syst Hum 33(4):451–465
11. Zhang N, Achaibou K, Mora-Camino F (2010) Fault detection using difference flatness and fuzzy logic. Eng Lett 18(2):199–205
12. Aljoumaa H, Soffker D (2011) Multi-class classification approach based on fuzzy-filtering for condition monitoring. Int J Comput Sci 38(1):66–73

Chapter 8
Balanced Performance/Robustness PID Design

Orlando Arrieta, Ramon Vilanova, and Víctor M. Alfaro

Abstract The design of the closed-loop control system must take into account the system performance to load-disturbance and set-point changes and its robustness to variation of the controlled process characteristics, preserving the well-known trade-off among all these variables. This work faces with the combined servo/regulation performance and robustness problem, in order to get an *intermediate* solution between the robustness increase and the consequent loss in the optimality degree of the performance. The proposed balanced Proportional-Integrative-Derivative (PID) control design is tested against other tuning methods.

Keywords PID control • Process control • Robustness/Performance balance

1 Introduction

Since their introduction in 1940 [11, 12] commercial *Proportional-Integrative-Derivative* (PID) controllers have been with no doubt the most extensive option that

O. Arrieta (✉)
Departament de Telecomunicació i d'Enginyeria de Sistemes, Escola d'Enginyeria, Universitat Autònoma de Barcelona, 08193 Bellaterra, Barcelona, Spain

Departamento de Automática, Escuela de Ingeniería Eléctrica, Universidad de Costa Rica, 11501-2060 San José, Costa Rica
e-mail: Orlando.Arrieta@uab.cat; Orlando.Arrieta@ucr.ac.cr

R. Vilanova
Departament de Telecomunicació i d'Enginyeria de Sistemes, Escola d'Enginyeria, Universitat Autònoma de Barcelona, 08193 Bellaterra, Barcelona, Spain
e-mail: Ramon.Vilanova@uab.cat

V.M. Alfaro
Departamento de Automática, Escuela de Ingeniería Eléctrica, Universidad de Costa Rica, 11501-2060 San José, Costa Rica
e-mail: Victor.Alfaro@ucr.ac.cr

S.I. Ao et al. (eds.), *Intelligent Control and Innovative Computing*, Lecture Notes in Electrical Engineering 110, DOI 10.1007/978-1-4614-1695-1_8,

can be found in industrial control applications [8]. Their success is mainly due to its simple structure and to the physical meaning of the corresponding three parameters (therefore making manual tuning possible). This fact makes PID control easier to understand by the control engineers than other most advanced control techniques. In addition, the PID controller provides satisfactory performance in a wide range of practical situations.

With regard to the design and tuning of PID controllers, there are many methods that can be found in the literature over the last sixty years. Special attention is made on the *IFAC workshop PID'00—Past, Present and Future of PID Control*, held in Terrassa, Spain, in April 2000, where a glimpse of the state-of-the-art in PID control was provided. Moreover, because of the widespread use of PID controllers, it is interesting to have simple but efficient methods for tuning the controller.

In fact, since the initial work of Ziegler and Nichols [31], an intensive research has been done, developing autotuning methods to determine the PID controller parameters [9, 19, 25]. It can be seen that most of them are concerned with feedback controllers which are tuned either with a view to the rejection of disturbances [14, 20] or for a well-damped fast response to a step change in the controller set-point [21, 23, 24].

Moreover, in some cases the methods considered only the system performance [17], or its robustness [5, 15, 16]. However, the most interesting cases are the ones that combine performance and robustness, because they face with all system's requirements [17, 18, 27–29]. O'Dwyer [22], presents a complete collection of tuning rules for PID controllers, which show their abundance.

Taking into account that in industrial process control applications, it is required a good load-disturbance rejection (usually known as *regulatory-control*), as well as, a good transient response to set-point changes (known as *servo-control* operation), the controller design should consider both possibilities of operation.

Despite the above, the servo and regulation demands cannot be simultaneously satisfied with a One-Degree-of-Freedom (1-DoF) controller, because the resulting dynamic for each operation mode is different and it is possible to choose just one for an optimal solution.

Considering the previous statement, the studies have focused only in fulfilling one of the two requirements, providing tuning methods that are optimal to servo-control or to regulation-control. However, it is well-known that if we optimize the closed-loop transfer function for a step-response specification, the performance with respect to load-disturbance attenuation can be very poor and vice-versa [2]. Therefore, it is desirable to get a compromise design, between servo/regulation, by using 1-DoF controller.

The proposed method consider 1-DoF PID controllers as an alternative when *explicit* Two-Degree-of-Freedom (2-DoF) PID controllers are not available. Therefore, it could be stated that the proposed tuning can be used when both operation modes may happen and it could be seen as an *implicit* 2-DoF approach (because the design takes into account both objectives, servo, and regulation modes) [4].

Moreover, it is important that every control system provides a certain degree of robustness, in order to preserve the closed-loop dynamics, to possible variations in the process. Therefore, the robustness issue should be included within the multiple *trade-offs* presented in the control design and it must be solved on a balanced way [3].

With respect to the robustness issue, during the last years, there has been a perspective change of how to include the robustness considerations. In this sense, there is a variation from the classical Gain and Phase Margin measures to a single and more general quantification of robustness, such as the Maximum of the Sensitivity function magnitude.

Taking also into account the importance of the explicit inclusion of robustness into the design, the aim is to select a certain degradation for the performance, that at the same time gives the largest robustness increase, measured in terms of the Maximum Sensitivity function.

The existing methods, cited previously, study the performance and robustness jointly in the control design. However, no one treats specifically the performance/ robustness *trade-off* problem, nor consider in the formulation the servo/regulation *trade-off* or the interacting between all of these variables. On this respect, we attempt to generate a tuning rule (in fact a simple tuning rule) that provides a balance between the degree of optimality and the robustness increase. Therefore, as a main contribution, it is presented a balanced performance/robustness PID tuning for the best *trade-off*, between the robustness increase and the consequent loss in the optimality degree of the performance.

The rest of the chapter is organized as follows. Next section introduces the control system configuration, the general framework, as well as some related concepts and methods. The problem design is stated and faced in Sect. 3, where there are defined the evaluation indexes and it is presented the proposed balanced PID tuning. Some examples are provided in Sect. 4 and the chapter ends in Sect. 5 with some conclusions.

2 Materials and Methods

2.1 Control System Configuration

We consider the feedback control system shown in Fig. 8.1, where $P(s)$ is the controlled process, $C(s)$ is the controller, $r(s)$ is the set-point, $u(s)$ is the controller output signal, $d(s)$ is the load-disturbance and $y(s)$ is the system output.

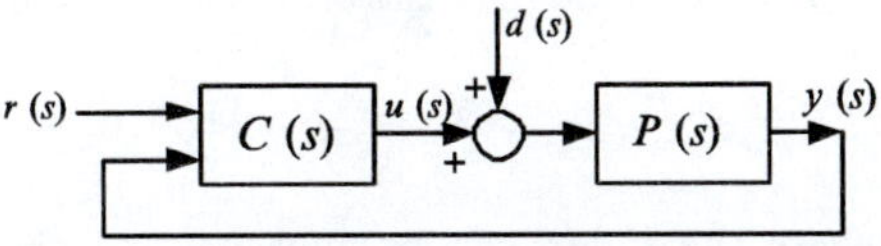

Fig. 8.1 Closed-loop control system

The output of the ISA-PID controller [10] is given by

$$u(s) = K_p \left(1 + \frac{1}{T_\mathrm{i} s}\right) e(s) - K_p \left(\frac{T_\mathrm{d} s}{1 + (T_\mathrm{d}/N)s}\right) y(s), \tag{8.1}$$

where $e(s) = r(s) - y(s)$ is the control error, K_p is the controller static gain, T_i the integral time constant, T_d the derivative time constant, and the derivative filter constant N is taken $N = 10$ as it is usual practice in industrial controllers.

Also, the process $P(s)$ is assumed to be modeled by a First-Order-Plus-Dead-Time (FOPDT) transfer function of the form

$$P(s) = \frac{K}{1 + Ts} \mathrm{e}^{-Ls}, \tag{8.2}$$

where K is the process gain, T is the time constant and L is the dead-time. This model is commonly used in process control because is simple and describes the dynamics of many industrial processes approximately [10].

The availability of FOPDT models in the process industry is a well-known fact. The generation of such model just needs for a very simple step-test experiment to be applied to the process. This can be considered as an advantage with respect to other methods that need a more *plant demanding* experiment such as methods based on more complex models or even data-driven methods where a sufficiently rich input needs to be applied to the plant. From this point of view, to maintain the need for plant experimentation to a minimum is a key point when considering industrial application of a technique.

2.2 *Performance*

One way to evaluate the performance of control systems is by calculating a cost function based on the error, i.e., the difference between the desired value (set-point) and the actual value of the controlled variable (system's output). Of course, as larger and longer in time is the error, the system's performance will be worse.

In this sense, a common reference for the evaluation of the controller performance, is a functional based on the integral of the error like: Integral-Square-Error (ISE), or Integral-Absolute-Error (IAE).

Some approaches had used the ISE criterion, because its definition allows an analytical calculation for the index [30]. However, nowadays can be found in the literature that IAE is the most useful and suitable index to quantify the performance of the system [10, 13, 19, 25, 26]. It can be used explicitly in the design stage or just as an evaluation measure.

The formulation of the criterion is stated as:

$$\mathrm{IAE} \doteq \int_0^\infty |e(t)|\,\mathrm{d}t = \int_0^\infty |r(t) - y(t)|\,\mathrm{d}t \tag{8.3}$$

where the index can be measure for changes in the set-point or in the load-disturbance.

2.3 Robustness

Robustness is an important attribute for control systems, because the design procedures are usually based on the use of low-order linear models identified at the closed-loop operation point. Due to the nonlinearity found in most of the industrial process, it is necessary to consider the expected changes in the process characteristics assuming certain relative stability margins, or robustness requirements, for the control system.

As an indication of the system *robustness* (relative stability) the Sensitivity Function peak value will be used. The control system Maximum Sensitivity is defined as:

$$M_s \doteq \max_{\omega} |S(j\omega)| = \max_{\omega} \frac{1}{|1 + C(j\omega)P(j\omega)|} \tag{8.4}$$

and recommended values for M_s are typically within the range 1.4–2.0 [10].

The use of the Maximum Sensitivity as a robustness measure, has the advantage that lower bounds to the Gain, A_m, and Phase, ϕ_m, margins [10] can be assured according to

$$A_m > \frac{M_s}{M_s - 1},$$

$$\phi_m > 2\sin^{-1}\left(\frac{1}{2M_s}\right).$$

Therefore, to assure $M_s = 2.0$ provides what is commonly considered minimum robustness requirement (that translates to $A_m > 2$ and $\phi_m > 29°$, for $M_s = 1.4$ we have $A_m > 3.5$ and $\phi_m > 41°$).

In many cases, robustness is specified as a target value of M_s, however the accomplishment of the resulting value is never checked.

2.4 Simple PID Tuning Rules for Arbitrary M_s-Based Robustness Achievement

In the work of Arrieta and Vilanova (A&V) [3], it is presented a joint criteria that faces with the *trade-off* between the performance for servo and regulation operation and also that takes into account the accomplishment of a robustness level.

A cost objective function is formulated, where J_x^z represents the criteria (8.3) taking into account the operation mode x, for a tuning mode z. The index is stated in order to get closer, as much as possible, the resulting point (J_r^{rd}, J_d^{rd}), to the "ideal" one, (J_r^o, J_d^o). Therefore,

$$J_{rd} = \sqrt{\left(J_r^{rd} - J_r^o\right)^2 + \left(J_d^{rd} - J_d^o\right)^2}, \tag{8.5}$$

Table 8.1 PID settings for servo/regulation tuning with robustness consideration

Constant	M_s^d Free	$M_s^d = 2.0$	$M_s^d = 1.8$	$M_s^d = 1.6$	$M_s^d = 1.4$
a_1	1.1410	0.7699	0.6825	0.5678	0.4306
b_1	−0.9664	−1.0270	−1.0240	−1.0250	−1.0190
c_1	0.1468	0.3490	0.3026	0.2601	0.1926
a_2	1.0860	0.7402	0.7821	0.8323	0.7894
b_2	0.4896	0.7309	0.6490	0.5382	0.4286
c_2	0.2775	0.5307	0.4511	0.3507	0.2557
a_3	0.3726	0.2750	0.2938	0.3111	0.3599
b_3	0.7098	0.9478	0.7956	0.8894	0.9592
c_3	−0.0409	0.0034	−0.0188	−0.0118	−0.0127

where J_r^o and J_d^o are the optimal values for servo and regulation control respectively, and J_r^{rd}, J_d^{rd} are the performance indexes for the *intermediate* tuning considering both operation modes.

The index (8.5) is minimized with the aim to achieve a balanced performance for both operation modes of the control system. Also, using (8.4) as a robustness measure, the optimization is subject to a constraint of the form

$$\left|M_s - M_s^d\right| = 0 \tag{8.6}$$

where M_s and M_s^d are the Maximum Sensitivity and the desired Maximum Sensitivity functions respectively. This constraint tries to guarantee the selected robustness value for the control system. See [3], for more details.

Results are expressed just in terms of the FOPDT process model parameters (8.2), in a tuning methodology for PID parameters, with the corresponding desired level robustness as:

$$\begin{aligned} K_pK &= a_1\tau^{b_1} + c_1 \\ \frac{T_i}{T} &= a_2\tau^{b_2} + c_2 \\ \frac{T_d}{T} &= a_3\tau^{b_3} + c_3 \end{aligned} \tag{8.7}$$

where the constants a_i, b_i and c_i are given in Table 8.1, according to the desired robustness level for the control system.

With the aim to give more completeness to the previous tuning method, an extension of the approach was proposed, allowing to determine the PID controller for any arbitrary value M_s^d in the range [1.4, 2.0]. Thus, tuning expressions (8.7) can be rewritten as:

$$K_pK = a_1\left(M_s^d\right)\tau^{b_1\left(M_s^d\right)} + c_1\left(M_s^d\right)$$
$$\frac{T_i}{T} = a_2\left(M_s^d\right)\tau^{b_2\left(M_s^d\right)} + c_2\left(M_s^d\right)$$
$$\frac{T_d}{T} = a_3\left(M_s^d\right)\tau^{b_3\left(M_s^d\right)} + c_3\left(M_s^d\right) \tag{8.8}$$

where the constants are expressed as functions of M_s^d. Therefore, from Table 8.1 each constant a_i, b_i and c_i was generated from a generic second order M_s^d dependent polynomial as:

$$a_1 = -0.3112\left(M_s^d\right)^2 + 1.6250\left(M_s^d\right) - 1.2340$$
$$b_1 = 0.0188\left(M_s^d\right)^2 - 0.0753\left(M_s^d\right) - 0.9509$$
$$c_1 = -0.1319\left(M_s^d\right)^2 + 0.7042\left(M_s^d\right) - 0.5334$$
$$a_2 = -0.5300\left(M_s^d\right)^2 + 1.7030\left(M_s^d\right) - 0.5511$$
$$b_2 = -0.1731\left(M_s^d\right)^2 + 1.0970\left(M_s^d\right) - 0.7700$$
$$c_2 = -0.0963\left(M_s^d\right)^2 + 0.7899\left(M_s^d\right) - 0.6629$$
$$a_3 = 0.1875\left(M_s^d\right)^2 - 0.7735\left(M_s^d\right) + 1.0740$$
$$b_3 = 0.3870\left(M_s^d\right)^2 - 4.7810\left(M_s^d\right) + 4.9470$$
$$c_3 = 0.1331\left(M_s^d\right)^2 - 0.4733\left(M_s^d\right) + 0.4032 \tag{8.9}$$

It is important to note that the tuning just depend of the system's model information and the design parameter M_s^d.

3 PID Balanced Controller Design

The analysis exposed here shows the interaction between performance and robustness in control systems. It is possible to say that, an increase of robustness implies an optimality loss in the performance (i.e., a degradation), with respect to the one that can be achieved without any robustness constraint.

3.1 Performance Optimality Index

It is possible to define the degree of optimality of the constrained case, with respect to the unconstrained one (that is the optimum).

In order to quantify the degree of optimality, the following index is proposed

$$I_{\mathrm{Perf}} \doteq \frac{J_{\mathrm{rd}}^{\mathrm{o}}}{J_{\mathrm{rd}}^{\mathrm{rd}M_{\mathrm{s}}}}, \tag{8.10}$$

where $J_{\mathrm{rd}}^{\mathrm{o}}$ is the optimal index value (8.5), using the tuning (8.7) for no constraint of M_{s} (first column of Table 8.1), that means the best one that can be achieved. Then, $J_{\mathrm{rd}}^{\mathrm{rd}M_{\mathrm{s}}}$ is the value of index (8.5) for the cases where the tuning has a robustness constraint.

Note that (8.10) is normalized in the [0, 1] range, where $I_{\mathrm{Perf}} = 1$ means a perfect optimality and, as much as the robustness is increased, the index $J_{\mathrm{rd}}^{\mathrm{rd}M_{\mathrm{s}}}$ will increase and, consequently, $I_{\mathrm{Perf}} < 1$, meaning an optimality reduction.

The degree of optimality that each control system has when a desirable value of M_{s} is stated, can be evaluated taking advantage of the generic tuning rule (8.8). For each value of $M_{\mathrm{s}}^{\mathrm{d}} \in [1.4 - 2.0]$, the optimality degree (8.10) can be obtained.

For each τ, we take advantage of the possibilities of the tuning (8.8) and (8.9), to get the PID parameters for any value of $M_{\mathrm{s}}^{\mathrm{d}} \in [1.4 - 2.0]$ and then, compute the degree of optimality using (8.10). Figure 8.2 shows the I_{Perf} variation, as a function of $M_{\mathrm{s}}^{\mathrm{d}}$, for some values of the normalized dead-time, τ.

Note that, as an example, the horizontal line indicates when the degree of optimality is 55%. With the intersection points between this line and the curves corresponding to the I_{Perf} variation for each τ, it is possible to determine a set of desired robustness that is related with this degree of optimality ($I_{\mathrm{Perf}} = 0.55$). This

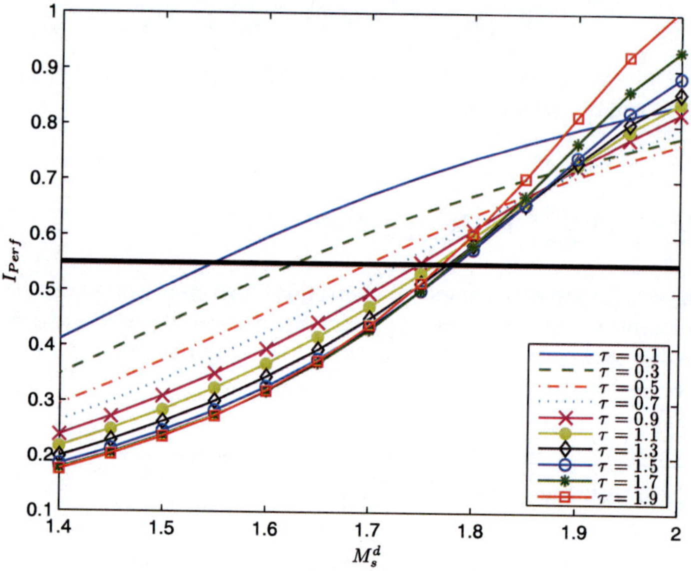

Fig. 8.2 Variation of the index I_{Perf}

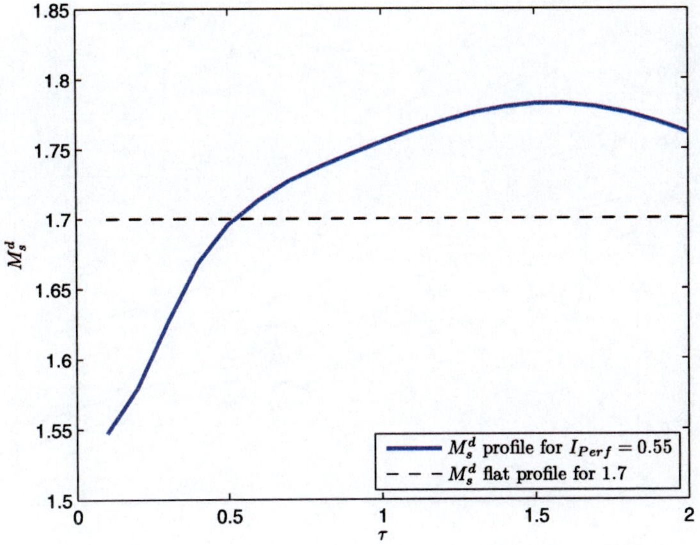

Fig. 8.3 M_s^d profile corresponding to a 55% optimality degree

set of $M_s^d(\tau)$ can be seen as a profile that the tuning should follow to satisfy the degree of optimality. Figure 8.3 shows the detail where also, just to clarify, there is the previous case of $M_s^d = 1.7$, that can be considered as a flat profile.

3.2 *Robustness Increase Measure*

From the above, index (8.10) provides an idea about the degree of optimality for the system's performance. Here, we need to state a similar concept but with the aim to quantify the robustness increase. Then, we compare the achieved robustness of the constrained case with respect to the unconstrained one, that is the optimum in performance but with lower robustness.

Therefore, the following index is proposed.

$$I_{\mathrm{Rob}} \doteq \left| \frac{M_s^o - M_s^{J_{\mathrm{rd}}}}{M_s^o} \right|, \tag{8.11}$$

where M_s^o represents the robustness value achieved by tuning (8.5) without any constraint for M_s (first column of Table 8.1), and $M_s^{J_{\mathrm{rd}}}$ is the robustness value for the cases where the robustness is constrained (i.e., increased). Then, (8.11) is a normalized index, where as much as it grows, the robustness increase will be larger.

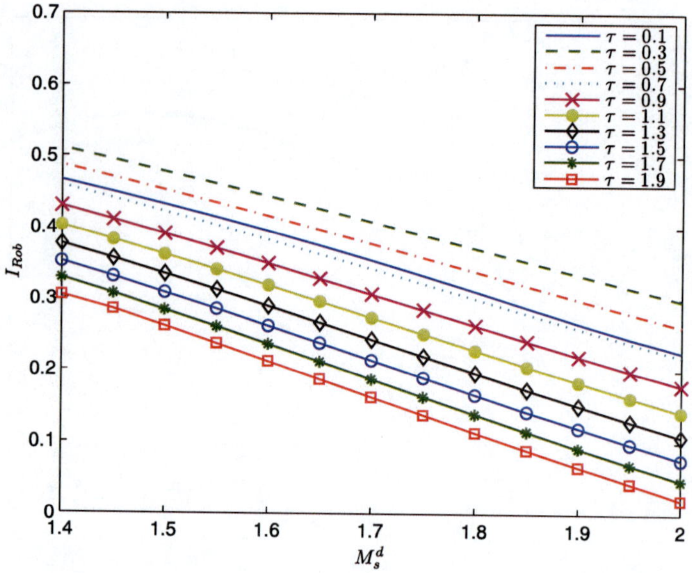

Fig. 8.4 Variation of the index I_{Rob}

Once more, we want to take advantage of the generic tuning rule (8.8), in order to evaluate the robustness increase for the whole range of family plants within $\tau \in [0.1, 2.0]$. Then, for each τ and each value $M_{\mathrm{s}}^{\mathrm{d}} \in [1.4, 2.0]$, it is possible to find the PID parameters and obtain index (8.11). This procedure is analogous to the presented previously for I_{Perf} index.

Figure 8.4 shows the I_{Rob} variation, as a function of $M_{\mathrm{s}}^{\mathrm{d}}$, for some cases of the normalized dead-time, τ. Each line represents the relative robustness increase for a particular process (value of τ).

Note that in Fig. 8.4, we could define a certain value (level), for the robustness increase and with the intersection points, determine a suitable robustness profile (analogous to the procedure for I_{Perf}). The idea is to increase the robustness of the system and provide at the same time, the best servo/regulation performance. Here, the aim is to look for a balanced tuning.

3.3 Robustness/Performance Balance

Taking into account that indexes (8.10) and (8.11) are both normalized into the range [0, 1], therefore having the same scale, if we put the information of Figs. 8.2 and 8.4 all together, the intersection points for each pair of curves, can be interpreted as equilibrium points. These, represent when the degree of optimality is consistent to the robustness increase, and vice-versa.

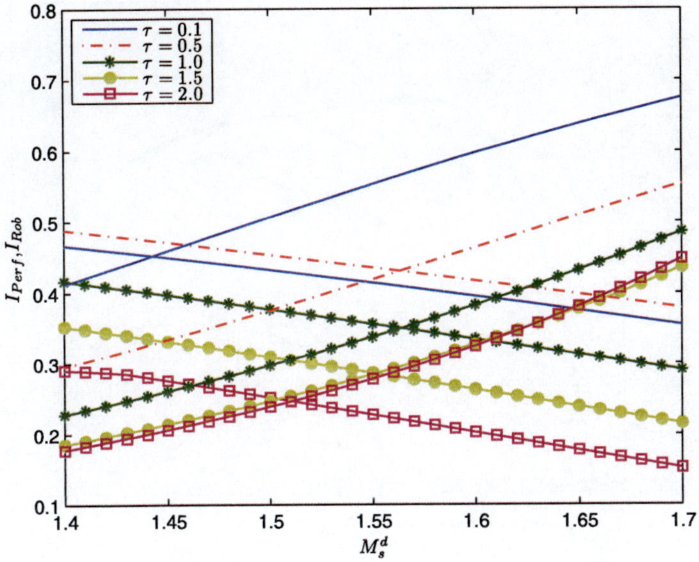

Fig. 8.5 I_{Perf} and I_{Rob} variation for some values of τ

In Fig. 8.5, there are just few cases for the model normalized dead-times τ, where it is possible to see the consideration exposed before.

With the intersection points between the corresponding pairs of curves, we can determine the suitable set of desired robustness, that provides the best balance between performance optimality degree and robustness increase. This set of $M_s^d(\tau)$ determines the robustness profile for the best robustness/performance compromise, for all the τ range. See Fig. 8.6.

3.4 Balanced PID Tuning

As it has been explained before, with information provided by indexes I_{Perf} and I_{Rob}, it is possible to obtain the corresponding M_s^d value, that achieves the best compromise between the loss of the optimality degree and the robustness increase.

The set of desired robustness values for all the plants within $\tau \in [0.1, 2.0]$, is a M_s^d profile that, as before, can be used in tuning (8.8) and (8.9), in order to get the PID controller parameters.

Then, following the same aim to propose a simple and homogeneous tuning, the set for each parameter is fitted to the general tuning expressions (8.7). Therefore,

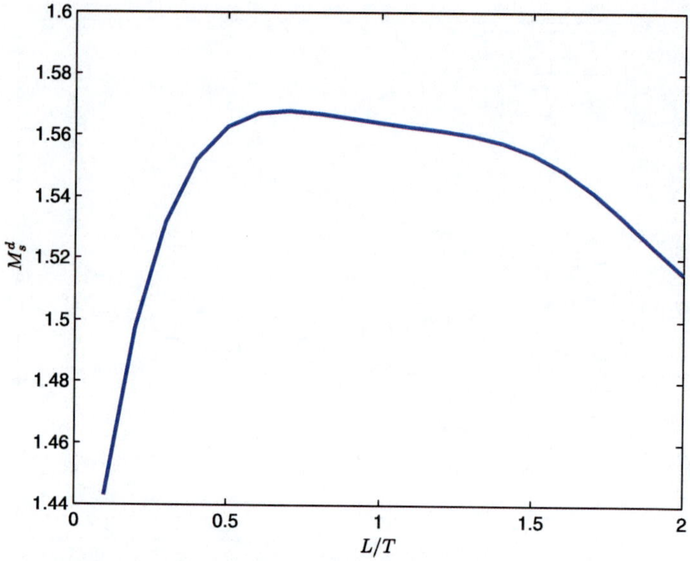

Fig. 8.6 M_s^d profile for the best robustness/performance balance

$$K_pK = 0.6776\tau^{-0.8630} + 0.1162$$
$$\frac{T_i}{T} = 0.9950\tau^{0.4016} + 0.1564$$
$$\frac{T_d}{T} = 0.2998\tau^{0.9760} + 0.0110 \qquad (8.12)$$

Tuning (8.12) is the one that provides the best compromise/balance between the robustness increase and the resulting loss of optimality degree for the system's performance. It is important to note that, due to the performance index formulation, the *trade-off* problem between servo and regulation control operation modes, is also included.

4 Comparative Examples

This section presents two kinds of examples in order to evaluate the characteristics of the proposed tuning rule. First example is an analysis not only for a specific process, but for the whole set of plants provided in their range of validity, $\tau \in [0.1, 2.0]$, in order to show the global advantages that the proposal can provide. Secondly, the other example is for a specific plant, providing the control system time responses and some evaluation data. It is supposed that the process output can vary in the 0–100% normalized range and that in the normal operation point, the controlled variable has a value close to 70%.

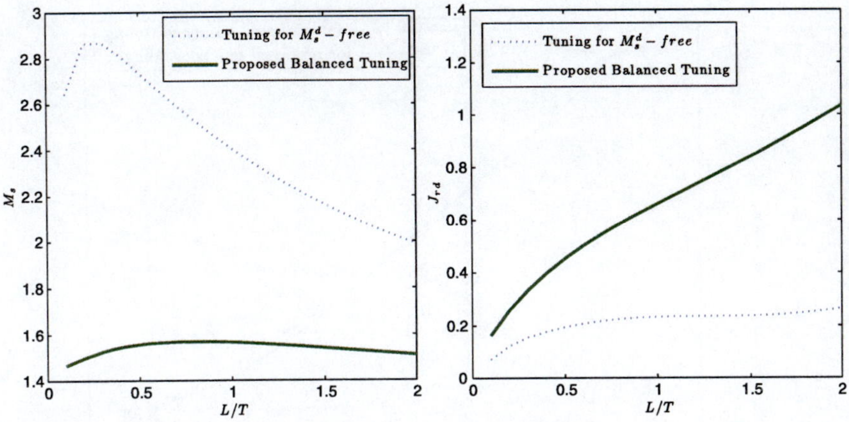

Fig. 8.7 Robustness and performance values for proposed balanced tuning

4.1 Complete Tuning Case

Here the intention is to evaluate the performance and robustness features for the proposed balanced tuning. Also, it is desirable to establish a more precise quantification of the balance concept, stated before. Therefore, we define the following index,

$$I_{\mathrm{B}} \doteq \int_{\tau_o}^{\tau_f} |I_{\mathrm{Perf}}(\tau) - I_{\mathrm{Rob}}(\tau)| \, \mathrm{d}\tau, \tag{8.13}$$

where the idea is to measure the difference between the robustness increase and the degree of optimality, for the whole range of plants $\tau \in [0.1, 2.0]$. Note that, when the index (8.13) is low, it means a good balance between the robustness increase and the corresponding price for performance optimality. A high value of I_{B} indicates a lack of balance for the studied indexes.

In Fig. 8.7, there are the robustness and servo/regulation performance evaluation for the proposed balanced tuning and compared with the unconstrained case, equivalent to M_s^d-free. Then, it can be seen the robustness increase, as well as the consequent loss of degree of optimality for the system's performance.

In addition, Fig. 8.8 shows I_{Perf} and I_{Rob} indexes for the proposed balanced tuning (8.12), where the similarity between the values indicates that, for the whole τ range, the robustness increase is equivalent to the loss of performance's optimality, therefore having a balance. It means a "fair price between what is paid and what is got".

Table 8.2 shows the balanced index I_{B}, for the proposed tuning rule and also compared with AMIGO [9] and Kappa–Tau [6] and also the methods by Arrieta and Vilanova (A&V) [3].

It is possible to see that, the case for proposed balanced tuning has obviously the best I_{B} index. Also, A&V tuning [3] with $M_s^d = 1.6$ provides a lower value for

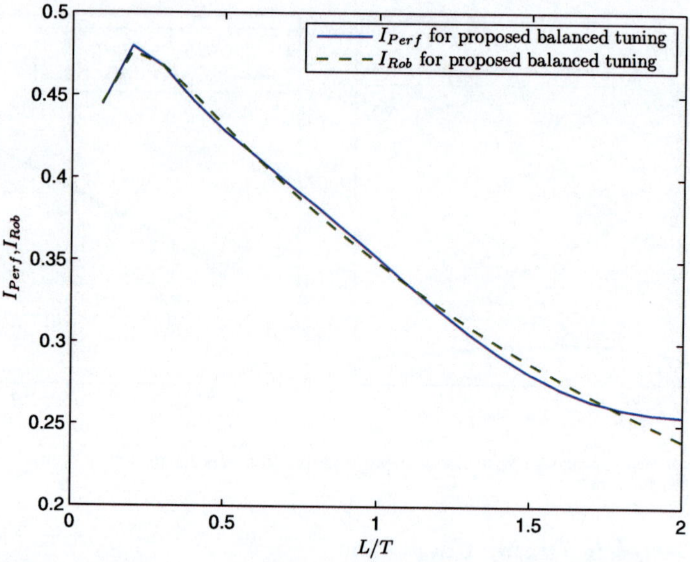

Fig. 8.8 I_{Perf} and I_{Rob} indexes for the proposed balanced tuning

Table 8.2 Balanced index (8.13) for different tuning rules

Tuning	Criteria design	I_{B}
AMIGO	$M_s^d = 1.4$	0.2900
κ–τ	$M_s^d = 1.4$	2.1373
κ–τ	$M_s^d = 2.0$	3.6303
A&V [3]	$M_s^d = 1.4$	0.3163
	$M_s^d = 1.6$	0.1416
	$M_s^d = 1.8$	0.7011
	$M_s^d = 2.0$	1.3228
Proposed	Balanced-I_{B}	0.0070

the balanced index (8.13). It has a lot of sense because, if we look in Fig. 8.6, the robustness profile in almost the cases are within the range [1.5, 1.6].

4.2 *Particular Process Case*

In order to evaluate the proposed balanced tuning, we will consider the following fourth order controlled process

$$P_1(s) = \frac{1}{\prod_{n=0}^{3}(\sigma^n s + 1)} \tag{8.14}$$

Table 8.3 Particular processes-PID controller parameters for P_1 ($\sigma =$ 0.50)

Tuning	Criteria design	K_p	T_i	T_d
AMIGO	$M_s^d = 1.4$	1.012	1.079	0.296
κ–τ	$M_s^d = 2.0$	1.719	1.151	0.285
A&V [3]	$M_s^d = 1.6$	1.300	1.193	0.215
Proposed	Balanced-I_B	1.244	1.174	0.224

Table 8.4 Particular process P_1 ($\sigma = 0.50$)-Controller tuning evaluation

Tuning	Criteria design	I_{Perf}	I_{Rob}	$\lvert I_{Perf} - I_{Rob} \rvert$
AMIGO	$M_s^d = 1.4$	0.2913	0.4651	0.1738
κ–τ	$M_s^d = 2.0$	0.6675	0.2359	0.4316
A&V [3]	$M_s^d = 1.6$	0.4457	0.4062	0.0495
Proposed	Balanced-I_B	0.4202	0.4211	0.0009

with $\sigma = 0.50$ taken from [7]. Using a two-point identification procedure [1], FOPDT models were obtained as: $K = 1.0$, $T = 1.247$, and $L = 0.691$.

In this case, we compare again with AMIGO and Kappa–Tau methods and also we choose the A&V tuning [3], for the $M_s^d = 1.6$ case (because as we said before, it is the one that provides good performance/robustness balance). From the model information and applying the corresponding tuning methods, it is possible to obtain the PID parameters as shown in Table 8.3.

In Table 8.4, the indexes (8.10) and (8.11) are shown, as well as the corresponding difference $\lvert I_{Perf} - I_{Rob} \rvert$, that indicates the system robustness/performance balance. Just to remember, index I_{Perf} indicates the degree of optimality for the system, so it is desired high values near to one. Then, for the system's robustness increase I_{Rob}, it must be as large as possible.

From Table 8.4, it is possible to see that Kappa–Tau has a good performance optimality but a small robustness increase, therefore having a bad balance. The opposite happens for AMIGO, providing a good robustness increase but a very low performance optimality.

Last two rows of Table 8.4 give the evaluation for the tuning by Arrieta and Vilanova [3] ($M_s^d = 1.6$) and for the proposed balanced tuning (8.12). Both cases show a good balance, meaning a similar values for I_{Perf} and I_{Rob}, that for this system is when the performance degree of optimality and the robustness increase are values around the 40%. As expected, proposed balanced tuning is that provides the best balance calculation.

In Fig. 8.9, there are the system's and controller's outputs for all the studied cases. It is possible to see that, responses for the balanced tuning are between the other ones, therefore providing an intermediate solution for the controller's tuning.

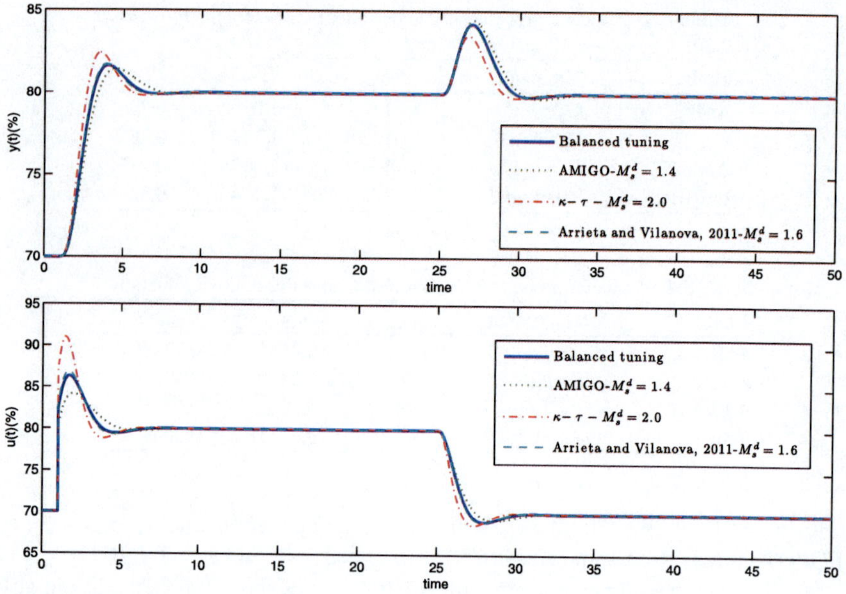

Fig. 8.9 Particular process P_1-Proposed methods ($\sigma = 0.50$)

5 Conclusions

In process control, it is very important to guarantee some degree of robustness, in order to preserve the closed-loop dynamics, to possible variations in the control system. Also, at the same time, it must be provided the best achievable performance for servo and regulation operation.

All of the above specifications, lead to have different *trade-offs*, between performance and robustness or between servo and regulation modes, that must be solved on a balanced way. Here, we looked for a PID controller tuning methodology, with the aim to achieve the best possible balance between robustness and performance issues.

The problem was faced using the proposed evaluation indexes, where just the FOPDT model information is needed to calculate the PID controller parameters. This balanced tuning looks for the best compromise between the robustness increase and the consequent reduction in the performance optimality for the system, therefore trying to get a fair price between what is paid and what is got.

Acknowledgments This work has received financial support from the Spanish CICYT program under grant DPI2010-15230. Also, the financial support from the University of Costa Rica and from the MICIT and CONICIT of the Government of the Republic of Costa Rica is greatly appreciated.

References

1. Alfaro VM (2006) Low-order models identification from process reaction curve. Ciencia y Tecnología (Costa Rica) 24(2):197–216 (in Spanish)
2. Arrieta O, Vilanova R (2010) Performance degradation analysis of controller tuning modes: application to an optimal PID tuning. Int J Innov Comput Inf Contr 6(10):4719–4729
3. Arrieta O, Vilanova R (2011) Simple PID tuning rules with guaranteed M_s robustness achievement. In: Proceedings of 18th IFAC world congress, August 28–September 2, Milano, Italy
4. Arrieta O, Visioli A, Vilanova R (2010) PID autotuning for weighted servo/regulation control operation. J Process Contr 20(4):472–480
5. Åström KJ, Hägglund T (1984) Automatic tuning of simple regulators with specifications on phase and amplitude margin. Automatica 20:645–651
6. Åström KJ, Hägglund T (1995) PID controllers: theory, design, and tuning. Instrument of Society of America, Research Triangle Park, North Carolina
7. Åström KJ, Hägglund T (2000) Benchmark systems for PID control. In: Proceedings of IFAC digital control: past, present and future of PID control, Terrassa, Spain
8. Åström KJ, Hägglund T (2001) The future of PID control. Contr Eng Pract 9:1163–1175
9. Åström KJ, Hägglund T (2004) Revisiting the Ziegler–Nichols step response method for PID control. J Process Contr 14:635–650
10. Åström KJ, Hägglund T (2006) Advanced PID control. ISA: The Instrumentation, Systems, and Automation Society
11. Babb M (1990) Pneumatic instruments gave birth to automatic control. Contr Eng 37(12):20–22
12. Bennett S (2000) The past of PID controllers. In: Proceedings of IFAC digital control: past, present and future of PID control, Terrassa, Spain
13. Chen D, Seborg DE (2002) PI/PID controller design based on direct synthesis and disturbance rejection. Ind Eng Chem Res 41(19):4807–4822
14. Cohen GH, Coon GA (1953) Theoretical considerations of retarded control. ASME Trans 75:827–834
15. Fung HW, Wang QG, Lee TH (1998) PI tuning in terms of gain and phase margins. Automatica 34:1145–1149
16. Ho WK, Hang CC, Cao LS (1995) Tuning PID controllers based on gain and phase margin specifications. Automatica 31(3):497–502
17. Ho WK, Lim KL, Hang CC, Ni LY (1999) Getting more phase margin and performance out of PID controllers. Automatica 35:1579–1585
18. Ingimundarson A, Hägglund T, Åström KJ (2004) Criteria for desing of PID controllers. Technical report, ESAII, Universitat Politecnica de Catalunya
19. Kristiansson B, Lennartson B (2006) Evaluation and simple tuning of PID controllers with high-frequency robustness. J Process Contr 16:91–102
20. López AM, Miller JA, Smith CL, Murrill PW (1967) Tuning controllers with Error-Integral criteria. Instru Technol 14:57–62
21. Martin J, Smith CL, Corripio AB (1975) Controller tuning from simple process models. Instru Technol 22(12):39–44
22. O'Dwyer A (2003). Handbook of PI and PID controller tuning rules. Imperial College Press, London, UK
23. Rivera DE, Morari M, Skogestad S (1986) Internal model control 4 PID controller design. Ind Eng Chem Res 25:252–265
24. Rovira A, Murrill PW, Smith CL (1969) Tuning controllers for setpoint changes. Instru Contr Syst 42:67–69
25. Skogestad S (2003) Simple analytic rules for model reduction and PID controller tuning. J Process Contr 13:291–309

26. Tan W, Liu J, Chen T, Marquez HJ (2006) Comparison of some well-known PID tuning formulas. Comput Chem Eng 30:1416–1423
27. Vilanova R (2008) IMC based robust PID design: tuning guidelines and automatic tuning. J Process Contr 18:61–70
28. Vilanova R, Alfaro VM, Arrieta O (2011) Ms based approach for simple robust PI controller tuning design. In: Proceedings of the international multiconference of engineers and computer scientists 2011, Hong Kong, 16–18 March 2011. Lecture notes in engineering and computer science, pp 767–771
29. Yaniv O, Nagurka M (2004) Design of PID controllers satisfying gain margin and sensitivity constrains on a set of plants. Automatica 40:111–116
30. Zhuang M, Atherton D (1993) Automatic tuning of optimum PID controllers. IEE Proc Part D 140(3):216–224
31. Ziegler JG, Nichols NB (1942) Optimum settings for automatic controllers. ASME Trans 64:759–768

Chapter 9
Hybrid Genetic Algorithm Optimisation of Distribution Networks—A Comparative Study

Romeo Marin Marian, Lee Luong, and Son Duy Dao

Abstract This chapter focuses on the second of a three-stage, integrated methodology for modeling and optimising distribution networks (DN) based on hybrid genetic algorithms (HGA). The methodology permits any combination of transportation and warehousing costs for deterministic/stochastic demand. This chapter analyses and compares the fluctuation of overall costs when the number of facilities varies and indicates how to minimize them. The chapter concentrates on capacitated location allocation of distribution centers, a large scale, highly constrained, NP-hard, combinatorial problem. The HGA used has a classical structure, but incorporates a special encoding of solutions as chromosomes and integrates linear programming/mixed integer programming modules in the genetic operators (GO). A complex and extensive case study is described, demonstrating the robustness of the HGA and the optimization approach.

Keywords Capacitated location-allocation problem • Distribution network • Hybrid genetic algorithms • Optimisation

1 Introduction

The distribution network (DN) critically affects the structure, complexity, costs and overall efficiency associated with operating a supply chain (SC), as well as the service level, the two critical factors in any SC [1]. This directly influences the aptitude of participants in a SC to enter or stay competitive in a market. An essential aspect is the escalating complexity of DN. As SC become increasingly large and

R.M. Marian (✉) • L. Luong • S.D. Dao
School of Advanced Manufacturing & Mechanical Engineering, University of South Australia, Mawson Lakes Campus, Mawson Lakes, SA 5095, Australia
e-mail: romeo.marian@unisa.du.au; lee.luong@unisa.edu.au; daosd001@mymail.unisa.edu.au

S.I. Ao et al. (eds.), *Intelligent Control and Innovative Computing*, Lecture Notes in Electrical Engineering 110, DOI 10.1007/978-1-4614-1695-1_9,

complex, a general trend today, due mainly to globalization [2], designing the DN becomes vital [3, 4]. Due to sheer size, capacity of classical tools to solve and optimise location allocation (LA) problems in real DN was exceeded and required development of new methodologies.

The chapter presents a framework developed for the optimisation of DN, especially as the LA of distribution centres (DC) or warehouses are concerned. The size of DC/warehouses is limited—thus, the problem becomes a capacitated location allocation (CLA) problem.

The CLA problem is a large scale, highly constrained, NP-hard, combinatorial problem [5]. Genetic algorithms (GA) have been chosen in this research for the design and optimisation of the DN for their remarkable capacity to successfully work with problems with huge solution spaces [6].

Various GA have been developed, as shown briefly in the next section, to solve various LA-like problems. Some of them were successful to tackle—in general—restricted variants of the problem (obvious when the run-time of a GA clocks hours of CPU indicating that particular methodology has reached its limits due to combinatorial explosion (CE)).

The approach presented in this paper to tackle the CLA problem differs from others through the use of an integrated methodology, flexible enough to accommodate most realistic assumptions and, at the same time, computationally frugal, avoiding CE. The methodology presented here is a development of algorithms of similar complexity presented in [7, 8] and an extension of the work in [3, 4].

A complex case study, of considerable size and a complex cost structure, achieved excellent run times. It demonstrates the robustness of the HGA, its capacity to tackle problems of considerable size and expandability to even larger and more complex problems.

2 Literature Review on Optimisation of Location-Allocation Problems

The LA problems received considerable attention in the last 50 years due to their immediate practical applicability. A wide variety of methodologies and techniques were employed, alone or in diverse combinations in attempting to find good solutions.

LA problem were treated in depth in recent years [9, 10]. The multitude of definitions of variants of LA problems make in fact difficult to compare results. The literature review presented here highlights the diversity of problems and approaches used to tackle them.

The initial LA problem has been generally further complicated to integrate different aspects relating to production and distribution [11], product returns [12], facilities restricted to being located along a line, while the destinations that will be served by them are generally located off the line [13], etc. It is, however, important to highlight that LA problems are combinatorial problem and any complication of the model generally has a detrimental effect on the size of the problem that can be, eventually, solved.

MIP has been used extensively, directly [14] or in combination with other techniques—e.g. GA [12]—in tackling LA problems with various degrees of success.

Artificial intelligence techniques have also been used in optimising LA problems, standalone, or in combination with other AI or mathematical optimisation techniques.

A hybrid heuristic based on the simulated annealing, Tabu list, and improvement procedures are proposed to resolve an un-capacitated single allocation hub location problem [15]. This algorithm is compared with a GA [16] using similar sets of data.

GA have been applied in the optimisation of LA problems. The extreme variety of LA and LA-like problems led to the development of various GA and combinations of GA with other mathematical tools.

A multi-phase mathematical approach for the design of a complex SC network is presented in [17]. The proposed approach is based on GA, the analytical hierarchy process (AHP), and the multi-attribute utility theory (MAUT) to satisfy simultaneously the preferences of the suppliers and the customers at each level in the network. Even if the model is very comprehensive, it is likely to be limited to DN of reduced size. A network of 3–4–4–2 facilities illustrates the application of the approach.

A LA problem involving reverse logistics is presented in [12]. It addresses the problem of determining the number and location of centralized return centres (i.e. reverse consolidation points) where returned products from retailers or end-customers are collected, sorted and consolidated into large shipments destined for manufacturers' or distributors' repair facilities. The paper proposes a nonlinear MIP model and a GA that can solve the reverse logistics problem involving product returns. The model and algorithm were validated by an example of 30 customers, 10 potential sites for collection and 5 potential sites of centralised return (although the actual number of sites for collection and centralisation is smaller). The solution times tend to indicate potential CE issues.

From this literature review it can be seen that the LA problems are of an extremely diverse nature. Also, GA are important candidates when it comes to select optimisation tools. Due to the extreme variety of the problems and implementation of algorithms, it is difficult to compare or classify them.

Most of the algorithms above are limited in either one or more of the following: size of solution space, realism of the modelling of the problem—this includes decoding chromosomes to meaningful solutions—versatility and capacity to be customised for more or less related problem, and, probably the most important, robustness [18].

For solving realistic and versatile CLA, avoiding simultaneously all limitations presented above, a new methodology and optimisation tool was necessary and are presented in the next sections.

3 Methodology to Optimise the Distribution Network

The complexity and contradictory nature of optimisation criteria requires the problem to be optimised in stages.

The main tool for the optimisation of the CLA problem in this research is a HGA, chosen for the versatility of GA to implement optimisation criteria, their flexibility to be adapted to extremely diverse types of problems (hybridised with LP/MIP tools) and their capacity to work in extremely vast solution spaces [19]. The methodology presented here is an extension of [20].

It is challenging working with solutions of variable size. In CLA, one of the parameters to be optimised is the number of DCs (complete definition of the problem in the next section). As a result, and to still optimise for the number of DCs as a variable, the optimisation is initially done for a set number of DCs. After a local optimum/near optimum set of solutions has been found for a set number of DCs (in the first stage), the algorithm is run for different numbers of DCs (second stage). This approach can quickly highlight the most promising zone of the solution space (as number of DCs is concerned) and permits to determine a global optimum.

The major steps of the first two stages are as follows:

3.1 First Stage

- Determine the optimal (local optimum—minimal cost to operate the DN) result using HGA for a set number of DCs and a deterministic demand.
- Set number of DCs to be opened and operated.
- Determine the location of DCs.
- Allocate PFs to each DC.
- Allocate flow of merchandise from each PF to each allocated DC.
- Allocate DCs to each R.
- Allocate flow of merchandise from each DC to each allocated R.

3.2 Second Stage

- Determine the global optimum (minimum cost) by varying the number of DCs, for a deterministic demand.
- Repeat the first stage for a different number of DCs.
- Determine optimal result in each case.
- Select the solutions that offer global minima of costs.

Note the stochastic character of many variables playing a role in any SC and in the DN in particular. To capture the stochastic nature of the SC in the final results of the optimisation, a third stage is proposed.

3.3 *Third Stage*

- Verify robustness of the solutions—by modelling and simulating the DN and by varying different parameters and using stochastic input values—so as to improve the realism of the results. Test when uncertainty is present in the various components of the SC and its actors, primarily:
    - Stochastic demand.
    - Variable/stochastic transportation costs.
    - Stochastic variation of different costs in opening/operating DCs, etc.
- Determine, in the new conditions, actual service levels, costs, potential bottlenecks, etc. in realistic settings;

If the final results of a realistic, stochastic SC, are not satisfactory and not within the accepted practical levels, different parameters and inputs of the problem can be used (e.g. multiply the demand, the storage capacity, transportation costs, etc. by a safety factor) and re-run the optimisation.

This approach offers maximum flexibility to adapt a method developed for deterministic data into a more realistic one involving stochastic data/settings.

This chapter only focuses on the second stage of the research, the first stage being extensively detailed in [3, 4].

4 The Model of the Capacitated Location Allocation Problem

In general location allocation problems require locating a set of new facilities—DCs in this case—such that transportation costs from facilities to customers are minimized.

The CLA problem, as considered in this work (Fig. 9.1 presents a case) is defined as follows [3, 4]:

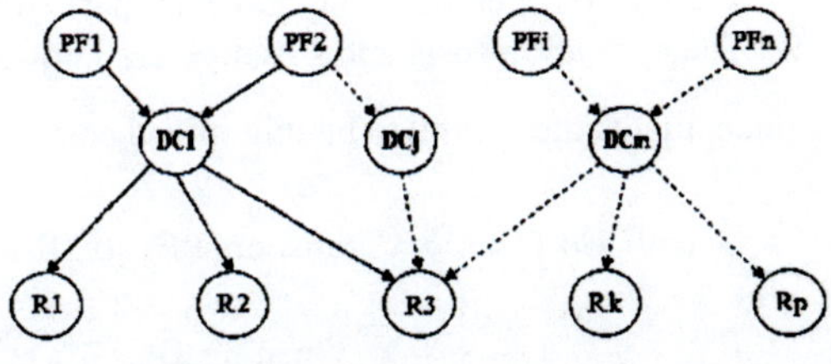

Fig. 9.1 Distribution network for n PF, m DC and p R

4.1 Considering Known

1. Number, location and production capacity of production facilities (PF).
2. Number, location and demand in a given time frame of product users or retailers (R).
3. Demand.

4.2 Determine

- Number, location and capacity of DCs (Note: the number of DCs is set in the first stage of optimisation and is varied/optimised in the second stage).
- Allocation of DCs to production facilities.
- Allocation of the flow of products from each PF to the relevant DC.
- Allocation of retailers to DCs.
- Allocation of the flow of products from each DC to the relevant Rs.

4.3 So that

- The overall operating costs are minimized.
- A pre-set level of service is maintained.

4.4 Conditions

1. It is assumed, initially, that demand can be fully satisfied and all production can be sold.
2. The capacity of DCs is limited, selected from a modular set of capacities.
3. Capacity conditions—capacity of DCs as selected cannot be exceeded.
4. Each DC is allocated to at least one PF and at least a R.
5. In a given time frame, flow of products IN equals the flow of products OUT of a DC.
6. A single type of product is considered.
7. The products move strictly from PF to a DC and from the DC to a R.
8. The cost of opening a DC for each particular size of DC is known.
9. The cost of operating a DC for each particular size of DC is known.
10. Transportation costs and structure are known.

Conceptually, the overall to be minimized cost can be modeled as follows: minimize

$$\begin{aligned}\text{Overall\_Cost} = &\left(\sum \text{Transport\_PFs\_to\_DCs} + \sum \text{Transport\_DCs\_to\_Rs}\right)\\ &+ \left(\sum \text{Opening\_DCs} + \sum \text{Operating\_DCs}\right)\end{aligned}$$

Figure 9.1 presents a typical graph of a DN, including facilities (as vertices) and connections between them (as edges), as considered in LA problems. In this case, PF1 sends products to DC1, PF2 sends products to DC1 and DCj, etc. DC1 receives products from PF1 and PF2 and distributes it to R1, R2 and R3, etc. (directed graph).

This model is general enough to cover most of real-life problems, and at the same time it permits the optimisation of the CLA problems with a single general methodology.

5 Structure of the Genetic Algorithm

The structure of the HGA is classic, but a special attention has to be attached to handling the numerous and varied constraints of the problem and between different genes of the chromosome.

The issue of constraints becomes critical for this problem and for the development of the HGA. The probability to obtain a legal and feasible chromosome by random generation of genes or simply by using any completely random operation (as part of a genetic operator (GO)) on genes is quasi-nil. If special measures are not implemented, any GO is very likely to produce an illegal chromosome or offspring.

Also, due to the size of the chromosome and complex relations between genes, the number of operations needed to be performed within any GO has to be minimized, while preserving the stochastic character of the HGA. Rejecting, repairing and penalty strategies to handle constraints were comprehensively explored and discarded as inadequate for the present problem. Instead, modified GO developed for the particularities of the CLA problem are used, as no suitable GO, could be found in the literature.

The HGA is designed to work only with feasible chromosomes, to avoid the need to repeatedly check the feasibility of the chromosome during the running of the algorithm (operation prone to CE). As a result, all operators are designed to filter the potential candidates in any operations on genes. To ensure feasibility of the solutions, the GOs are hybridised with LP/MIP tools—essentially, a low-level optimisation is done using LP/MIP. Actual selection to use either of them depends on the particular conditions of the problem. LP/MIP filter potential candidates and retain, at each step, only the ones guaranteed to lead to a gene that satisfies all constraints. Thus, any random operation within the chromosome is applied only on suitable candidates—guided search using LP/MIP during the application of any operation [19], and the resulting gene/chromosome is guaranteed feasible. Three major GO are hybridised with LP/MIP: random generation of chromosomes, the crossover and the pseudo-mutation (PM).

A major, difference between a classic, complete GA and the HGA used to optimise the CLA problem in this case, is the absence of a mutation operator per se, replaced with a PM [4], which successfully replaces the mutation operator in its function. The PM is applied to the population after evolution, i.e. after the selection process has taken place

6 Case Study

The case study mainly focuses on the second stages of the methodology to optimise the CLA problem.

The DN is composed of 25 PFs, 10 DCs and 25 R. All components of the network are placed on an array of 100 × 100 cells [20]. The *x* and *y* coordinates of the cells in which PFs and Rs are located are given in Table 9.1, where PFX(i), PFY(i)—*x* and *y* coordinates of PFi and RX(i), RY(i)—*x* and *y* coordinates of Ri, the annual output of PFs (PPi) and also the annual demand of Rs (PRi).All output can be sold, all demand can be satisfied, so the sum of the annual output of PFs equals the sum of annual demand of Rs. Additionally, the following assumptions are considered:

- The output of each PF is to be transported in five equal installments (i.e. monthly installments) to DCs. This demonstrates the HGA can be applied also when production is seasonal.
- Each R has a demand of products to be delivered as requested—i.e. the capacity of the DC has to accommodate the peak production of PFs.
- All products are produced, delivered and sold within a standard period (e.g. one year).
- The structure of the costs takes into consideration economies of scale. DC capacities and costs are presented in Table 9.2. The opening costs of a DC are compounded and allocated as annual costs depending on the capacity of the DC. On the other hand, the operating costs for a DC are allocated as costs per actual unit transiting the DC.
- Transportation is considered by trucks, and the cost is allocated as $0.2 × 10^{-3} per actual unit transported and per unit of distance (Manhattan distance).

7 Implementation

The algorithm has been successfully implemented in a computer program coded in MS Visual Basic 6, not compiled and run on a notebook with a Core 2 Duo 2.4 GHz processor and 4 GB DDR RAM.

The costs are implemented as separate subroutines and can be easily modified and adapted to suit any particular practical cost structure and allocation and actual transportation rates. This makes the case study and its implementation, even if developed for a scenario and not a real case from practice, easily transferable and customisable for problems encountered in practice.

The whole HGA has been tested for the settings presented above (25 PFs, 10 DCs and 25 Rs), and the complex cost structure. For a population of 100 chromosomes evolving through 150 generations:

- The initial population ($\leq$100 chromosomes) is generated in less than 1 s.
- The evolution through one generation takes typically under 2 s.

Table 9.1 Case study data

	Production facilities positions		25 Retailers positions		Output of production facilities	Demand of retailers
I	PFX(i)	PFY(i)	RX(i)	RY(i)	PP(i)	PR(i)
1	59	85	41	83	6,420	2,820
2	96	24	66	75	7,140	6,240
3	94	59	49	50	3,840	8,160
4	25	22	62	17	7,260	4,080
5	41	26	62	72	4,080	7,920
6	42	42	59	69	7,860	7,260
7	27	75	46	18	4,920	2,700
8	8	67	65	36	4,080	4,500
9	32	58	49	48	7,740	7,680
10	75	15	60	100	3,180	6,060
11	95	20	17	52	5,700	8,040
12	98	2	96	44	8,160	6,480
13	55	51	82	13	5,220	3,420
14	85	7	56	12	2,400	6,360
15	34	74	70	11	5,220	7,680
16	61	12	98	76	5,040	2,400
17	89	68	5	75	7,500	7,680
18	89	79	39	61	3,780	2,880
19	71	28	85	94	5,340	2,520
20	29	52	88	41	4,020	5,640
21	15	56	4	45	7,980	4,920
22	92	66	38	35	7,080	3,540
23	37	11	16	75	5,040	7,260
24	63	66	54	13	4,080	7,920
25	7	78	3	47	6,540	5,460

Table 9.2 Structure of costs

DC capacities (units)	Opening costs per unit of capacity	Operating costs per actual unit through DC
10,000	$1	$0.5/actual unit
20,000	$0.75	$0.4/actual unit
50,000	$0.6	$0.3/actual unit
100,000	$0.4	$0.2/actual unit

- The evolution of a population of 100 chromosomes over 150 generations takes typically under 5 min.
- The PM rate is 8%.
- The output of the algorithm is recorded in a*.txt* file, from which the information is processed further—the size of the output.txt file is in the order of 200 MB, which corresponds to about 750,000 pages of raw data.
- The output of the HGA gets stabilised after about 100 generations.

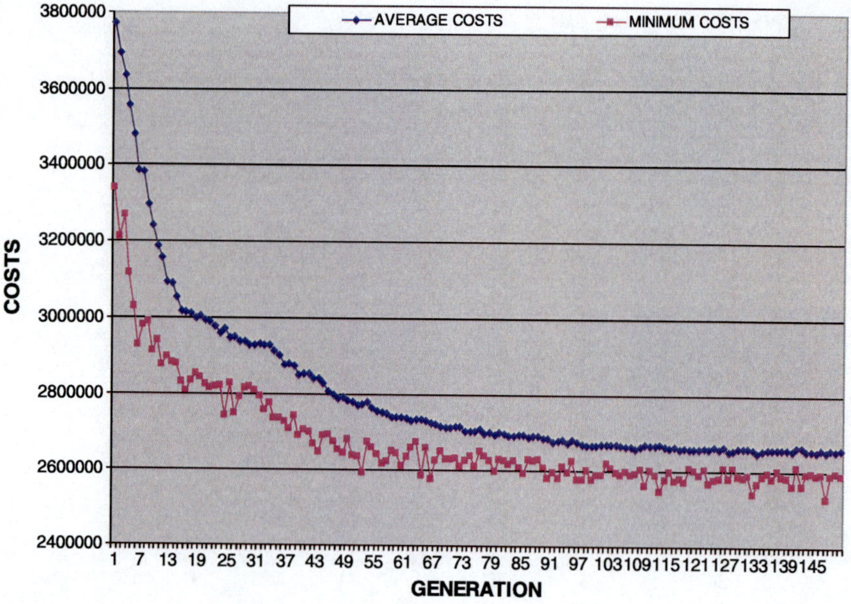

Fig. 9.2 Typical evolution of the results of the HGA

The implementation proves to be very fast and with plenty of potential for expansion—i.e. the possibility to tackle significantly larger problems.

A typical evolution of the results of the HGA is presented in Fig. 9.2: evolution of the average and minimum values of costs for each generation.

8 A Comparative Study—Application of the HGA to Various Configurations of the Distribution Network

The HGA permits testing of different scenarios and variations applied to the DN. The structure of the costs (including different costing policies), the number of DCs, different sets of constraints applied to the allocation of DCs to PFs and Rs to DCs, etc.

Table 9.3 presents the results of series of runs of the HGA for a different number of DCs in the network. All parameters were kept constant—a population of 100 chromosomes evolving over 150 generations. The only change involved the number of DCs.

Specifically, for each number of DCs, set from 10 to 5, the HGA has been run 10 times and the minimum values of the costs over the DN have been recorded in Table 9.3. (minimum of minima—minim minimorum MM—and average of the minimums, FA—final generation). It also contains the average of start cost averages

Table 9.3 Comparative study of distribution network overall costs

	Cost					
	Number of DCs					
Run no.	10 DCs	9 DCs	8 DCs	7 DCs	DC6	5 DCs
1	2,582,699	2,650,427	2,709,175	2,621,674	2,555,249	2,661,262
2	2,671,802	2,600,410	2,596,027	2,639,483	2,621,090	2,629,925
3	2,595,235	2,548,232	2,725,879	2,664,019	2,645,709	2,576,142
4	2,643,025	2,631,391	2,564,001	2,517,838	2,592,724	2,583,852
5	2,668,349	2,611,887	2,538,922	2,629,772	2,634,023	2,561,923
6	2,592,083	2,594,487	2,591,196	2,591,397	2,651,871	2,608,955
7	2,609,241	2,693,317	2,509,232	2,598,378	2,626,243	2,777,139
8	2,609,381	2,605,224	2,551,041	2,568,115	2,754,417	2,678,564
9	2,628,396	2,585,734	2,699,229	2,563,010	2,617,379	2,652,160
10	2,642,547	2,717,553	2,679,737	2,616,710	2,618,132	2,667,405
(SA)	**3,764,934**	**3,771,158**	**3,764,992**	**3,755,449**	**3,745,062**	**3,720,276**
(MM)	**2,582,699**	**2,548,232**	**2,509,232**	**2,517,838**	**2,555,249**	**2,561,923**
(FA)	**2,624,276**	**2,623,866**	**2,616,444**	**2,601,039**	**2,631,683**	**2,639,732**
(FA/SA)	**0.697031**	**0.695772**	**0.69494**	**0.692604**	**0.702708**	**0.709553**

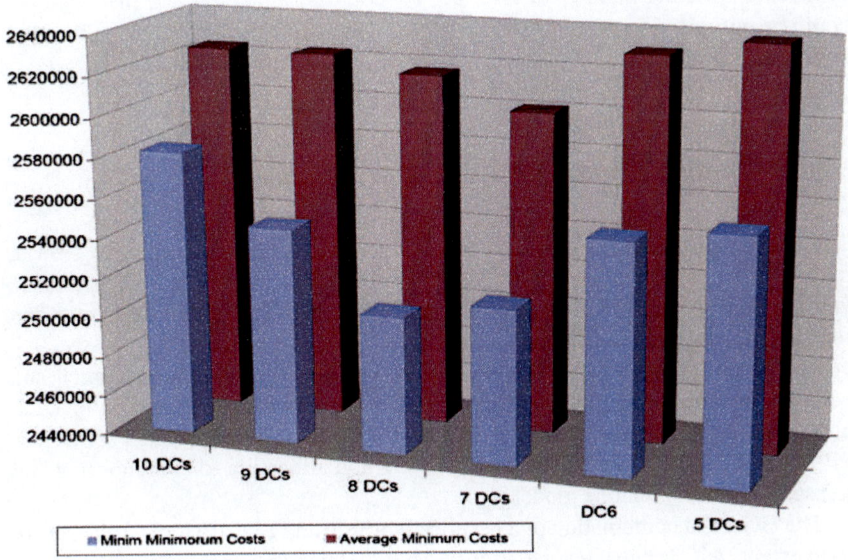

Fig. 9.3 Minimum and average values across various numbers of DCs

(SA)—defined as average over the 10 runs of the averages of costs of the population of solutions in the first generation of each run. The minimum and average values across various numbers of DCs in the network have been represented in Fig. 9.3.

The ratio between FA and SA (optimisation ratio FA/SA) is consistently around 30% for the number of DCs from 10 to 5. This means that, by modelling the network and running the algorithm, the totality of costs in the SC can be reduced by about 30%. It is worth considering these figures in light of some statistics: grocery industry in US can save about $30 billion, or 10% of the annual operating costs by using more efficient SC strategies [1]. In case this algorithm is applied, savings can be of the order of 30% for the portion of SC where results are applied.

A number of conclusions can be drawn from examining Table 9.3 and Fig. 9.3, and the context in which results were obtained:

- The MM and FA values are very close to each other (less than 3% and 1.5%, respectively). This can be attributed to the number of DCs per PF, which has been kept constant over the experiment (max 5), the capacity of the algorithm to find consistently good solutions, just slight variation of the costs if the number of DCs vary and the structure of the opening/operating costs for DCs.
- Even if the difference is small, it seems that the optimum number of DCs for this problem is 7 or 8. The final number should be determined after additional constraints are considered (robustness of DN).

The runtime is proportional with the number of DCs.

Considering the observations above, significant savings could be achieved by further tweaking variables and constraints and their adaptation to real situations. As an example, the number of DCs per PF could be modified and their number could be optimised in this context. Different structures of costs and cost allocation could be explored.

9 Concluding Remarks and Further Work

The chapter presented an integrated methodology to model and optimise DNs in SC using a hybrid genetic algorithm. This methodology works for deterministic demand and is being developed for stochastic demand. The optimisation criterion for the problem is the cost, considering the service level set to 100%.

The proposed model of the CLA problem is able to encode, in an explicit and compact manner, in an array, all relevant information about the location of DCs, allocation of DCs to PFs and Rs to DCs and flow of products for each active pair PF-DC and DC-R. Any realistic cost structure and allocation for warehousing and transportation costs can be modeled.

The HGA is a combination of a genetic algorithm and LP/MIP tools. LP/MIP are used extensively during the generation of the initial population of chromosomes and during crossover and pseudo mutation. They are embedded in the structure of GO and filter the generation of genes of the chromosomes to ensure feasibility. LP/MIP modules ensure all constraints are satisfied, whereas the stochastic character of any operator is preserved.

The HGA has been extensively tested, for various combinations of the input parameters. The evolution of the output is consistently convergent towards the optimum.

It is possible to adapt the HGA for any realistic situation. The flexibility to define costs permits exploration of different business models, allocation of costs, economies of scale, seasonal production, inverse logistics, multi-products, etc.

Further work is being conducted in:

- Generalisation of the model and HGA for multi-echelon SCs.
- Possibility to add the time dimension to the encoding, so scheduling can be incorporated.
- Testing of a yet more complex cost structure, to balance the economies of scale, characteristic to having fewer and larger DCs with the associated risks.
- Improvement of the crossover operator, to preserve more of the parents' genetic information, especially as the second part of the chromosome is involved.

After the full implementation of the HGA and after obtaining the results from running the algorithm, the robustness of the DN will be assessed using a simulation software package (Automod/Arena/Simprocess), for different, preset levels of uncertainty and the level of service will be determined in each case.

The computer code developed to implement the HGA, managed to achieve excellent times for the optimisation of the CLA problem. The capacity of the algorithm to work and consistently produce good results in those conditions proves its robustness. The results lead to the justified assumption that the technique can be safely extrapolated for most practical-size CLA problems with complex cost structures.

References

1. Simchi-Levi D, Kaminsky P, Simchi-Levi E (2008) Designing and managing the supply chain: concepts, strategies and case studies. McGraw-Hill, New York
2. Fahimnia B, Marian R, Motevallian B (2009) Analysing the hindrances to the reduction of manufacturing lead-time and their associated environmental pollution. Int J Environ Technol Manage 10(1):16–25
3. Marian RM, Luong LHS, Akararungruangkul R (2008) Optimisation of distribution networks using genetic algorithms - part 1 - Problem modeling and automatic generation of solutions. Int J Manufact Technol Manage 15(Special issue on: Latest concepts and approaches in logistics engineering and management):64–83
4. Marian RM, Luong LHS, Akararungruangkul R (2008) Optimisation of distribution networks using genetic algorithms - Part 2 - The genetic algorithm and genetic operators. Int J Manufact Technol Manage 15(Special issue on: Latest concepts and approaches in logistics engineering and management):84–101
5. Perez M, Almeida F, Moreno-Vega JM (1998) Genetic algorithm with multistart search for the p-Hub median problem. In: 24th Euromicro Conference, Vasteras
6. Fahimnia B, Luong L, Marian R (2009) Optimization of a two-Echelon supply network using multi-objective genetic algorithms in 2009 World Congress on Computer Science and Information Engineering, Los Angeles, USA, March 31–April 2, 2009
7. Lai KT et al (2008) An optimisation framework for the design of multi-echelon production-distribution networks using genetic algorithms. Asian Int J Sci Technol (AI-JSTPME) Product Manufact Eng 1(2):17–29

8. Marian RM, Luong LHS, Abhary K (2006) A genetic algorithm for the optimisation of assembly sequences. Comput Ind Eng 50:503–527
9. Gen M, Cheng R (1997) Genetic algorithms and engineering design. In: Parsaei H (ed.) Wiley series in engineering design and automation. Wiley, New York
10. Gen M, Cheng R (2000) Genetic algorithms and engineering optimisation. Wiley, New York
11. Jayaraman V, Pirkul H (2001) Planning and coordination of production and distribution facilities for multiple commodities. Eur J Oper Res 133:394–408
12. Eben-Chaime M, Mehrez A, Markovich G (2002) Capacitated location-allocation problems on a line. Comput Oper Res 29(5):459–470
13. Silva MR, Cunha CB (2009) New simple and efficient heuristics for the uncapacitated single allocation hub location problem. Comput Oper Res 36(12):3152–3165
14. Min H, Ko HJ, Ko CS (2006) A genetic algorithm approach to developing the multi-echelon reverse logistics network for product returns. Omega 34(1):56–69
15. Chen J-F (2007) A hybrid heuristic for the uncapacitated single allocation hub location problem. Omega 35(2):211–220
16. Topcuoglu H et al (2005) Solving the uncapacitated hub location problem using genetic algorithms. Comput Oper Res 32:967–984
17. Sha DY, Che ZH (2006) Supply chain network design: partner selection and production/distribution planning using a systematic model. J Oper Res Soc 57(1):52–62
18. Hoffmann CM (1989) The problems of accuracy and robustness in geometric computation. IEEE Comput 22(3):31–41
19. Marian RM, Luong LHS, Abhary K (2003) Assembly sequence planning and optimisation using genetic algorithms. Part I: Automatic generation of feasible assembly sequences. Appl Soft Comput (2/3F):223–253
20. Marian RM, Luong LHS, Dao SD (2011) Modeling and optimisation of distribution networks using hybrid genetic algorithms. A comparative study – The 2011 IAENG international conference on artificial intelligence and applications, Hong Kong, 16–18 March 2011

Chapter 10
Cooking-Step Scheduling Algorithm for Simultaneous Cooking of Multiple Dishes

Nobuo Funabiki and Yukiko Matsushima

Abstract Nowadays, cooking everyday is hard for busy persons such as workers, students, and child-rearing families, because the time is limited on weekdays. At the same time, it is important for a healthy diet not to rely on eating-out and instant foods heavily. One solution to this situation is to cook dishes for the whole week on the weekend and eat them on the following weekdays. Then, the way of cooking with a good cooking-step schedule for multiple dishes is very important to save the time. In this chapter, we first propose a *cooking model* to estimate the cooking time accurately under various conditions including the kitchen layout and the number of cooks. Then, using this model, we propose a *cooking-step scheduling algorithm* for simultaneous cooking of multiple dishes under the cooking time constraint. Through experiments, we verify the effectiveness of our model and algorithm, where the cooking time difference between the model and the real cooking is only 2 min.

Keywords Homemade cooking • Cooking model • Cooking-step scheduling • Algorithm

1 Introduction

Nowadays, cooking everyday is hard for busy persons such as workers, students, and child-rearing families, because the time is limited for this activity. Actually, a lot of people are eating at restaurants or buying lunch boxes at stores. As a result, they may fail to keep the well-balanced nutritional diet that has been recommended in the *Food Balance Guide* by the government [3]. On the other hand, there has been

N. Funabiki (✉) • Y. Matsushima
Department of Information and Communication Systems, Okayama University, 3-1-1 Tsushimanaka, Okayama, 700-8530, Japan
e-mail: funabiki@cne.okayama-u.ac.jp; matsushima@sec.cne.okayama-u.ac.jp

S.I. Ao et al. (eds.), *Intelligent Control and Innovative Computing*, Lecture Notes in Electrical Engineering 110, DOI 10.1007/978-1-4614-1695-1_10,

a great deal of public fears about the metabolic syndrome [4]. Thus, more and more people have desired to have a healthy and cost-effective diet by cooking foods at home.

One method to solve this situation can be to cook multiple dishes on a weekend, and to eat them on the following weekdays by keeping them in a refrigerator. However, this task of efficiently cooking various dishes at the same time on a weekend is often difficult for ordinary persons who have no professional skills because of the limited availability of pots, stoves, and microwave ovens in their kitchens. If they cook these dishes without a proper schedule, it may consume an inhibitory long time before completing to cook all the dishes. Besides, if they want to eat them at a dinner, some dishes completed first may become too cold when the others are completed. Thus, the weekend cooking of multiple dishes is actually hard for conventional persons.

In this chapter, in order to help such busy persons to cook multiple dishes efficiently, we first propose a *cooking model* to estimate the cooking time accurately under various conditions including the kitchen layout and the number of cooks [2]. Then, using this model, we propose a *cooking-step scheduling algorithm* to help ordinary busy persons to cook multiple dishes at the same time under the cooking time limitation. Through experiments, we verify the effectiveness of our proposal, where the cooking time difference between the model and the real cooking is only 2 min.

In the cooking model, we first define the *kitchen layout* that is composed of cooks, pans and pots, cutting boards, stoves, microwave ovens, and a sink. This model becomes flexible in terms of the number of kitchen utensils and cooks. Then, we define six *cooking-steps* in the cooking process, namely, *Cut-step*, *Mix-step*, *Fry-step*, *Boil-step*, *Nuke-step*, and *Wash-step*. Lastly, we define state transition diagrams of these items in the kitchen layout to determine the constraints of applying the cooking-steps.

For a given cooking order of multiple dishes, the cooking model calculates the minimum cooking time by executing their cooking-steps as early as possible such that the constraints are satisfied. For example, *Boil-step* and *Nuke-step* can be executed with other cooking-steps simultaneously, because they do not need manual operations by a cook. Besides, two types of cooks are considered in the model, namely a *main-cook* and a *sub-cook*. The main-cook can execute any cooking-step, and the sub-cook may execute a subset of the cooking-steps. The sub-cook is regarded as a helper for the main-cook, and can be a partner or a child. Actually, the projects by the Japanese government such as "Equal employment and work and family harmonization" [5] and "Ikumen (child-rearing men) project" [6] have expected to increase the opportunity of cooking together with such family members.

Our cooking-step scheduling algorithm seeks an optimal cooking-step schedule for a given set of multiple dishes based on the *simulated annealing* (*SA*) [7]. It repeats the random update of the dish order and the calculation of the cooking time using the cooking model to seek the optimal solution. The estimated cooking time of the algorithm can be used to plan menus for homemade cooking by busy persons under the limited cooking time [1].

2 Cooking Model

In this section, we present the cooking model to optimize the arrangement of the cooking-steps for multiple dishes and to estimate the cooking time accurately.

2.1 Kitchen Layout

Figure 10.1 illustrates the kitchen layout in the cooking model. The kitchen consists of cooks, pots, cutting boards, stoves, microwave ovens, and one sink. The number of cooks may be one or two. When two cooks are considered, we designate them as the *main-cook* and the *sub-cook*. The main-cook, assuming a housewife, can execute any cooking-step and complete any dish by himself (herself). On the other hand, the sub-cook, assuming a partner or child, can execute a part of the cooking-steps to help the main-cook. The number of pots, cutting boards, stoves, and microwave ovens can be specified by the model user, where they have the same performance. For simplicity, this model assumes that for any ingredient, the cleaning and the division into each dish are finished beforehand, and that there is enough space to put ingredients in the kitchen between cooking-steps.

We regard the cooking process for any dish as a sequence of *cooking-steps*. We define six different cooking-steps depending on the nature of the procedure, namely, *Cut-step*, *Mix-step*, *Fry-step*, *Boil-step*, *Nuke-step*, and *Wash-step*. The following four cooking-steps always require a cook load. In *Cut-step*, a cook processes

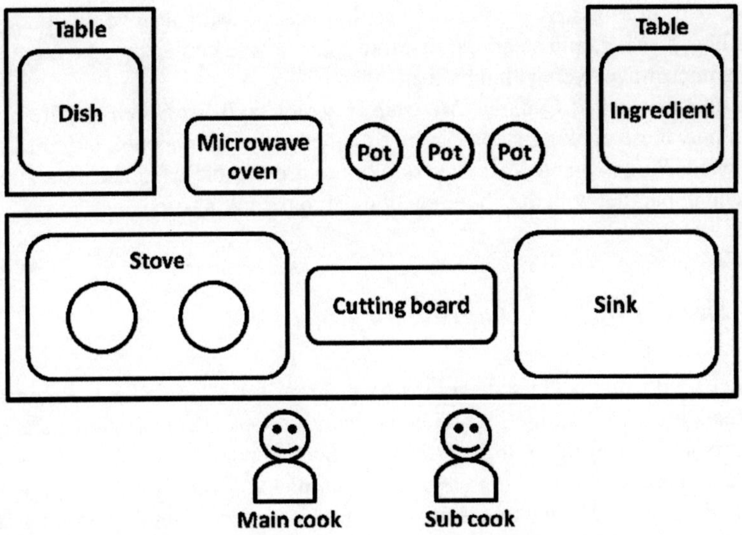

Fig. 10.1 Kitchen layout

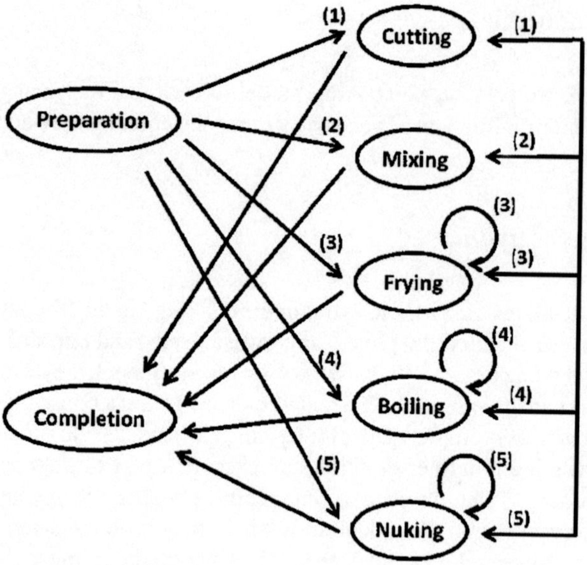

Fig. 10.2 State transition diagram for ingredient

ingredients by peeling, cutting, and slicing them on a cutting board. In *Mix-step*, a cook mixes ingredients by blending, kneading, and wrapping them. In *Fry-step*, a cook heats ingredients by frying, deep-frying, or grilling them using a pan on a stove where a cook is always engaged there to avoid scorching. In *Wash-step*, a cook cleans pans and pots in a sink. The following two cooking-steps do not require a cook load. In *Boil-step*, a cook can heat ingredients using a stove without a load by boiling, stewing, and steaming them. In *Nuke-step*, a cook can nuke ingredients using a microwave oven without a load.

Here, we note that *Cut-step*, *Mix-step*, *Fry-step*, and *Wash-step* require a cook load. Thus, these cooking-steps can be executed only when a cook is free. On the other hand, *Boil-step* and *Nuke-step* do not require a cook load. Thus, they can be executed in parallel with the above-mentioned four cooking-steps.

2.2 *State Transition Diagram for Ingredient*

Figure 10.2 illustrates the state transition diagram for an ingredient. *Preparation* represents the state that the ingredient is ready to cook. *Pause* does the state that the ingredient is waiting for the transition to another cooking-step. *Completion* does the state where all the cooking-steps have completed for a dish. *Cutting*, *Mixing*, *Frying*, *Boiling* and *Nuking* represent the states for the corresponding cooking-steps.

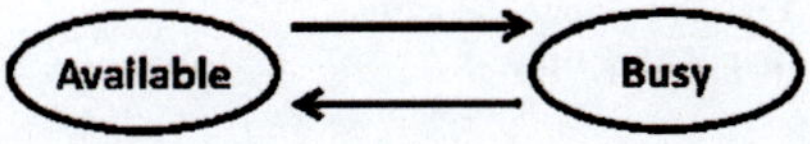

Fig. 10.3 State transition diagram for cook

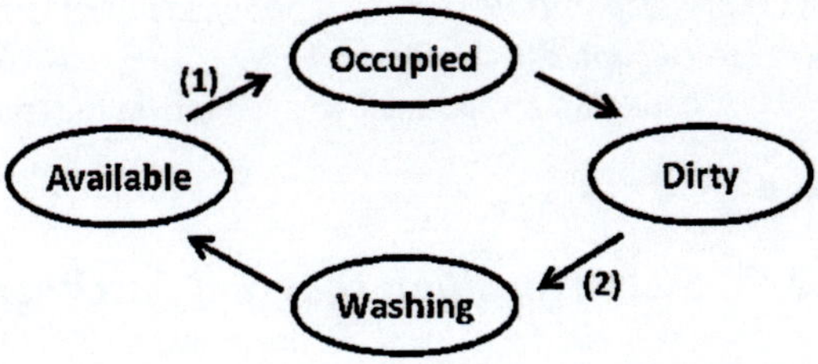

Fig. 10.4 State transition diagram for pot

Fry-step, *Boil-step*, and *Nuke-step* can transit to the same state to deal with the recipe like "first boiling, then stewing". Hence, the following five conditions must be satisfied at the transitions from (1) to (5) in Fig. 10.2:

(1) A cook and a cutting board must be available for the transition to *Cut-step*.
(2) A cook must be available for the transition to *Mix-step*.
(3) A cook, a stove, and a pan (a pot) must be available for the transition to *Fry-step*.
(4) A stove and a pan (a pot) must be available for the transition to *Boil-step*.
(5) A microwave oven must be available for the transition to *Nuke-step*.

2.3 *State Transition Diagram for Cook*

Figure 10.3 illustrates the state transition diagram for a cook. *Available* represents the state where the cook can execute a cooking-step requiring the load such as *Cut-step*, *Mix-step*, *Fry-step*, and *Wash-step*. *Busy* represents the state where a cook is executing a cooking-step and cannot perform another one. The cooking model can change the number of cooks and the roles of the main-cook and the sub-cook.

2.4 *State Transition Diagram for Pot*

Figure 10.4 illustrates the state transition diagram for a pot. *Available* represents the state where a pot has been washed and can be used for cooking. *Occupied* represents the state where it is currently used for cooking in *Fry-step* or *Boil-step*. *Dirty* represents the state where the cooking-step using the pot has finished. *Washing* represents the state where a cook is actually washing the pot in a sink. When *Dirty* happens, a cook stops the current cooking-step to wash the dirty pot, because the number of pots is limited. Hence, the following two conditions must be satisfied at the transitions of (1) and (2) in Fig. 10.4:

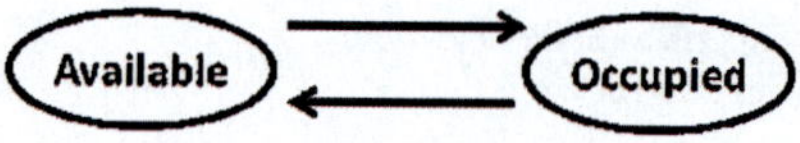

Fig. 10.5 State transition diagram for others

(1) *Fry-step* or *Boil-step* must be selected as the next cooking-step at the transition to *Occupied*.
(2) A cook and a sink must be available at the transition to *Wash-step*.

2.5 State Transition Diagram for Others

Figure 10.5 illustrates the state transition diagram for a cutting board, a stove, a microwave oven, and a sink. *Available* represents the state where the corresponding item can be used for cooking. *Occupied* represents the state where it is currently used.

2.6 Cooking Process Simulation

The cooking model estimates the completion time of cooking each dish by simulating the cooking processes of the multiple dishes, where the starting order of the dishes for the first cooking-steps is given. This starting order of the dishes is called the *dish order*. In this simulation, at every minute T, the following procedure is executed until all the dishes are completed.

(1) T is updated by $T{+}{+}$, where T is initialized by 0.
(2) The completion time of a dish is recorded, when all the states of the corresponding ingredients become *Completion*.
(3) The starting time of washing a pot is recorded, when the state of the pot becomes *Dirty*.
(4) The next cooking-step for a cook is selected and the starting time is recorded, when the state of a cook becomes *Available*.
(5) The next cooking-step for a stove or a microwave oven is selected and the starting time is recorded, when the state of the corresponding one becomes *Available*.

In the following subsections, we describe the details in Steps (3), (4), and (5) in the procedure.

2.6.1 Washing Pot

As soon as the state of a pot becomes *Dirty* and the state of a sink is *Available*, a cook washes the pot, assuming the pot washing has the first priority to other cooking-steps. Even if a cook is executing a different cooking-step, the cook interrupts it and starts washing the pot. Note that when the remaining number of cooking-steps using pots does not exceed the number of available pots, a cook does not execute *Wash-step*. Besides, a sub-cook is selected preferentially here, when the states of both cooks are *Available*.

2.6.2 Next Cooking-Step Selection for Cook

The cooking model selects the next cooking-step of a dish for an available cook by the following procedure.

(1) The first dish in the *dish order* that satisfies the following two conditions is found:

  (a) The state is *Preparation* or *Pause*.
  (b) The next cooking-step of the dish is neither *Boil-step* nor *Nuke-step* because they do not need a cook.

(2) When the next cooking-step of the found dish is *Cut-step*, the state of a cutting board is checked:

  (a) When the state of the cutting board is *Available*, the next cooking-step of the dish is selected for the cook, and the state of the cutting board is changed to *Occupied*.
  (b) Otherwise, this dish is given up, and (1) is repeated to check the next dish candidate.

(3) When the next cooking-step of the dish is *Mix-step*, this cooking-step is selected for the cook.
(4) When the next cooking-step of the dish is *Fry-step*, the state of a stove is checked:

  (a) When the state of the stove is *Available*, the next cooking-step of the dish is selected for the cook, and the state of the stove is changed to *Occupied*.
  (b) Otherwise, this dish is given up, and (1) is repeated to check the next dish candidate.

2.6.3 Next Cooking-Step Selection for Stove or Microwave Oven

The cooking model selects the next cooking-step of a dish for an available stove or microwave oven by the following procedure.

(1) The first dish in the *dish order* that satisfies the following two conditions is found:

  (a) The state is *Preparation* or *Pause*.

(b) The next cooking-step of the dish is not *Cut-step*, *Mix-step*, or *Fry-step* because they need a cook.

(2) When the next cooking-step of the found dish is *Boil-step*, the state of a stove is changed to *Occupied*.
(3) When the next cooking-step of the found dish is *Nuke-step*, the state of a microwave oven is changed to *Occupied*.

3 Cooking-Step Scheduling Problem and Algorithm

In this section, we formulate the cooking-step scheduling problem for multiple dishes, and propose its algorithm using the cooking model.

3.1 Cooking-Step Scheduling Problem

First, we formulate the cooking-step scheduling problem for multiple dishes as a combinatorial optimization problem.

3.1.1 Input

The inputs of this problem are as follows:

- the number of servings for cooking: m
- the number of pots, cutting boards, stoves, and microwave ovens
- the role of a sub-cook
- the washing time of a pot (min)
- the list of n dishes for cooking: $V = \{1, \ldots, n\}$
  - the name of dish i $(i \in V)$
  - the cooking-step of dish i at step j: *Cut-step*, *Mix-step*, *Fry-step*, *Boil-step*, and *Nuke-step*
  - the cooking time of dish i at step j (min)

3.1.2 Output

The output of the problem is the cooking-step schedule for all the dishes.

3.1.3 Constraint

The cooking-step schedule must satisfy the constraints in the cooking model in Sect. 2.

3.1.4 Objective

Two objective functions are defined for this problem. The first one represents the *maximum cooking time* to complete all the dishes in (10.1). The second one represents the *cooking time difference* between the first dish and the last one in (10.2). C_i represents the cooking time (min) to complete dish i.

$$f_1(\sigma) = \max_i \{C_i\}. \tag{10.1}$$

$$f_2(\sigma) = \max_i \{C_i\} - \min_i \{C_i\}. \tag{10.2}$$

3.2 Cooking-Step Scheduling Algorithm

A heuristic algorithm for the cooking-step scheduling problem is proposed based on the SA. The objective function is minimized by repeating a random small change of the *dish order* and the cooking time calculation using the cooking model.

3.2.1 Preprocessing Stage

In the preprocessing stage, the cooking time of each dish at each cooking-step is calculated according to the number of servings m. The duration of *Cut-step* is usually proportional to the quantity of the ingredients. Thus, the time is given by the product of the time for one person and m. The duration of *Fry-step*, *Boil-step*, *Mix-step*, and *Nuke-step* is not proportional to the quantity, although the rise of temperature is slightly delayed due to the increase of ingredients. Thus, the time is given by product of the time for one person and $(1+0.1\cdot m)$. Finally, the number of cooking-steps using pots in *Boil-step* or *Fry-step* is calculated to judge the necessity of *Wash-step*.

3.2.2 Dish Order Generation Stage

For a given set of dishes, our algorithm seeks the *dish order* σ to minimize the objective function, where the cooking model generates the cooking-step schedule from σ and calculates the cooking completion time. The initial value of σ is randomly generated. Then, the algorithm improves σ by swapping randomly selected two adjacent dishes.

3.2.3 Cooking Model Simulation Stage

By feeding σ to the cooking model, the cooking-step schedule is generated and the completion time of each dish is calculated. Then, the objective functions are calculated.

3.2.4 Dish Order Update Stage

The adoption of the newly generated *dish order* is judged based on SA. When σ_{old} denotes the previous solution and Δ denotes the difference between two objective function values ($= f(\sigma) - f(\sigma_{\text{old}})$), then σ is adopted with the probability of 1 for $\Delta \leq 0$. Otherwise, σ is adopted with the probability $e^{-\frac{\Delta}{t}}$ where t is the SA temperature.

3.2.5 Algorithm Parameter

In our algorithm, the temperature t is changed in constant times before termination, while the *dish order* is generated in constant times at each temperature. At each temperature change, t is multiplied by 0.95 [8]. Thus, the following three parameters, k_1, k_2, and k_3, should be adjusted in our algorithm [9–11].

(1) Initial temperature: $k_1 \times n$
(2) Number of temperature changes: $k_2 \times n$
(3) Number of dish order updates at each temperature: $k_3 \times n$

4 Evaluation

In order to evaluate our proposal, we implemented the cooking model and the cooking-step scheduling algorithm as Java applications.

4.1 Algorithm Parameter Adjustment

First, we adjust the three algorithm parameters, k_1, k_2, and k_3, through applying our algorithm to an instance of randomly generated 10 dishes. Either of 5, 10, and 20 is

Table 10.1 Algorithm parameter adjustment

k_1	k_2	k_3	f_1	k_1	k_2	k_3	f_1	k_1	k_2	k_3	f_1
5	5	5	81	10	5	5	85	20	5	5	87
5	5	10	84	10	5	10	85	20	5	10	88
5	5	20	83	10	5	20	86	20	5	20	89
5	10	5	72	10	10	5	70	20	10	5	77
5	10	10	66	10	10	10	74	20	10	10	76
5	10	20	72	10	10	20	75	20	10	20	77
5	20	5	72	10	20	5	65	20	20	5	67
5	20	10	64	10	20	10	70	20	20	10	72
5	20	20	68	10	20	20	66	20	20	20	69

Table 10.2 Dishes and cooking time in simulation instance for accuracy evaluation

i	Dish	Process	$j=1$	2	3	4
		Type	Cut	Mix	Boil	
1	Stuffed sweet pepper	min	2	5	15	
		Type	Cut	Fry	Boil	
2	Fried clams	min	2	4	3	
		Type	Boil	Nuke	Cut	Fry
3	Fried rice & salmon	min	10	2	3	5
		Type	Cut	Boil		
4	Miso soup	min	1	10		
		Type	Cut	Nuke	Cut	
5	Eggplant salad	min	1	3	1	
		Type	Cut	Mix		
6	Cabbage pickles	min	2	2		

used for them to find the best choice. As the cooking environment, the number of cooks is set two, where the main-cook can execute any cooking-step and the sub-cook can execute *Mix-step*, *Fry-step*, and *Wash-step*. The number of pots, cutting boards, stoves, and microwaves ovens are set three, one, two, and one, respectively. The number of servings m is four. The washing time for a pot is set three minutes. Table 10.1 shows the best value of f_1 among 30 executions using different random numbers, where the parameter set of $k_1 = 5$, $k_2 = 20$, and $k_3 = 10$ provides the best result of $f_1 = 64$.

4.2 Cooking Time Accuracy Evaluation

Then, we select the six dishes in Table 10.2 as a simulation instance to evaluate the accuracy of the estimated cooking time in the model. Table 10.3 shows the best values for the two objective functions among 20 executions by our algorithm in the same cooking environment for the parameter adjustment. Table 10.4 shows the cooking-step schedule using the best solution for f_1.

Table 10.3 Objective function values

	f_1(min)	f_2(min)
Initial state	68	55
Final state	52	25

Table 10.4 Cooking-step schedule for f_1 optimization

Start	Who	Dish	Process	Minutes
0	Main	Eggplant salad	Cut	4
	Stove0	Fried rice & salmon	Boil	14
4	Main	Stuffed sweet pepper	Cut	8
	M-oven	Eggplant salad	Nuke	4
12	Main	Eggplant salad	Cut	4
	Sub	Stuffed sweet pepper	Mix	7
14	Sub	Pot	Wash	3
	M-oven	Fried rice & salmon	Nuke	2
16	Main	Miso soup	Cut	4
20	Main	Fried clam	Cut	8
	Stove0	Miso soup	Boil	14
22	Stove1	Stuffed sweet pepper	Boil	21
28	Main	Fried rice & salmon	Cut	12
34	Sub	Pot	Wash	3
37	Sub	Fried clam	Fry	5
40	Main	Cabbage pickles	Cut	8
42	Sub	Pot	Wash	3
	Stove0	Fried clam	Boil	4
45	Sub	Fried rice & salmon	Fry	7
48	Main	Cabbage pickles	Mix	2

To evaluate the accuracy of the cooking time, we actually cooked the six dishes in the same environment with the cooking-step schedule in Table 10.4. Then, the difference between the estimated cooking time by the model and the actual cooking time by real cooking is only 2 min, where the estimated time is 52 min and the actual time is 54 min.

4.3 Cooking Time Change by Sub-cook Role Change

Finally, we evaluate the scalability of our proposal when the number of dishes increases and the role of the sub-cook is changed. Figure 10.6 shows the changes of f_1, where C, M, F, and W represent *Cut-step*, *Mix-step*, *Fry-step*, and *Wash-step* that can be executed by the sub-cook respectively. As the number of roles increases for a sub-cook, the cooking time decreases. When a sub-cook can execute all of the cooking-steps, the cooking time becomes about the half of the time by a single cook.

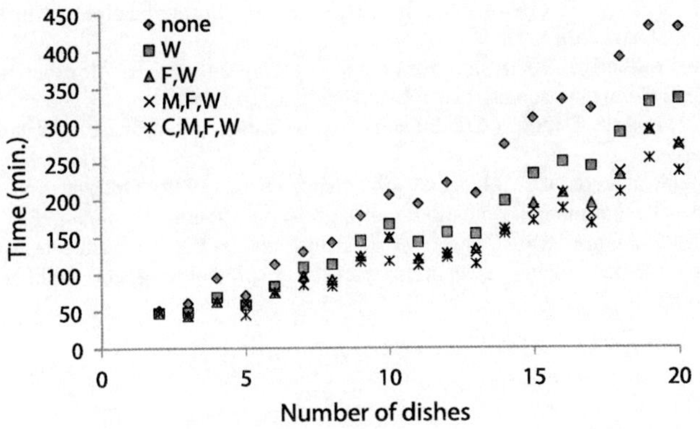

Fig. 10.6 Change of f_1 by role changes of sub-cook

5 Conclusion

In this chapter, we proposed a cooking model and a cooking-step scheduling algorithm for simultaneous cooking of multiple dishes by busy persons. We verified the accuracy of the estimated cooking time by the model and the effectiveness of our algorithm through the simulation using six dishes. In addition, we confirmed that the increase of roles by a sub-cook can reduce the cooking time. Our future works may include further improvements of the cooking model and algorithm, the construction of a cooking-step database for our proposal, and the development of a total cooking support system using a Web application.

References

1. Funabiki N, Taniguchi S, Matsushima Y, Nakanishi T (2011) A proposal of a menu planning algorithm for two-phase cooking by busy persons. In: Proceedings of 3rd international workshop on virtual environment and network-oriented applications (VENOA 2011), pp. 668–673.
2. Matsushima Y, Funabiki N, Nakanishi T (2011) A proposal of cooking model and cooking step scheduling algorithm for multiple dishes. In: Proceedings of international multiconference of engineers and computer scientists 2011 (IMECS 2011), Hong Kong, 16–18 March. Lecture notes in engineering and computer science, pp. 127–131.
3. Ministry of Health, Labour, and Welfare (MHLW) (2005) Food balance guide. *http://www.mhlw.go.jp/bunya/kenkou/eiyou-syokuji.html*. Accessed on November 14, 2011.
4. Ministry of Health, Labour, and Welfare (MHLW) (2007) Prevention of metabolic syndrome. *http://www.mhlw.go.jp/bunya/kenkou/metabo02/*. Accessed on November 14, 2011.
5. Ministry of Health, Labour, and Welfare (MHLW) (2011a) Equal employment. *http://www.mhlw.go.jp/bunya/koyoukintou/index.html*. Accessed on November 14, 2011.
6. Ministry of Health, Labour, and Welfare (MHLW) (2011b) Ikumen (child-rearing men) project. *http://www.ikumen-project.jp/index.html*. Accessed on November 14, 2011.

7. Osman IH, Kelly JP (1996) Meta-heuristics: theory & applications. Kluwer Academic Publishers, Norwell, MA, USA.
8. Sadiq MS, Youssef H (2000) Iterative computer algorithms with applications in engineering. Wiley-IEEE Computer Society Press, Los Alamitos, CA, USA.
9. Yagiura M, Ibaraki T (2001) Combinatorial optimization (in Japanese). Asakura Publishers, Tokyo, Japan.
10. Johnson DS, Aragon CR, Mcgeoch LA, Schevom C (1989) Optimization by simulated annealing: an experimental evaluation; part I, graph partitioning, pp. 865–892.
11. Johnson DS, Aragon CR, Mcgeoch LA, Schevom C (1991) Optimization by simulated annealing: an experimental evaluation; part II, graph coloring and number partitioning, pp. 378–406.

Chapter 11
Estimation of Distributed Generation Using Complex-Valued Network Inversion with Regularization

Takehiko Ogawa, Kyosuke Nakamura, and Hajime Kanada

Abstract Network inversion has been studied as a neural network based solution of inverse problems. Complex-valued network inversion has been proposed as the extension of this inversion to the complex domain. Further, regularization is considered for solving ill-posed inverse problems. On the other hand, the estimation of the parameters of a distributed generation from observed data is a complex-valued inverse problem with ill-posedness. In this chapter, we propose the application of a complex-valued network inversion with regularization to the inverse estimation of a distributed generation.

Keywords Complex-valued neural networks • Distributed generation • Ill-posed inverse problems • Regularization

1 Introduction

The inverse problem of estimating causes from observed results has been studied in various engineering fields [1, 14]. Network inversion has been studied as a neural network based solution of inverse problems [2]. While the original network inversion method has been applied to usual real-valued neural networks, the complex-valued network inversion method has been proposed to solve inverse problems on a complex-valued neural network [3]. Inverse problems are generally ill-posed,

T. Ogawa (✉) • H. Kanada
Department of Electronics and Computer Systems, Takushoku University, 815–1 Tatemachi, Hachioji-shi, Tokyo 193–0985 Japan
e-mail: togawa@es.takushoku-u.ac.jp; khajime@es.takushoku-u.ac.jp

K. Nakamura
Electronics and Information Science Course, Graduate School of Engineering, Takushoku University, 815–1 Tatemachi, Hachioji-shi, Tokyo 193–0985 Japan
e-mail: y0m317@st.takushoku-u.ac.jp

S.I. Ao et al. (eds.), *Intelligent Control and Innovative Computing*, Lecture Notes in Electrical Engineering 110, DOI 10.1007/978-1-4614-1695-1_11,

which implies that the existence, uniqueness, and stability of their solution are not guaranteed. Regularization imposes specific conditions on an ill-posed inverse problem to convert it into a well-posed problem [4]. In the case of complex-valued network inversion, regularization has been examined on simple inverse problems [5].

Distributed generation is an important technique that involves natural power sources or fuel cells. A distributed generation system is an electric power network that connects a number of small-scale power supplies [612]. In general, a small-scale power supply is unstable because of the characteristics of natural energy and the required cost reduction. A stable operation of the electric power network is important. Hence, it is necessary to control a small-scale power supply for the stabilization of the electric power. In distributed generation, it is important to estimate the parameters of several power supplies to control them. The problem of estimating the parameters from a large amount of observed data is an ill-posed inverse problem featuring complex numbers [6].

In this chapter, we propose the application of complex-valued network inversion with regularization to the ill-posed inverse estimation of the distributed generation. Specifically, we consider the problem of estimating the voltage of the power supply from the observed output voltage and current of a circuit with two power supplies. The problem of ill-posedness concerning the uniqueness of the solution appears in the inverse estimation of the distributed generation. To show the effect of the proposed method, we carry out a simulation using the complex-valued network inversion with regularization.

2 Inverse Problems and Network Inversion

The inverse problem refers to the problem of estimating the cause from the observed phenomenon. The cause is estimated from the fixed model and given result in the inverse problem. The solution of inverse problems is important in various science and engineering fields [1], such as astronomy [7], and inverse kinematics [8].

Network inversion is a method for solving inverse problems using multilayer neural networks. Network inversion has been applied to image restoration [9] and the inverse kinematics of robot arms [10]. In this method, we estimate the corresponding input from a given output using a trained network. The network is typically trained using the error back-propagation method. In the trained network, we provide the observed output with fixed trained weights. The input can then be updated according to the calculated input correction signal. Essentially, the input is estimated from the output using an iterative update of the input based on the output error, as shown in Fig. 11.1. By doing so, the inverse problem of estimating input from the given output is solved using the multilayer neural network.

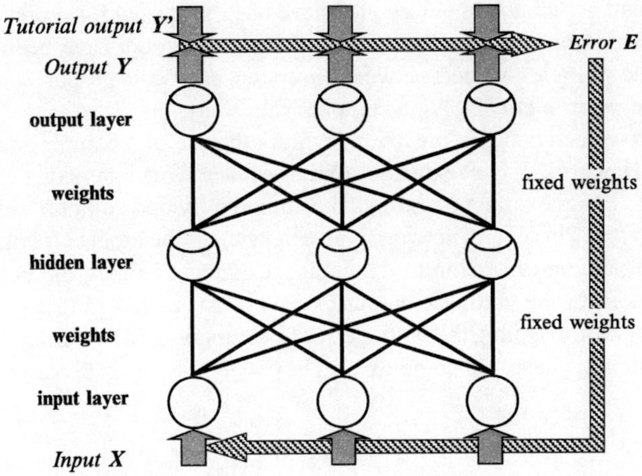

Fig. 11.1 Iterative update of input using network inversion

To solve the inverse problem using network inversion, the network is used in two phases: forward training and inverse estimating. In the training phase, the weights w are updated using

$$w(n+1) = w(n) - \varepsilon_t \frac{\partial E}{\partial w} \tag{11.1}$$

where x, y, E, and ε_t are the training input, training output, output error, and training gain, respectively. It is assumed that the output error is caused by the inaccurate adjustments of the weights in the training phase. This is the procedure employed in the usual error back-propagation method. In the inverse estimation phase, the input x is updated using

$$x(n+1) = x(n) - \varepsilon_e \frac{\partial E}{\partial x} \tag{11.2}$$

where x, y, E, and ε_e are the random input, provided output, output error, and input update gain, respectively. It is assumed that the output error is caused by the maladjustments of the input during inverse estimation. By repeating this updating procedure, the input is estimated from the provided output [2].

2.1 Complex-Valued Network Inversion

The original network inversion procedure is a solution for inverse problems using a typical real-valued multilayer neural network. Recently, an extension of the multi-layer neural network to the complex domain has been studied. These networks learn the complex input–output relation using complex weights and complex neurons. Various models such as the multilayer-type neural network [11], the self-organizing map [12], and associative memory [13] have been proposed for representing these networks, and a number of applications of these networks have been studied. In addition, the complex-valued network inversion has been proposed to solve the general inverse problem involving complex values [3].

Complex-valued network inversion involves the use of a complex-valued multi-layer neural network. It is an extension of the usual network inversion to the complex domain. The complex input is estimated from the provided complex output using a trained network. This neural network is an extension of the input correction principle of a typical network inversion to the complex domain. In fact, the complex input is estimated from the complex output by giving the random input to the trained network, back-propagating the output error to the input, and repeating the input.

In the training phase, the complex weight $w = w_R + iw_I$ is updated using

$$w_R\,(n+1) = w_R\,(n) - \varepsilon_t \left(\frac{\partial E_R}{\partial w_R} + \frac{\partial E_I}{\partial w_R} \right)$$

$$w_I\,(n+1) = w_I\,(n) - \varepsilon_t \left(\frac{\partial E_I}{\partial w_I} - \frac{\partial E_R}{\partial w_I} \right) \qquad (11.3)$$

where $x = x_R + ix_I$, $y = y_R + iy_I$, $E = E_R + iE_I$, and ε_t are the complex training input, training output, output error, and training gain, respectively. By repeating this updating procedure, a forward relation is obtained. This is the usual complex error back-propagation method. In the inverse estimation phase, the complex input $x = x_R + ix_I$ is updated using

$$x_R\,(n+1) = x_R\,(n) - \varepsilon_e \left(\frac{\partial E_R}{\partial x_R} + \frac{\partial E_I}{\partial x_R} \right)$$

$$x_I\,(n+1) = x_I\,(n) - \varepsilon_e \left(\frac{\partial E_I}{\partial x_I} - \frac{\partial E_R}{\partial x_I} \right) \qquad (11.4)$$

where $x = x_R + ix_I$, $y = y_R + iy_I$, $E = E_R + iE_I$, and ε_e are the complex random input, provided output, output error, and input update gain, respectively. By repeating this procedure, the complex input approaches the corresponding value of the provided output. When the error becomes sufficiently small, the input correction is completed and the obtained complex input becomes a solution. As a result, the complex input can be inversely estimated from the complex output using the trained complex weights [3,5].

2.2 *Ill-Posedness and Regularization*

In the inverse problem, the existence, uniqueness, and stability of solution are not guaranteed. The problem that does not satisfy these three conditions is referred to as being ill-posed. The ill-posedness of a problem is an important issue in complex-valued network inversion.

In this chapter, we consider the regularization for complex-valued network inversion. The method is based on Tikhonov regularization [4]. We use a regularization functional that is minimized in accordance with the output error in the inverse estimation phase in order to impose the constraint condition. For real-valued network inversion, we define the energy function E with the regularization functional $K(x)$ as

$$E = \frac{1}{2}\sum_{r}(y'_r - y_r)^2 + \lambda \sum_{k} K(x_k) \tag{11.5}$$

where $y'_r = y'_{rR} + i\ y'_{rI}$, $y_r = y_{rR} + i\ y_{rI}$, and $x_k = x_{kR} + i\ x_{kI}$ are the r-th tutorial output, r-th network output, and k-th input, respectively. The first and second terms represent the output error and regularization functional, respectively. The parameter λ is the regularization coefficient. The objective of the regularization term is to express the restraint condition that the input should satisfy and to be minimized simultaneously with the error term. By using the regularization term, we aim to obtain a feasible solution that minimizes the error and satisfies the restraint condition.

Next, we extend the regularization to complex-valued network inversion as

$$E_R = \frac{1}{2}\sum_{r}(y'_{rR} - y_{rR})^2 + \lambda \sum_{k} K(x_{kR})$$

$$E_I = \frac{1}{2}\sum_{r}(y'_{rI} - y_{rI})^2 + \lambda \sum_{k} K(x_{kI}) \tag{11.6}$$

where $y'_r = y'_{rR} + i\ y'_{rI}$, $y_r = y_{rR} + i\ y_{rI}$, and $x_k = x_{kR} + i\ x_{kI}$ are the r-th complex provided output, r-th complex network output, and k-th complex input, respectively [5].

The coefficient λ is usually fixed and determined empirically or by trial and error. However, the balance between the output error minimization and regularization is often uncertain. The regularization coefficient may actually be reduced with a decrease in the output error. We refer to this as dynamic regularization. Dynamic regularization is also applicable to the complex-valued network inversion. We consider the decay of the regularization coefficient $\lambda(t)$ from $\lambda(0)$ to zero with the epoch number *t*.

3 Inverse Estimation of Distributed Generation

One example of distributed generation is the small-scale power supply allocated near a consumer. The distributed generation system can be expected to decrease the environmental load and power cost and to increase the stability of the power supplies. However, it is difficult to maintain the quality of the electric power and to detect the incidents in case of numerous power supplies.

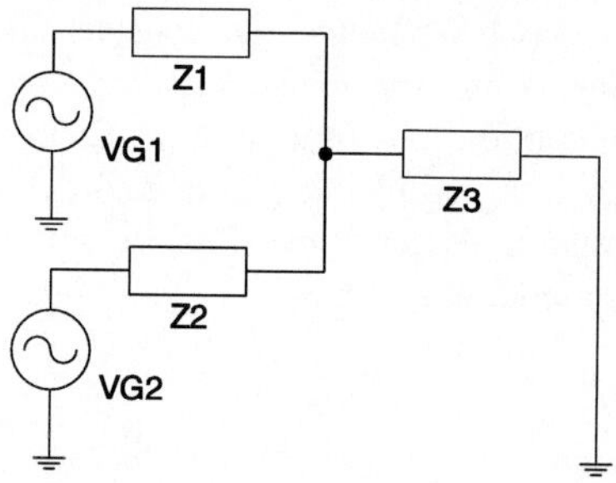

Fig. 11.2 Circuit model of distributed generation for simulation

In this study, we consider the inverse problem of estimating the voltage of the power supplies from the observed voltage and current data. We compose an inverse estimation problem using a simple AC electric circuit model that consists of several power supplies. In this circuit, we assume that the impedance element is driven by many AC power supplies. Concretely, we use a model that includes two power supplies with complex impedance and output complex impedance, as shown in Fig. 11.2. The circuit parameters are as follows. The values of each impedance are $Z_1 = Z_2 = Z_3 = 1 + i\,[\Omega]$. The parameters in the power supplies VG_1 and VG_2 are varied to obtain the training data. In addition, we prepare the test data by changing the output voltage and current. The amplitudes of VG_1 and VG_2 are varied from 30 to 180 [V] in steps of 30 [V]. Their phase is also varied from $-150°$ to $180°$ in $30°$ steps.

The problem of ill-posedness, which concerns the uniqueness of the solution, arises in the inverse estimation of the power supplies VG_1 and VG_2. This problem is ill-posed because we cannot distinguish between VG_1 and VG_2 and hence treat them as the same performances. In this study, we examine the ill-posed inverse estimation of the parameters of the power supply using complex-valued network inversion under conditioning using regularization. We use the following regularization term

$$K(x_{kR}) = \frac{1}{2}\sum_{k}\left(x'_{kR} - x_{kR}\right)^2$$
$$K(x_{kI}) = \frac{1}{2}\sum_{k}\left(x'_{kI} - x_{kI}\right)^2 \tag{11.7}$$

where k is the number of neurons that we impose on a specific condition. Here, we provide the correct value of VG_1 or VG_2 as $x_k' = x'_{kR} + i\,x'_{kI}$. This implies that the restraint condition that VG_1 or VG_2 is correct is considered to be the regularization. The input of the complex-valued network inversion with regularization is also updated based on (11.4). The input has to be updated in the real and imaginary parts of the complex-valued network inversion. The ill-posedness in the complex domain can be reduced by iterative correction of the input.

4 Simulation

We carry out the inverse estimation of the power supply parameters of the distributed generation circuit shown in Fig. 11.2. We use a complex-valued neural network with two input neurons and two output neurons, which correspond to the complex voltages of the two sources and the measured complex current and voltage, respectively. These values are normalized by each maximum and minimum value and used as an input and output value for the network. Here, we consider a neuron that independently applies the sigmoid function to the real and imaginary parts of the weighted sum of inputs. This neuron applies a complex sigmoid function to each real part and imaginary part independently and can be defined as

$$f_C(s) = f(s_R) + if(s_I), \quad f(u) = \frac{1 - e^{-u}}{1 + e^{-u}} \tag{11.8}$$

where i and $s = s_R + is_I$ indicate the imaginary unit and the weighted sum of the neuron input, respectively. The network architecture and network parameters are shown in Fig. 11.3 and Table 11.1, respectively.

To confirm the regularization, we perform the following simulation. First, we examine the inverse estimation without the regularization. The network learns the input/output relation using the training data. We inversely estimate the corresponding input to the provided output by using the trained network. Next, we perform

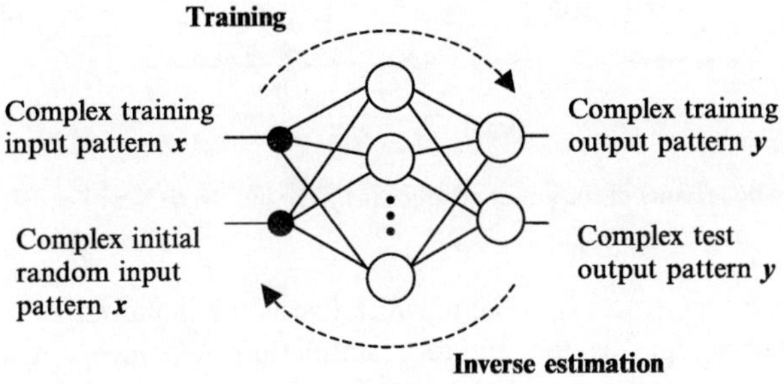

Fig. 11.3 The network architecture used in simulation

Table 11.1 Network parameters

Number of input neurons	2
Number of hidden neurons	10
Number of output neurons	2
Training rate ε_t	0.0001
Input correcting rate ε_e	0.0001
Maximum number of training epochs	10,000
Maximum number of estimating epochs	10,000
Training error to be attained	0.0001
Estimation error to be attained	0.0001

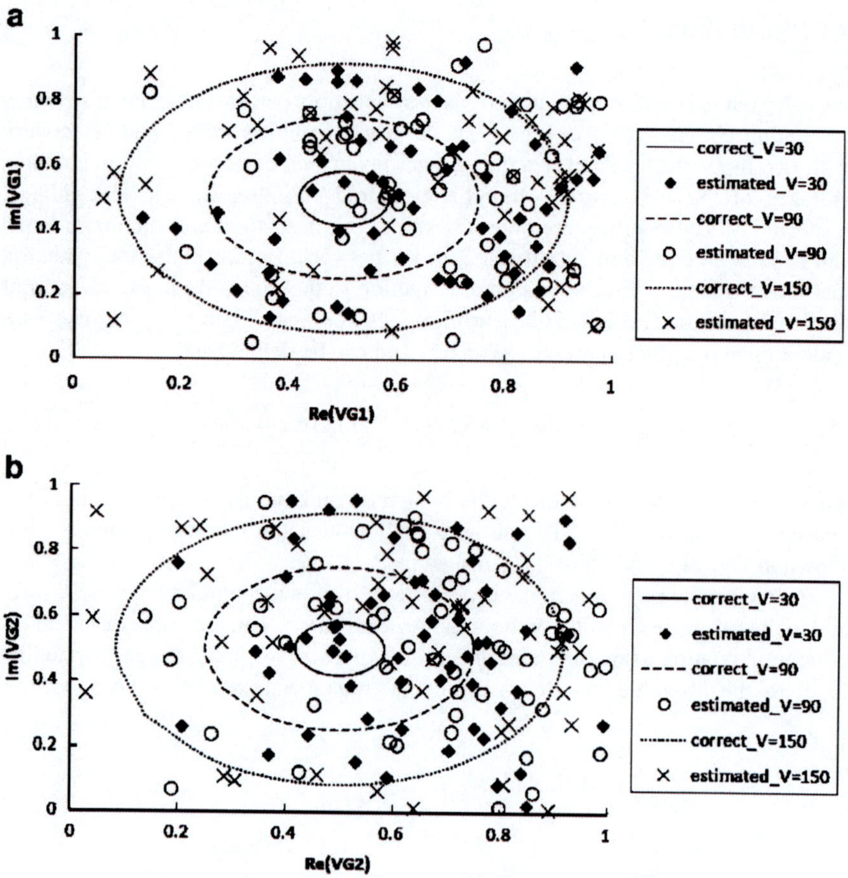

Fig. 11.4 Estimated results of the power supplies (**a**) VG_1 and (**b**) VG_2 without regularization

regularization for VG_1 or VG_2. The network learns the input/output relation using the same training data as the previous simulation. We inversely estimate the corresponding input to the provided output by using the trained network with regularization for VG_1 or VG_2. The dynamic regularization that is explained in previous section is applied to the complex-valued network inversion. The regularization coefficient $\lambda(t)$ in (11.6) is reduced for each correction of input as

$$\lambda(t) = \lambda(0)\left(1 - \frac{t}{2T}\right) \tag{11.9}$$

where $\lambda(0)$, t, and T are the initial value of λ, training epoch t, and maximum number of training epochs, respectively. The initial value $\lambda(0)$ is set to 0.5. We carry out each simulation five times and show the plots of the five estimated results.

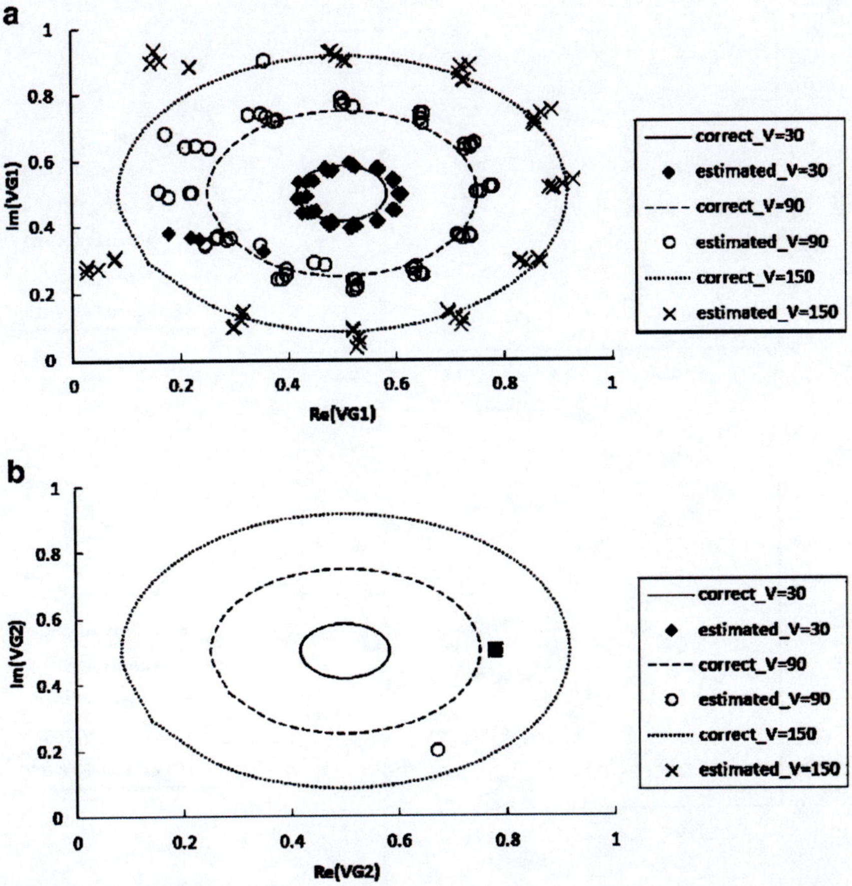

Fig. 11.5 Estimated results of the power supplies **(a)** VG_1 and **(b)** VG_2 with regularization for VG_2

4.1 Results

The inversely estimated inputs without regularization are shown in Fig. 11.4. Figure 11.4a and b shows the estimated results of VG_1 and VG_2, respectively. We found that both the estimated inputs are not the correct voltages of VG_1 and VG_2. This is because the solution cannot be computed in the correct manner because of the ill-posedness of the problem with respect to the uniqueness of the solution.

The estimated inputs with regularization are shown in Figs. 11.5 and 11.6. Figure 11.5(a) and (b) shows the estimated results of VG_1 and VG_2, respectively, when the limitation is imposed to VG_2 by the regularization. Similarly, Fig. 11.6a and b shows the estimated results of VG_1 and VG_2, respectively, when the limitation is imposed on VG_1. The results showed that the input correction is sufficiently

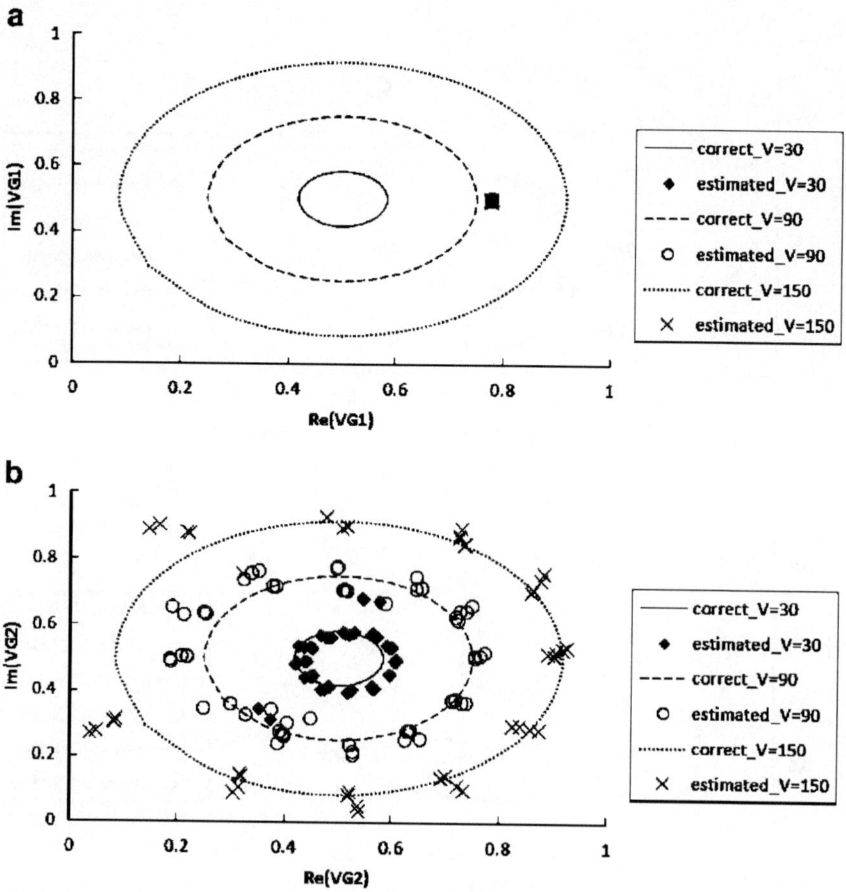

Fig. 11.6 Estimated results of the power supplies (**a**) VG_1 and (**b**) VG_2 with regularization for VG_1

suitable and the input is correctly estimated when the limitation is imposed on one of the inputs. Therefore, the problem of ill-posedness is not the initial-value problem, and we confirmed the effectiveness of the regularization method.

5 Conclusion

In this study, we proposed the application of a complex-valued network inversion to the ill-posed inverse estimation of the distributed generation. We carried out the simulation of the inverse estimation of the voltage source using a simple distributed generation model. To investigate the effect of ill-posedness, we examined the complex-valued network inversion with regularization. Consequently, we confirmed that the complex-valued network inversion method with regularization was effective in solving the ill-posed inverse estimation problem of distributed generation.

In future studies, we will attempt to increase the number of voltage sources and improve their estimation accuracy. In addition, we intend to search for a more effective regularization method for actual problems by examining different types of regularization methods.

Acknowledgement This work was supported in part by a Grant-in-Aid for Scientific Research #21700260 from the Japan Society for the Promotion of Science.

References

1. Grötsch CW (1993) Inverse problems in the mathematical sciences. Informatica International
2. Linden A, Kindermann J (1989) Inversion of multilayer nets. Proceedings of the International Joint Conference on Neural Networks II, pp. 425–430
3. Ogawa T, Kanada H (2005) Network inversion for complex-valued neural networks. Proceedings of the IEEE International Symposium on Signal Processing and Information Technology, pp. 850–855
4. Tikhonov AN, Arsenin VY (1977) Solutions of ill-posed problems. Winston
5. Fukami S, Ogawa T, Kanada H (2008) Regularization for complex-valued network inversion. Proceedings of the SICE Annual Conference, 1237–1242
6. Korovkin NV, Chechurin VL, Hayakawa M (2007) Inverse problems in electric circuits and electromagnetics. Springer
7. Craig IJD, Brown JC (1986) Inverse problems in astronomy. Adam Hilger
8. Craig JJ (1989) Introduction to robotics mechanics and control. Addison-Wesley
9. Valova I, Kameyama K, Kosugi Y (1995) Image decomposition by answer-in-weights neural network. IEICE Trans Inform Syst E78-D-9:1221–1224
10. Lu B-L, Ito K (1999) Regularization of inverse kinematics for redundant manipulators using neural network inversion. Proceedings of the IEEE international conference on neural networks, pp.2726–2731
11. Nitta T (1997) An extension of the backpropagation algorithm to complex numbers. Neural Networks 10(8):1392–1415
12. Hirose A, Hara T (2003) Complex-valued self-organizing map dealing with multi-frequency interferometeric data for radar imaging systems. Proceedings of WSOM 2003, pp. 255–260
13. Nemoto I, Kubono M (1996) Complex associative memory. Neural Networks 9:253–261
14. Ogawa T, Nakamura K, Kanada H (2011) Solution of ill-posed inverse problem of distributed generation using complex-valued network inversion. Lecture notes in engineering and computer science: proceedings of the international multiconference of engineers and computer scientists 2011, IMECS 2011, Hong Kong, 16–18 March 2011

Chapter 12
A New Method of Cooperative PSO: Multiple Particle Swarm Optimizers with Inertia Weight with Diversive Curiosity

Hong Zhang

Abstract This chapter presents a new method of cooperative PSO—multiple particle swarm optimizers with inertia weight with diversive curiosity (MPSOIWα/DC). Compared to a plain MPSOIW, it has the following outstanding features: (1) Decentralization in multi-swarm exploration with hybrid search, (2) Concentration in evaluation and behavior control with diversive curiosity (DC), (3) Practical use of the results of an evolutionary PSOIW, and (4) Their effective combination. The actualization of its overall composition expands the applied object of cooperative PSO, and effectually alleviates stagnation in optimization with the multi-swarm's decision-making. To demonstrate the effectiveness of the proposal, computer experiments on a suite of multi-dimensional benchmark problems are carried out. We examine its intrinsic characteristics, and compare the search performance with other methods. The obtained experimental results clearly indicate that the search performance of the MPSOIWα/DC is superior to that of the PSOIW, OPSO, and RGA/E, and is better than that of the MPSOα/DC, and MCPSOα/DC except for the Rosenbrock Problem.

Keywords Cooperative particle swarm optimization • Hybrid search • Localized random search • Diversive and specific curiosity • Swarm intelligence

1 Introduction

In comparison with some traditional optimization methods, e.g., steepest descent method, conjugate gradient method and quasi-Newton method, which may be good at solution accuracy and exact computation but have brittle operations and necessary

H. Zhang (✉)
Department of Brain Science and Engineering, Kyushu Institute of Technology, 2-4 Hibikino, Wakamatsu, Kitakyushu 808-0196, Japan
e-mail: zhang@brain.kyutech.ac.jp

S.I. Ao et al. (eds.), *Intelligent Control and Innovative Computing*, Lecture Notes in Electrical Engineering 110, DOI 10.1007/978-1-4614-1695-1_12,

information to search environment, the methods of genetic and evolutionary computation (GEC) generally provide a more robust, efficient, and expandable approach in treating with high-grade nonlinear, multi-modal optimization problems, and complex practical problems in the real-world without prior knowledge [13, 17, 24].

As a new member of GEC, particle swarm optimization (PSO) [8, 15] has been successfully applied in different fields of science, technology, engineering, and applications [21]. This is because the technique has distinctive features: information exchange, intrinsic memory, and directional search in the mechanism and composition compared to the other members such as genetic algorithms [12] and evolutionary programming [11].

For improving the convergence, solution accuracy, and search efficiency of a plain particle swarm optimizer (the PSO), many basic variants such as a particle swarm optimizer with inertia weight [22], and a canonical particle swarm optimizer [5, 6], etc., were proposed. Especially, in recent years, the approach of multi-swarm search is rapidly developing from the basis of the approach of single-swarm search. A large number of studies and investigations on cooperative PSO in relation to symbiosis, interaction, and synergy are in the researcher's spotlight. Various kinds of the algorithms on cooperative PSO, for instance, hybrid PSO, multi-population PSO, multiple PSO with decision-making strategy, etc., were published [1, 10, 19, 26].

Compared to those methods that only operate a particle swarm, it is an indisputable fact that different attempts, schemes and strategies to reinforcement of multi-swarm search can be fulfilled well, which mainly focus on the rationality of information propagation, cooperation, optimization, and intelligent control within these particle swarms for efficiently finding an optimal solution or near-optimal solutions [4, 14, 28].

Due to great requests to enlarge search performance in reality, utilizing the techniques of group searching, parallel processing with intelligent strategy has become one of extremely important approaches to optimization. For magnifying cooperative PSO research and improving the search performance of a plain multiple particle swarm optimizers with inertia weight (MPSOIW), a new method of cooperative PSO: multiple particle swarm optimizers with inertia weight with diversive curiosity (MPSOIWα/DC) [29] is described.

In comparison with the plain MPSOIW, the proposal has the following outstanding features: (1) Decentralization in multi-swarm exploration with hybrid search (MPSOIWα), (2) Concentration in evaluation and behavior control with diversive curiosity (DC), (3) Practical use of the results of an evolutionary PSOIW (PSOIW*), and (4) Their effective combination. According to these features, the MPSOIWα/DC could be expected to enhance search ability by alleviating stagnation in optimization with group decision-making, and to improve solution accuracy by enforcement of managing the trade-off between exploitation and exploration in the multi-swarm's heuristics.

From the viewpoint of methodology, the MPSOIWα/DC [31] is an analogue of the method of multiple particle swarm optimization with DC [30], which has been successfully applied to the plain multiple particle swarm optimizers (MPSO) and multiple canonical particle swarm optimizers (MCPSO). Nevertheless, the creation of the proposal is not only to improve the search performance of the plain MPSOIW, but also to expand the applied object and area of the curiosity-driven multi-swarm search, further to develop various methods of cooperative PSO by an integration manner.

2 Algorithms

In this section, the algorithms of the PSO, PSOIW, and PSOIWα/DC are briefly described. For the convenience of the description, let the search space be N-dimensional, $\mathbf{\Omega} \in \Re^N$, the number of particles in a swarm be P, the position and velocity of the ith particle be $\vec{x}^i = \left(x^i_1, x^i_2, \ldots, x^i_N\right)^T$ and $\vec{v}^i = \left(v^i_1, v^i_2, \ldots, v^i_N\right)^T$, respectively.

2.1 *The PSO*

In the beginning of the PSO search, the position and velocity of the ith particle are generated in random, then they are updated by

$$\begin{cases} \vec{x}^i_{k+1} = \vec{x}^i_k + \vec{v}^i_{k+1} \\ \vec{v}^i_{k+1} = w_0 \vec{v}^i_k + w_1 \vec{r}_1 \otimes \left(\vec{p}^i_k - \vec{x}^i_k\right) + w_2 \vec{r}_2 \otimes \left(\vec{q}_k - \vec{x}^i_k\right) \end{cases}, \tag{12.1}$$

where w_0 is an inertial coefficient, w_1 is a coefficient for individual confidence, w_2 is a coefficient for swarm confidence. $\vec{r}_1, \vec{r}_2 \in \Re^N$ are two random vectors in which each element is uniformly distributed over [0, 1], and the symbol $\otimes$ is an element-wise operator for vector multiplication. $\vec{p}^i_k \left(= \arg \max_{j=1,\ldots,k} \left\{ g\left(\vec{x}^i_j\right) \right\} \right)$, where $g(\cdot)$ is the criterion value of the ith particle at time-step k is the local best position of the ith particle up to now, and $\vec{q}_k \left(= \arg \max_{i=1,2,\ldots} \left\{ g\left(\vec{p}^i_k\right) \right\} \right)$ is the global best position found by the whole particle swarm. In the original PSO, the parameter values, $w_0 = 1.0$ and $w_1 = w_2 = 2.0$, are used [15].

2.2 *The PSOIW*

As to be generally known, the PSO has weak convergence [2, 5]. For improving the convergence of the PSO, Shi et al. modified the update rule of the particle's velocity in (12.1) by constant reduction of the inertia coefficient over time-step [9, 22] as follows:

$$\vec{v}^i_{k+1} = w(k)\, \vec{v}^i_k + w_1 \vec{r}_1 \otimes \left(\vec{p}^i_k - \vec{x}^i_k\right) + w_2 \vec{r}_2 \otimes \left(\vec{q}_k - \vec{x}^i_k\right), \tag{12.2}$$

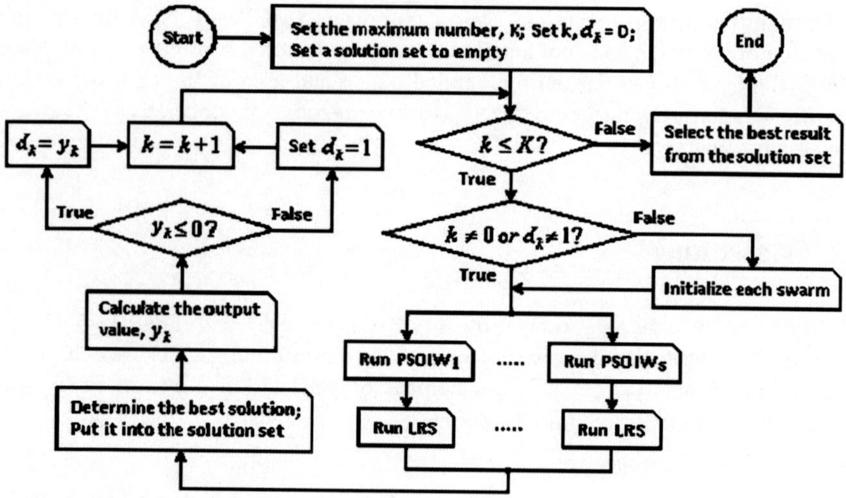

Fig. 12.1 A flowchart of the MPSOIWα/DC

where $w(k)$ is the variable inertia weight which is linearly reduced from a starting value, w_s, to a terminal value, w_e, with the increment of time-step k.

$$w(k) = w_s + \frac{w_e - w_s}{K} \times k \tag{12.3}$$

where K is the number of iteration for the PSOIW search. In the original PSOIW, the boundary values are adopted to $w_s = 0.9$ and $w_e = 0.4$, respectively.

Owing to the effect of the linearly damped inertia weight, the weakness of the PSO in convergence is overcome, and the solution accuracy is also improved. However, the search results of the PSOIW often converge to local solutions in treating with multi-modal problems. The phenomenon called premature convergence or stagnation in optimization is easily caused as much as the small terminal value w_e. Consequently, this fault is fatal for the use of the PSOIW to handle complex optimization problems.

2.3 *The MPSOIWα/DC*

In order to thoroughly conquer the above mentioned shortcoming of the PSOIW, Zhang proposed to use the method of multiple particle swarm optimizers with inertia weight with DC, called MPSOIWα/DC [31]. This is an integration way which mixes different approaches including the practical use of parameter selection, and intelligent, hybrid and multi-swarm search.

Figure 12.1 illustrates a flowchart of the MPSOIWα/DC for showing the data processing and information control in optimization. In the MPSOIWα/DC, plural

PSOIWs are executed in parallel, and the localized random search (LRS) [31] is implemented to find the most suitable solution from a small limited space surrounding the solution found by each PSOIW. Here, the continuous action of the PSOIW and LRS constitutes a hybrid search. Subsequently, the best solution, $\vec{q}_k^{\,b}$, is determined with maximum selection from the whole solutions found by the multi-swarm search (i.e., parallel hybrid search). Then it is put in a solution set being a storage memory for information processing later.

The role of the internal indicator is to monitor whether the status of the best solution $\vec{q}_k^{\,b}$ continues to change or not over time-step. It makes up the concentration in evaluation and behavior control. While the output y_k is zero, this means that the multi-swarm concentrates on exploring the surroundings of the solution $\vec{q}_k^{\,b}$ for "cognition". If once the output y_k become positive, this means that the multi-swarm has lost interest, i.e., feeling of boredom, to search the region around the solution $\vec{q}_k^{\,b}$ for "motivation". The concepts of psychology, "cognition" and "motivation", are further explained later.

Due to the big reduction of boredom behavior in the curiosity-driven multi-swarm search, the search efficiency finding an optimal solution or near-optimal solutions will be drastically improved. Here, it is to be noted that the re-initialization decided by the signal d_k =1 in Fig. 12.1 is a mere operation style which implements the mechanism of DC for realizing a positive search. Of course, the operation style is not an unique one, it also can be performed by other operation ways to show the inquisitive behavior of the multi-swarm in practice.

2.3.1 The LRS

As a resource to reinforce search ability, the LRS [29] is introduced into a MPSOIW/DC to make up the MPSOIWα/DC, which can find the most suitable solution from a limited space surrounding the solution found by each PSOIW.

Concretely, the procedure of the LRS is implemented as follows:

Step-1: Let $\vec{q}_k^{\,s}$ be a solution found by the sth particle swarm at time-step k, and set $\vec{q}_{\text{now}}^{\,s} = \vec{q}_k^{\,s}$. Give the terminating condition, J (the total number of the LRS run), and set $j = 1$.

Step-2: Generate a random data, $\vec{d}_j \in \Re^N \sim N(0, \sigma_N^2)$ (where σ_N is a small positive value given by user, which determines the small limited space). Check whether $\vec{q}_k^{\,s} + \vec{d}_j \in \mathbf{\Omega}$ is satisfied or not. If $\vec{q}_k^{\,s} + \vec{d}_j \notin \mathbf{\Omega}$ then adjust $\vec{d}_j$ for moving $\vec{q}_k^{\,s} + \vec{d}_j$ to the nearest valid point within $\mathbf{\Omega}$. Set $\vec{q}_{\text{new}} = \vec{q}_k^{\,s} + \vec{d}_j$.

Step-3: If $g(\vec{q}_{\text{new}}) > g(\vec{q}_{\text{now}}^{\,s})$ then set $\vec{q}_{\text{now}}^{\,s} = \vec{q}_{\text{new}}$.

Step-4: Set $j = j + 1$. If $j \le J$ then go to the Step-**2**.

Step-5: Set $\vec{q}_k^{\,s} = \vec{q}_{\text{now}}^{\,s}$ to correct the solution found by the sth particle swarm at time-step k. Stop the search.

Due to the complementary characteristics of the used hybrid search (i.e., PSOIW + LRS, a kind of memetic algorithm [18]), the correctional function seems to be close to the HGAPSO [14] in performance effect, which implements a plain GA and the PSO with the mixed operations for improving the adaptation to treat with various blended distribution problems.

2.3.2 Internal Indicator

Curiosity, as a general concept in psychology, is an emotion related to natural inquisitive behavior for humans and animals, and its importance and effect in motivating search cannot be ignored [7, 20]. Berlyne categorized it into two types: DC and specific curiosity (SC) [3]. In the matter of the mechanism of the former, Loewenstein insisted that "diversive curiosity occupies a critical position at the crossroad of cognition and motivation" in [16].

Based on the assumption of the "cognition" is the act of exploitation, and the "motivation" is the intention to exploration, Zhang et al. proposed the following internal indicator for distinguishing and detecting the above two behavior patterns in the multi-swarm search [25–27].

$$y_k(L,\varepsilon) = \max\left(\varepsilon - \sum_{l=1}^{L} \frac{|g(\vec{q}_k^{\,b}) - g(\vec{q}_{k-l}^{\,b})|}{L},\ 0\right), \tag{12.4}$$

where $\vec{q}_k^{\,b}\left(= \arg\max_{s=1,\ldots,S} \{g(\vec{q}_k^{\,s})\}\right.$, where S is the number of plural particleswarms$\left.\right)$ denotes the best solution found by the whole particle swarms at time-step k. As two adjustable parameters of the internal indicator, L is duration of judgment, and ε is the positive tolerance coefficient (sensitivity).

3 Computer Experiments

To facilitate comparison and analysis of the search performance of the proposed method, we use a suite of multi-dimensional benchmark problems [23] and the corresponding criteria in Table 12.1 to identification.

Table 12.2 gives the major parameters in the MPSOIWα/DC employed for the next computer experiments.

3.1 Preliminaries

To ensure higher search performance of the MPSOIWα/DC, the optimized PSOIWs corresponding to each given problem are adopted in it. For obtaining these optimal PSOIWs, the method of meta-optimization of the PSOIW heuristics, called evolutionary particle swarm optimizer with inertia weight (EPSOIW) [32] is used.

Table 12.1 A suite of benchmark optimization problems. The search space for each problem is limited to the search space $\mathbf{\Omega} \in (-5.12, 5.12)^N$

Problem	Function	Criterion ($N = 2$)
Sphere	$f_{\mathrm{Sp}}(\vec{x}) = \sum_{d=1}^{N} x_d^2$	$g_{\mathrm{Sp}}(\vec{x}) = \frac{1}{f_{\mathrm{Sp}}(\vec{x})+1}$
Griewank	$f_{\mathrm{Gr}}(\vec{x}) = \frac{1}{4{,}000}\sum_{d=1}^{N} x_d^2 - \prod_{d=1}^{N} \cos\left(\frac{x_d}{\sqrt{d}}\right) + 1$	$g_{\mathrm{Gr}}(\vec{x}) = \frac{1}{f_{\mathrm{Gr}}(\vec{x})+1}$
Rastrigin	$f_{\mathrm{Ra}}(\vec{x}) = \sum_{d=1}^{N} \left(x_d^2 - 10\cos(2\pi x_d) + 10\right)$	$g_{\mathrm{Ra}}(\vec{x}) = \frac{1}{f_{\mathrm{Ra}}(\vec{x})+1}$
Rosenbrock	$f_{\mathrm{Ro}}(\vec{x}) = \sum_{d=1}^{N-1} \left[100(x_{d+1} - x_d^2)^2 + (1 - x_d)^2\right]$	$g_{\mathrm{Ro}}(\vec{x}) = \frac{1}{f_{\mathrm{Ro}}(\vec{x})+1}$
Schwefel	$f_{\mathrm{Sw}}(\vec{x}) = \sum_{d=1}^{N} \left(\sum_{j=1}^{d} x_j\right)^2$	$g_{\mathrm{Sw}}(\vec{x}) = \frac{1}{f_{\mathrm{Sw}}(\vec{x})+1}$
Hybrid	$f_{\mathrm{Hy}}(\vec{x}) = f_{\mathrm{Ra}}(\vec{x}) + 2f_{\mathrm{Sw}}(\vec{x}) + \frac{1}{12} f_{\mathrm{Gr}}(\vec{x}) + \frac{1}{20} f_{\mathrm{Sp}}(\vec{x})$	$g_{\mathrm{Hy}}(\vec{x}) = \frac{1}{f_{\mathrm{Hy}}(\vec{x})+1}$

Table 12.2 The major parameters used in the MPSOIWα/DC

Parameter	Value	Parameter	Value
The number of individuals, M	10	The number of iterations, K	400
The number of generation, G	20	The maximum velocity, $v_{\max}$	5.12
The number of particles, P	10	The range of the LRS, σ_N^2	0.05
The number of particle swarms, S	3	The number of the LRS run, J	10
The duration of judgment, L	$10 \sim 90$	The tolerance coefficient, ε	$10^{-6} \sim 10^{-2}$

Table 12.3 The resulting appropriate values of parameters in the PSOIW for each given 5D benchmark problem

	Parameters			
Problem	$\hat{w}_s$	$\hat{w}_e$	$\hat{w}_1$	$\hat{w}_2$
Sphere	0.7267 ± 0.0521	0.1449 ± 0.1574	1.2116 ± 0.6069	1.9266 ± 0.0651
Griewank	0.7761 ± 0.0822	0.2379 ± 0.2008	1.2661 ± 0.6332	0.2827 ± 0.0708
Rastrigin	2.0716 ± 0.9143	0.8816 ± 0.6678	12.942 ± 7.9204	5.0663 ± 1.5421
Rosenbrock	0.7702 ± 0.1660	0.5776 ± 0.2137	1.9274 ± 0.3406	1.9333 ± 0.4541
Schwefel	0.8552 ± 0.3210	0.1253 ± 0.2236	1.6106 ± 1.3345	1.9610 ± 1.3754
Hybrid	1.4767 ± 0.2669	0.6101 ± 0.5335	5.0348 ± 1.7687	9.2134 ± 5.0915

These estimated PSOIWs in Table 12.3 as the optimal PSOIWs (i.e., PSOIW*) are used in the MPSOIWα/DC for each given problem to ensure better convergence and search accuracy.

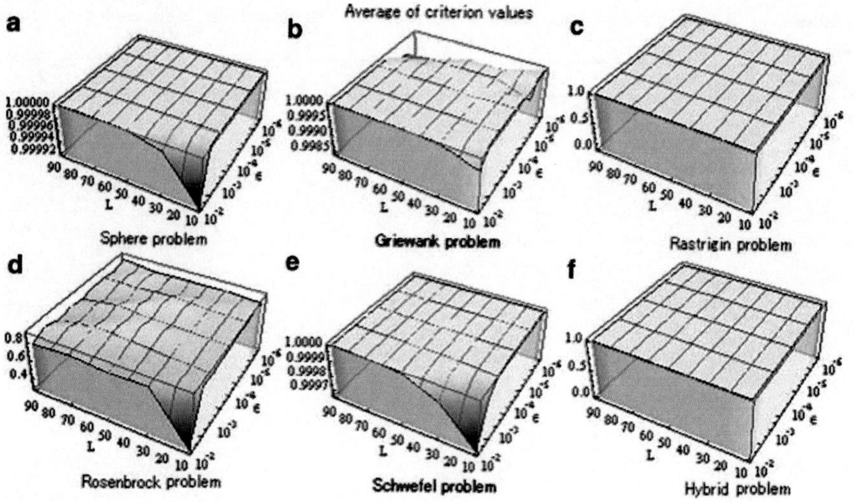

Fig. 12.2 The average of criterion values obtained by MPSOIW*a/DC with tuning the parameters, L and ε, for each problem

3.2 *Results of the MPSOIW*α/DC*

Figure 12.2 gives the search performance of the MPSOIW*α/DC for each benchmark problem by tuning the parameters of the internal indicator, L and ε with 20 trials. For obtaining superior search performance, the recommended range of parameters of the MPSOIW*α/DC are available: $L^*_{\mathrm{Sp}} \in (10\text{–}90)$ and $\varepsilon^*_{\mathrm{Sp}} \in (10^{-6}\text{–}10^{-2})$ for the Sphere problem; $L^*_{\mathrm{Gr}} \in (10\text{–}90)$ and $\varepsilon^*_{\mathrm{Gr}} \in (10^{-5}\text{–}10^{-3})$ for the Griewank problem; $L^*_{\mathrm{Ra}} \in (10\text{–}90)$ and $\varepsilon^*_{\mathrm{Ra}} \in (10^{-6}\text{–}10^{-2})$ for the Rastrigin problem; $L^*_{\mathrm{Ro}} \in (10\text{–}90)$ and $\varepsilon^*_{\mathrm{Ro}} \in (10^{-5}\text{–}10^{-3})$ for the Rosenbrock problem; $L^*_{\mathrm{Ri}} \in (10\text{–}90)$ and $\varepsilon^*_{\mathrm{Sw}} \in (10^{-6}\text{–}10^{-2})$ for the Schwefel problem; and $L^*_{\mathrm{Hy}} \in (10\text{–}90)$ and $\varepsilon^*_{\mathrm{Hy}} \in (10^{-6}\text{–}10^{-2})$ for the Hybrid problem.

We can see that the average of criterion values of the PSOIW*α/DC do not change at all with tuning the parameters, L and ε, for the Rastrigin and Hybrid problems. This phenomenon indicates that the MPSOIW*α/DC using the optimized PSOIWs has powerful search ability to handle the given multi-modal problems.

On the other hand, due to stochastic factor in the PSOIW search and complexity of the Rosenbrock problem, some irregular change of the experimental results can be discovered in Fig. 12.2d. Moreover, because the effect of the used hybrid search, the fundamental finding, "the zone of curiosity", in psychology [7] is not distinguished except for the Rosenbrock problem.

The average fitness values is very small in the case of the parameters, i.e., $L = 10$ and $\varepsilon = 10^{-2}$, for the Rosenbrock problem in Figure 12.2d. Since the average of the criterion values is the lowest compared to that in the other cases, this result shows that the behavior of the curiosity-driven multi-swarm search seems to have entered "the zone of anxiety", [7] which leads the search performance of the MPSOIWα/DC to be lower.

3.2.1 Optimized Swarms Versus Original Swarms

For verifying the effectiveness of the optimized PSOIW, Fig. 12.3 shows the resulting difference, $\Delta_{ON} = \bar{g}^*_O - \bar{g}^*_N$ ($\bar{g}^*_O$: the average of criterion values of the MPSOIW*α/DC, $\bar{g}^*_N$: the average of criterion values of the MPSOIWα/DC). The obtained results indicate that the search performance of the MPSOIWα/DC are greatly improved by introduction of the optimized PSOIW especially for the Rastrigin, Rosenbrock, and Hybrid problems. Since all of the differences are positive, the effect of the optimal PSOIW used in the MPSOIWα/DC run is remarkable.

3.2.2 Effect of the LRS

For confirming the effectiveness of the LRS, Fig. 12.4 shows the obtained difference, $\Delta_{PN} = \bar{g}^*_P - \bar{g}^*_N$ ($\bar{g}^*_N$: the average of criterion values of the MPSOIW*/DC). We can see that the search performance of the MPSOIW*α/DC is better than that of the MPSOIW*/DC in the most cases for the Griewank and Rosenbrock problems. It is confirmed that the LRS plays an important role in drastically improving the search performance of the MPSOIW*/DC.

On the other hand, the effect of the LRS is not remarkable for the Sphere, Rastrign, Schwefel, and Hybrid problems in many cases. It suggests that (1) the MPSOIW*/DC has powerful search ability to deal with these problems, (2) the effect of the LRS closely depends on the object of search, which related to how to set the parameter values for the running number, J, and the search range, σ^2_N, and the inherent feature of the given benchmark problems.

3.2.3 Comparison with Other Methods

Firstly, for further illuminating the effectiveness of the proposed method, Table 12.4 shows the search performance of the PSOIW, OPSO (optimized particle swarm optimization) [17], and RGA/E. In comparison with the mean of criterion values, we can see that the search performance of the MPSOIW*α/DC is better than that of the PSOIW, OPSO, and RGA/E except for the Sphere and Schwefel problems that are relatively simple. The results sufficiently reflect that the merging of both parallel hybrid search and the mechanism of DC takes the active role in efficiently handling the given problems.

Although the search capability of the MPSOIW*α/DC is the best among them, but its computational cost is relative high compared to the used simple-swarm or

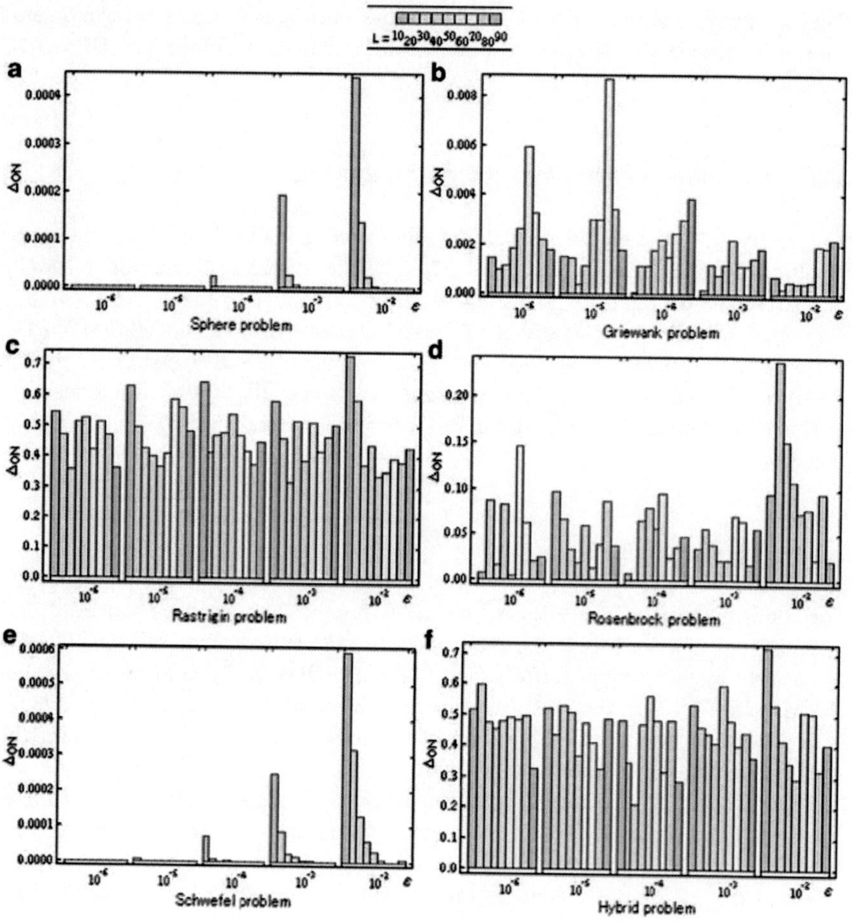

Fig. 12.3 Performance comparison between the MPSOIW*α/DC and MPSOIWα/DC

simple-population methods. As an example, Fig. 12.5 shows the computational times of implementing each method used in Table 12.4 for the Sphere problem. It is appropriate for the PSOIWα/DC to handle the given problems with the parallel hybrid search and intelligent judgment.

Secondly, for comparing three methods of the curiosity-driven multi-swarm search, Table 12.5 gives the resulting search performance of the MPSOIW*α/DC, MPSO*α/DC, and MCPSO*α/DC for handling each given 5D benchmark problems.

By comparing the mean of the criterion values obtained by each method, we can see that the search performance of the MPSOIW*α/DC is superior to the others. Although the use of the mechanism of DC, the search performance of the

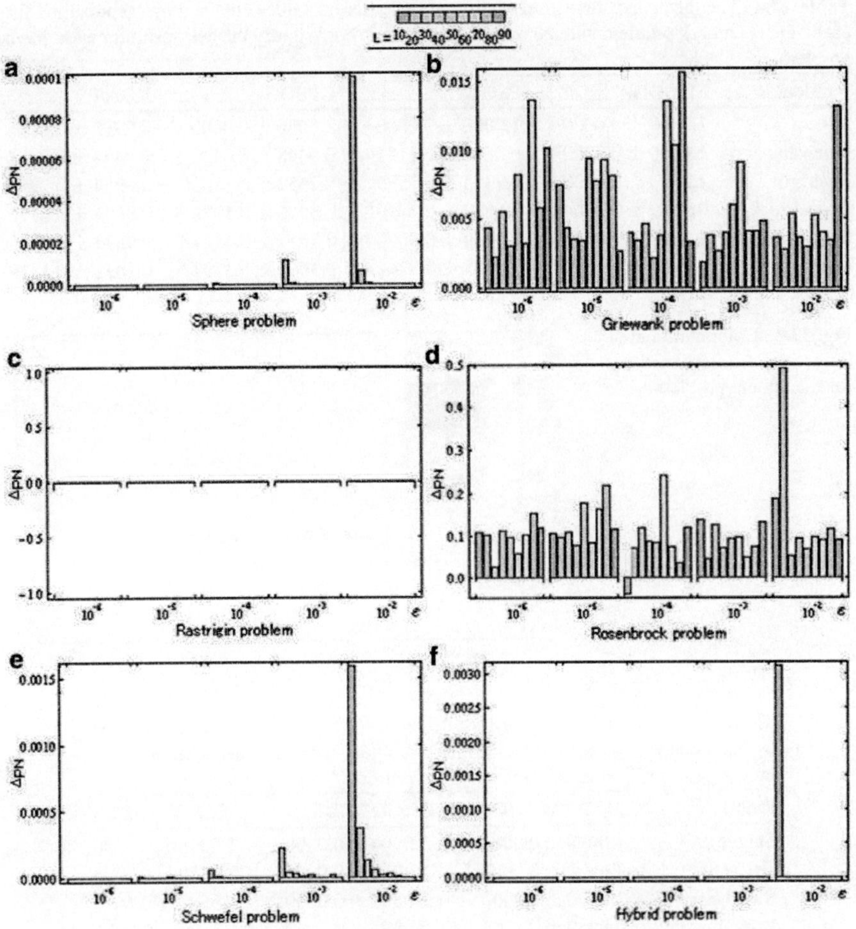

Fig. 12.4 Performance comparison between the MPSOIW*α/DC and MPSOIW*/DC

MCPSO*α/DC is not so good special for the Rastrigin and Hybrid problems. It is considered that the optimized CPSO has strong convergence to easily fall into local solutions.

4 Conclusion

A new method of multiple particle swarm optimizers with inertia weight with DC, MPSOIWα/DC, has been presented. Owing to the essential strategies of decentralization in search and concentration in evaluation and behavior control, the

Table 12.4 The mean and standard deviation of the criterion values obtained by each method for each 5D benchmark problem with 20 trials. The values in bold signify the best result for each given problem

Problem	MPSOIW*α/DC	PSOIW	OPSO	RGA/E
Sphere	**1.0000** ± 0.0000	**1.0000** ± 0.0000	**1.0000** ± 0.0000	0.9990 ± 0.0005
Griewank	**1.0000** ± 0.0000	0.8505 ± 0.1196	0.9448 ± 0.0439	0.9452 ± 0.0784
Rastrigin	**1.0000** ± 0.0000	0.2325 ± 0.1595	0.2652 ± 0.1185	0.9064 ± 0.2256
Rosenbrock	**0.9959** ± 0.0064	0.5650 ± 0.1796	0.3926 ± 0.1976	0.9893 ± 0.0120
Schwefel	**1.0000** ± 0.0000	**1.0000** ± 0.0000	0.7677 ± 0.4127	0.9875 ± 0.2145
Hybrid	**1.0000** ± 0.0000	0.3905 ± 0.3746	0.3061 ± 0.3591	0.1531 ± 0.1339

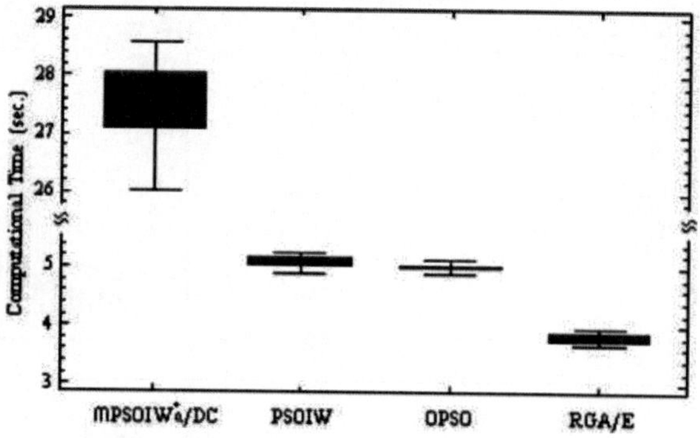

Fig. 12.5 Computational time (s) of different methods for the Sphere problem

Table 12.5 The mean and standard deviation of the criterion values obtained by each method for each 5D benchmark problem with 20 trials. The values in bold signify the best result for each given problem

Problem	MPSOIW*α/DC	MPSO*α/DC	MCPSO*α/DC
Sphere	**1.0000** ± 0.0000	**1.0000** ± 0.0000	**1.0000** ± 0.0000
Griewank	**1.0000** ± 0.0000	**1.0000** ± 0.0000	**1.0000** ± 0.0000
Rastrigin	**1.0000** ± 0.0000	**1.0000** ± 0.0000	0.5591 ± 0.4686
Rosenbrock	0.9959 ± 0.0064	0.9893 ± 0.0120	**0.9999** ± 0.0001
Schwefel	**1.0000** ± 0.0000	**1.0000** ± 0.0000	0.9956 ± 0.0059
Hybrid	**1.0000** ± 0.0000	**1.0000** ± 0.0000	0.5418 ± 0.5823

combination of the adopted hybrid search and the mechanism of DC, theoretically, it has good capability to improve search efficiency and to alleviate stagnation in handling complex optimization problems.

Applications of the MPSOIWα/DC to a suite of the 5D benchmark problems well demonstrated its effectiveness. The obtained experimental results verified that unifying the both characteristics of multi-swarm search and the LRS is successful and effective. By comparing the search performance of the PSOIW, OPSO, RGA/E, MPSOα/DC, and MCPSOα/DC, the MPSOIWα/DC has an enormous latent capability in handling the given benchmark problems.

It is left for further study to apply the MPSOIWα/DC to data mining, system identification, multi-objective optimization, and practical problems in the real-world.

Acknowledgment This research was supported by Grant-in-Aid Scientific Research(C) (22500132) from the Ministry of Education, Culture, Sports, Science and Technology, Japan.

References

1. Van den Bergh F, Engelbrecht AP (2004) A cooperative approach to particle swarm optimization. IEEE Trans Evolut Comput 8(3):225–239.
2. Van den Bergh F (2002) An analysis of particle swarm optimizers, Ph.D thesis, Department of Computer Science, University of Pretoria, Pretoria, South Africa.
3. Berlyne D (1960) Conflict, arousal, and curiosity. McGraw-Hill Book Co, New York.
4. Chang JF, Chu SC, Roddick JF, Pan JS (2005) A parallel particle swarm optimization algorithm with communication strategies. J Inf Sci Eng 21:809–818.
5. Clerc M, Kennedy J (2000) The particle swarm-explosion, stability, and convergence in a multidimensional complex space. IEEE Trans Evolut Comput 6(1):58–73.
6. Clerc M (2006) Particle swarm optimization. ISTE Ltd., London.
7. Day H (1982) Curiosity and the interested explorer. Perform Instr 21(4):19–22.
8. Eberhart RC, Kennedy J (1995) A new optimizer using particle swarm theory. In: Proceedings of the sixth international symposium on micro machine and human science, Nagoya, Japan, 4–6 October, pp 39–43.
9. Eberhart RC, Shi Y (2000) Comparing inertia weights and constriction factors in particleswarm optimization. In: Proceedings of IEEE congress on evolutionary computation (CEC2000), San Diego, CA, USA, 16–19 July, vol 1, pp 84–88.
10. El-Abd E, Kamel MS (2008) A taxonomy of cooperative particle swarm optimizers. Int J Comput Intell Res 4(2):137–144.
11. Fogel LJ (1999) Intelligence through simulated evolution: forty years of evolutionary programming. John Wiley & Sons, Inc. New York.
12. Goldberg DE (1989) Genetic algorithm in search optimization and machine learning, reading. Addison-Wesley, MA.
13. Hema CR, Paulraj MP, Nagarajan R, Yaacob S, Adom AH (2008) Application of particle swarm optimization for EEG signal classification. Biomed Soft Comput Human Sci 13(1):79–84.
14. Juang C-F (2004) A hybrid of genetic algorithm and particle swarm optimization for recurrent network design. IEEE Trans Syst Man Cyb B 34(2):997–1006.
15. Kennedy J, Eberhart RC (1995) Particle swarm optimization. In: Proceedings of the 1995 IEEE international conference on neural networks, Perth, Australia, 27 November–1 December, pp 1942–1948.
16. Loewenstein G (1994) The psychology of curiosity: a review and reinterpretation. Psychol Bull 116(1):75–98.
17. Meissner M, Schmuker M, Schneider G (2006) Optimized particle swarm optimization (OPSO) and its application to artificial neural network training. BMC Bioinformatics, 7(125), p 11. http://www.boimedcentral.com/content/pdf/147-2105-7-125.pdf.
18. Moscato P (1989) On evolution, search optimization, genetic algorithms and martial arts: towards memetic algorithms. Technical report caltech concurrent computation program, Report 826, California Institute of Technology, Pasadena, CA 91125.

19. Niu B, Zhu Y, He X (2005) Multi-population Cooperation Particle Swarm Optimization. In: Proceedings of VIII european conference on artificial life (ECAL2005), Canterbury, UK, 5–9 September. LNCS 3630, pp 874–883.
20. Opdal PM (2001) Curiosity, wonder and education seen as perspective development. Studies in philosophy and education, 20(4). Springer, Netherlands, pp 331–344.
21. Poli P, Kennedy J, Blackwell T (2007) Particle swarm optimization—an overview. Swarm Intell 1:33–57.
22. Shi Y, Eberhart RC (1998) A modified particle swarm optimiser, In: Proceedings of IEEE international conference on evolutionary computation, Anchorage, AK, USA, 4–9 May, pp 69–73.
23. Suganthan PN, Hansen N, Liang JJ, Deb K, Chen Y-P, Auger A, Tiwari S (2005) Problem definitions and evaluation criteria for the CEC 2005. http//:www.ntu.edu.sg /home /epnsugan/index_files/ CEC-05/Tech-Report-May-30-05.pdf
24. Valle YD, Venayagamoorthy GK, Mohagheghi S, Hernandez J-C, Harley RG (2008) Particle swarm optimization: basic concepts, variants and applications in power systems. IEEE Trans Evolut Comput 12(2):171–195.
25. Zhang H, Ishikawa M (2008) Improving the performance of particle swarm optimization with diversive curiosity. In: Proceedings of international multiconference of engineers and computer scientists 2008 (IMECS 2008), Hong Kong, 19–21 March, pp 1–6.
26. Zhang H, Ishikawa M (2008) Particle swarm optimization with diversive curiosity—an endeavor to enhance swarm intelligence. IAENG Int J Comput Sci 35(3):275–284.
27. Zhang H, Ishikawa M (2009) Characterization of particle swarm optimization with diversive curiosity. J Neural Comput Appl 18(5):409–415.
28. Zhang H, Ishikawa M (2010) The performance verification of an evolutionary canonical particle swarm optimizers. Neural Networks 23(4):510–516.
29. Zhang H (2010) Multiple particle swarm optimizers with diversive curiosity. In: Proceedings of international multiconference of engineers and computersScientists 2010 (IMECS 2010), Hong Kong, 17–19 March. Lecture notes in engineering and computer Science, pp 174–179
30. Zhang H (2010) A new expansion of cooperative particle swarm optimization. In: Proceedings of 17th international conference on neural information processing (ICONIP2010), Sydney, Australia, 22–25 November 2010. Neural information processing—theory and algorithms, Part I (LNCS 6443), pp 593–600.
31. Zhang H (2011) Multiple particle swarm optimizers with inertia weight with diversive curiosity. In: Proceedings of international multiconference of engineers and computer scientists 2011 (IMECS 2011), Hong Kong, 16–18 March. Lecture notes in engineering and computer science, pp 21–26.
32. Zhang H (2011) Assessment of an evolutionary particle swarm optimizer with inertia weight. In: Proceedings of IEEE congress on evolutionary computation (CEC2011), (ISBN 978-1-4244-7834-7), pp 1746–1753, June 5–8, 2011, New Orleans, LA, USA.

Chapter 13
A New Design Method for Train Marshaling Evaluating the Transfer Distance of Locomotive

Yoichi Hirashima

Abstract In this paper a new reinforcement learning system for generating marshaling plan of freight cars in a train is designed. In the proposed method, the total transfer distance of a locomotive is minimized to obtain the desired layout of freight cars for an outbound train. The order of movements of freight cars, the position for each removed car, the layout of cars in a train and the number of cars to be moved are simultaneously optimized to achieve the minimization of the transfer distance of locomotive. Initially, freight cars are located in a freight yard by the random layout, and they are removed and lined into a main track in a certain desired order in order to assemble an outbound train. Q-Learning is applied to reflect the transfer distance of the locomotive that is used to achieve one of the desired layouts in the main track. After adequate autonomous learning, the optimum schedule can be obtained by selecting a series of movements of freight cars that has the best evaluation.

Keywords Scheduling • Freight train • Marshaling • Q-Learning • Container transfer problem

1 Introduction

In transporting, goods are carried by containers that are loaded on freight cars in a freight yard. A freight train consists of several freight cars, and each freight car has its own destination. Thus, the train driven by a locomotive travels several destinations disjointing corresponding freight cars at each freight station. In addition, since freight trains can transport goods only between railway stations, modal

Y. Hirashima (✉)
Osaka Institute of Technology, 1-79-1, Kitayama, Hirakata, Osaka 573-0196, Japan
e-mail: hirash-y@is.oit.ac.jp

S.I. Ao et al. (eds.), *Intelligent Control and Innovative Computing*, Lecture Notes in Electrical Engineering 110, DOI 10.1007/978-1-4614-1695-1_13,

shifts are required for delivering them to area that has no railway. In intermodal transports from the road and the rail, containers carried into the station are loaded on freight cars in the arriving order. The initial layout of freight cars is thus random. Shifting operation at freight yard is required to joint several rail transports, or different modes of transportation including rail. For efficient shift, the desirable layout should be determined by considering destinations of freight cars. Then, freight cars must be rearranged before assembling the freight train. In general, the rearrangement process is conducted in a freight yard that consists of a main track and several sub tracks. Freight cars are initially placed on sub tracks, rearranged, and lined into the main track. This series of operation is called marshaling, and several methods to solve the marshaling problem have been proposed [1, 7]. Also, many different types of problem are treated by mathematical programming and genetic algorithm [2, 4, 8, 10], and some analyses are conducted for computational complexities [2, 3]. However, models assumed in these methods are different from the one in the addressed problem, and thus, they cannot apply directly to the problem. Recently, two reinforcement learning methods have been proposed in order to solve marshaling problems that have randomly defined initial layout in the fixed number of sub tracks. The one is derived based on the number of movements of locomotive [5], and the other one is based on the transfer distance of locomotive [6].

In this paper, a unified design method of reinforcement learning system for marshaling based on movement counts and transfer distance of locomotive is proposed. In the proposed method, the focus is centered on to reduce the total transfer distance of a locomotive required to achieve desirable layout on the main track. The optimal layout of freight cars in the main track is derived based on the destination of freight cars. This yields several desirable layouts of freight cars in the main track, and the optimal layout that can achieve the smallest transfer distance of the locomotive is obtained by autonomous learning. Simultaneously, the optimal sequence of car-movements as well as the number of freight cars that can achieve the desired layout is obtained by autonomous learning. Also, the feature is considered in the learning algorithm, so that, at each arrangement on sub track, an evaluation value represents the smallest transfer distance of the locomotive to achieve the best layout on the main track. The learning algorithm is derived based on the Q-Learning [11], which is known as one of the well established realization algorithm of the reinforcement learning.

In the learning algorithm, the state is defined by using a layout of freight cars, the car to be moved, the number of cars to be moved, and the destination of the removed car. An evaluation value called Q-value is assigned to each state, and the evaluation value is calculated by several update rules based on the Q-Learning algorithm. In the learning process, a Q-value in a certain update rule is referred from another update rule, in accordance with the state transition. Then, the Q-value is discounted according to the transfer distance of the locomotive. Consequently, Q-values at each state represent the total transfer distance of the locomotive required to achieve the desired layout from the state. Moreover, in the proposed method, only referred Q-values are stored by using table look-up technique, and the table is dynamically constructed by binary tree in order to obtain the best solution with feasible memory space [5]. In order to show effectiveness of the proposed method, computer simulations are conducted for several methods.

2 Problem Description

The yard consists of one main track and m sub tracks. Define k as the number of freight cars placed on the sub tracks, and they are carried to the main track by the desirable order based on their destination. In the yard, a locomotive moves freight cars from sub track to sub track or from sub track to main track. The movement of freight cars from sub track to sub track is called removal, and the car-movement from sub track to main track is called rearrangement. For simplicity, the maximum number of freight cars that each sub track can have is assumed to be n, the ith car is recognized by an unique symbol c_i $(i = 1, \cdots, k)$. Figure 13.1 shows the outline of freight yard in the case $k = 30, m = n = 6$. In the figure, track T_m denotes the main track, and other tracks [1], [2], [3], [4], [5], [6] are sub tracks. The main track is linked with sub tracks by a joint track, which is used for moving cars between sub tracks, or for moving them from a sub track to the main track. In the figure, freight cars are moved from sub tracks, and lined in the main track by the descending order, that is, rearrangement starts with c_{30} and finishes with c_1. When the locomotive L moves a certain car, other cars locating between the locomotive and the car to be moved must be removed to other sub tracks. This operation is called removal. Then, if $k \le n \cdot m - (n - 1)$ is satisfied for keeping adequate space to conduct removal process, every car can be rearranged to the main track.

In each sub track, positions of cars are defined by n rows. Every position has unique position number represented by $m \cdot n$ integers, and the position number for cars at main track is 0. Figure 13.2 shows an example of position index for $k = 30, m = n = 6$ and the layout of cars for Fig. 13.1.

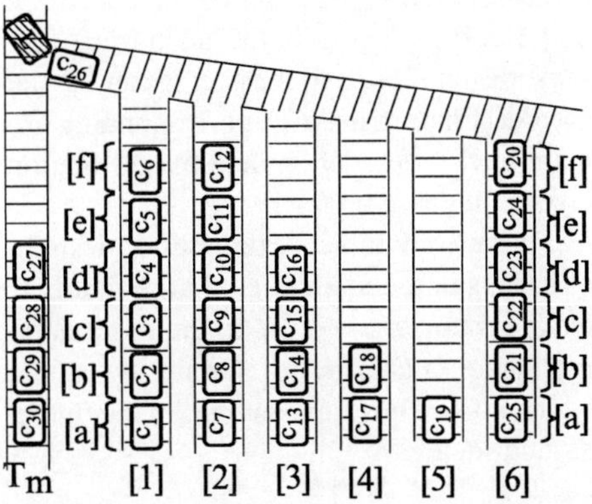

Fig. 13.1 Freight yard

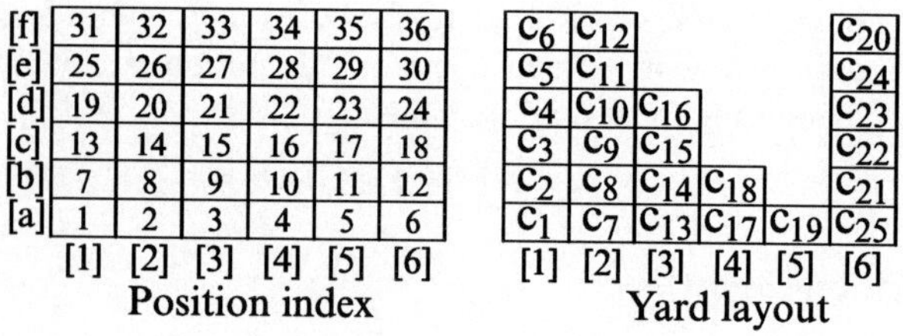

Fig. 13.2 Example of position index and yard state

In Fig. 13.2, the position "[a][1]" that is located at row "[a]" in the sub track "[1]" has the position number 1, and the position "[f][6]" has the position number 36. For unified representation of layout of car in sub tracks, cars are placed from the row "[a]" in every track, and newly placed car is jointed with the adjacent freight car. In the figure, in order to rearrange c_{25}, cars $c_{24}, c_{23}, c_{22}, c_{21}$ and c_{20} have to be removed to other sub tracks. Then, since $k \leq n \cdot m - (n-1)$ is satisfied, c_{25} can be moved even when all the other cars are placed in sub tracks.

In the freight yard, define $x_i (1 \leq x_i \leq n \cdot m, i = 1, \cdots, k)$ as the position number of the car c_i, and $\boldsymbol{s} = [x_1, \cdots, x_k]$ as the state vector of the sub tracks. For example, in Fig. 13.2, the state is represented by $\boldsymbol{s} = [1, 7, 13, 19, 25, 31, 2, 8, 14, 20, 26, 32, 3, 9, 15, 21, 4, 10, 5, 36, 12, 18, 24, 30, 6, 0, 0, 0, 0, 0]$. A trial of the rearrange process starts with the initial layout, rearranging freight cars according to the desirable layout in the main track, and finishes when all the cars are rearranged to the main track.

3 Desired Layout in the Main Track

In the main track, freight cars that have the same destination are placed at the neighboring positions. In this case, removal operations of these cars are not required at the destination regardless of layouts of these cars. In order to consider this feature in the desired layout in the main track, a group is organized by cars that have the same destination, and these cars can be placed at any positions in the group. Then, each destination makes a corresponding group, and the order of groups lined in the main track is predetermined by destinations. This feature yields several desirable layouts in the main track.

Figure 13.3 depicts examples of desirable layouts of cars and the desired layout of groups in the main track. In the figure, freight cars $c_1, \cdots, c_6$ to the destination$_1$ make group$_1$, $c_7, \cdots, c_{18}$ to the destination$_2$ make group$_2$, and $c_{19}, \cdots, c_{25}$ to the destination$_3$ make group$_3$. Groups$_{1,2,3}$ are lined by ascending order in the main track, which make a desirable layout. In the figure, examples of layout in group$_1$ are in the dashed square.

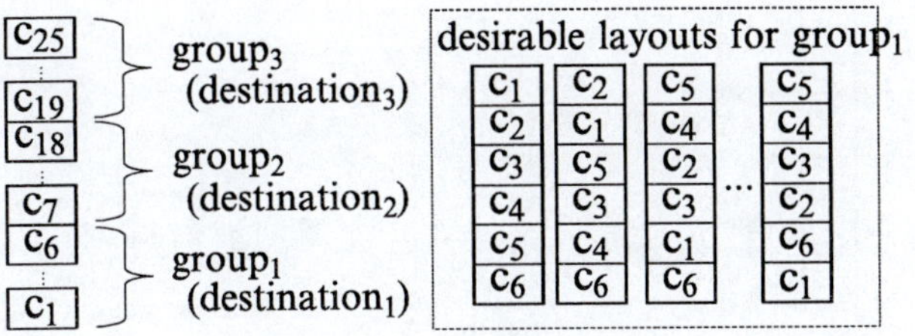

Fig. 13.3 Example of groups

4 Direct Rearrangement

When a car can be rearranged to the main track without a removal, the rearrangement precedes any removals. In the case that several cars can be rearranged without a removal, rearrangements are repeated until all the candidates for rearrangement requires at least one removal. If several candidates for rearrangement require no removal, the order of selection is random, because any orders satisfy the desirable layout of groups in the main track. In this case, the arrangement of cars in sub tracks obtained after rearrangements is unique, so that the movement counts of cars has no correlation with rearrangement orders of cars that require no removal. This operation is called direct rearrangement. When a car in a certain sub track can be rearrange directly to the main track and when several cars located adjacent positions in the same sub track satisfy the layout of groups in the main track, they are jointed and applied direct rearrangement.

Figure 13.4 shows an example of arrangement in sub tracks existing candidates for rearranging cars that require no removal. At the top of figure, from the left side, a desired layout of cars and groups, the initial layout of cars in sub tracks, and the position index in sub tracks are depicted for $m = n = 4, k = 9$. c_1, c_2, c_3, c_4 are in group$_1$, c_5, c_6, c_7, c_8 are in group$_2$, and group$_1$ must be rearranged first to the main track. In each group, any layouts of cars can be acceptable. In both cases, c_2 in step1 and c_3 in step3 are applied the direct rearrangement. Also, in step4, three cars c_1, c_4, c_5 located adjacent positions are jointed and moved to the main track by a direct rearrangement operation. In addition, at step5 in Case2, cars in group$_2$ and group$_3$ are moved by a direct rearrangement, since the positions of c_7, c_8, c_6, c_9 are satisfied the desired layout of groups in the main track.

5 Rearrangement Process

The rearrangement process for cars consists of following 6 operations:

(1) Rearrangement for all the cars that can apply the direct rearrangement into the main track.
(2) Selection of a freight car to be rearranged into the main track.

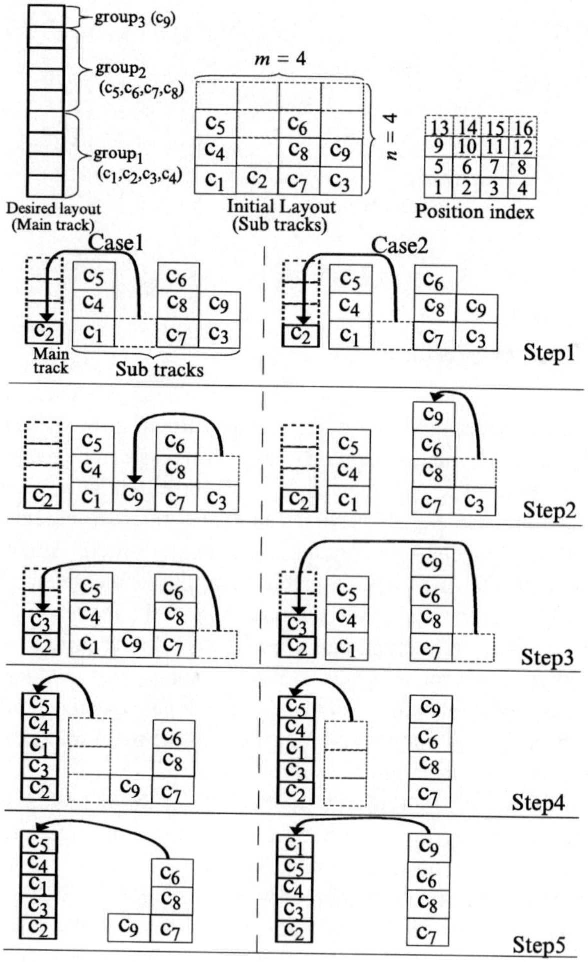

Fig. 13.4 Direct rearrangements

(3) Selection of a removal destination of cars located between the locomotive and the freight car selected in (2).
(4) Selection of the number of cars in the removal operation.
(5) Removal of the cars to the selected sub track.
(6) Rearrangement of the selected car to the main track.

These operations are repeated until one of desirable layouts is achieved in the main track, and a series of operations from the initial state to the desirable layout is defined as a trial.

In the operation (2), each group has the predetermined position in the main track. The car to be rearranged is defined as c_T, and candidates of c_T can be determined by excluding freight cars that have already rearranged to the main track. These candidates must belong to the same group that is determined uniquely by the desired layout of groups in the main track and the number of rearranged cars.

Now, define r as the number of groups, g_l as the number of freight cars in group$_l$ $(1 \leq l \leq r)$, and u_{j_1} $(1 \leq j_1 \leq g_l)$ as candidates of c_T.

In the operation (3), the removal destination of cars located in front of the car to be rearranged is defined as r_M. Then, defining $u_{j_2}(g_l + 1 \leq j_2 \leq g_l + m - 1)$ as candidates of r_M, excluding the sub track that has the car to be removed, and the number of candidates is $m - 1$.

In the operation (4), defining p_S as the number of removal cars required to rearrange c_T, and defining p_d as the number of removal cars that can be located on the sub track selected in the operation (3), the candidate numbers of cars to be moved are determined by $u_{j_3}, 2m \leq j_3 \leq 2m + \min\{p_S, p_d\} - 1$.

In both cases of Fig. 13.4, the direct rearrangement is conducted for c_2 at step1, and the selection of c_T conducted at step2, candidates are $u_1 = [1], u_2 = [4]$, that is, sub tracks where cars in group$_1$ are located at the top. u_3, u_4 are excluded from candidates. Then, $u_2 = [4]$ is selected as c_T. Candidates for the location of c_T are $u_5 = [1], u_6 = [2], u_7 = [3]$, sub tracks [1],[2], and [3]. In Case1, $u_6 = [2]$ is selected as r_M, and in Case2, $u_7 = [3]$ is selected. After direct rearrangements of c_3 at step3 and c_1, c_4, c_5 at step4, the marshaling process is finished at step5 in Case2, whereas Case1 requires one more step in order to finish the process. Therefore, the layout of cars and groups in the main track, the number of cars to be moved, the location of the car to be rearranged and the order of rearrangement affect the total movement counts of cars as well as the total transfer distance of locomotive.

5.1 *Transfer Distance of Locomotive*

A locomotive starts without freight cars, directs to the target car to be moved, and locates it at the corresponding destination. The distance D where the locomotive travels from the start location to the destination of the target car is defined as the transfer distance of the locomotive in a movement operation. Then, the location of the locomotive at the end of above process is the start location of the next movement process of the selected car. Also, the initial position of the locomotive is located on the joint track nearest to the main track.

Define the unit distance in a movement of cars in each sub track as $D_{\min_v}$, the length of joint track between adjacent sub tracks, or, sub track and main track as $D_{\min_h}$. Then, the transfer distance of the locomotive is D, and the maximum of D is $D_{\max} = 2(mD_{\min_v} + nD_{\min_h} + kD_{\min_v})$.

Figure 13.5 shows an example of transfer distance. In the figure, $m = n = 6$, $D_{\min_v} = D_{\min_h} = 1, k = 18$, (a) is position index, and (b) depicts movements of locomotive and freight cars. Also, the locomotive starts from position 8, the target is located on the position 18, the destination of the target is 4, and the number of cars to be moved is 2. Since the locomotive moves without freight cars from 8 to 24, the transfer distance is 12, whereas it moves from 24 to 16 with two freight cars, and the transfer distance is 13. Then $D = 25$ and $D_{\max} = 60$.

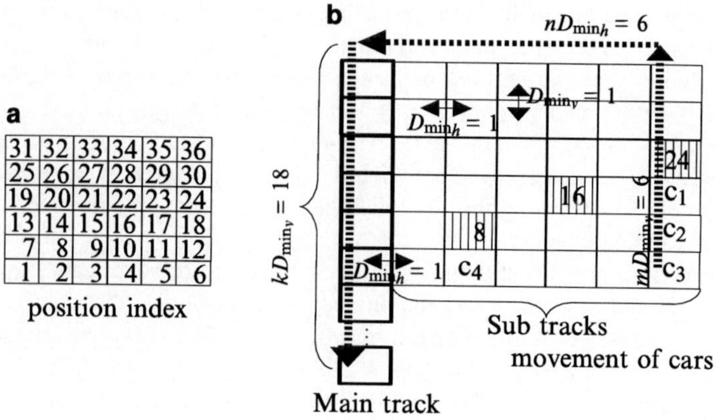

Fig. 13.5 Calculation of transfer distance

6 Learning Algorithm

6.1 Update Rules for Rearrange Order and Removal Destination

Define p_{M} as the number of removed cars, q as the movement counts of freight cars by direct rearrangement, and $\boldsymbol{s}'$ as the state that follows $\boldsymbol{s}$. Also, Q_1, Q_2 are defined as evaluation values for $(\boldsymbol{s}_1, u_{j_1})$, $(\boldsymbol{s}_2, u_{j_2})$, respectively, where $\boldsymbol{s}_1 = \boldsymbol{s}, \boldsymbol{s}_2 = [\boldsymbol{s}, c_T]$. $Q_1(\boldsymbol{s}_1, u_{j_1})$ and $Q_2(\boldsymbol{s}_2, u_{j_2})$ are updated by following rules:

$$Q_1 \ \ (\boldsymbol{s}_1, c_T) \leftarrow \max_{u_{j_2}} Q_2(\boldsymbol{s}'_2, u_{j_2}) \tag{13.1}$$

$$Q_2 \ \ (\boldsymbol{s}_2, r_{\mathrm{M}}) \leftarrow \begin{cases} (1-\alpha)Q_2(\boldsymbol{s}^{\dagger}(t), u_{j_2}) + \alpha[R + \gamma^{q+1} V_1] \\ \quad \text{(next operation is rearrangement)} \\ (1-\alpha)Q_2(\boldsymbol{s}_2, u_{j_2}) + \alpha[R + \gamma V_2] \\ \quad \text{(repetitive removal)} \end{cases} \tag{13.2}$$

$$V_1 = \max_{u_{j_1}} Q_1(\boldsymbol{s}'_1, u_{j_1}),$$

$$V_2 = \max_{u_{j_2}} Q_2(\boldsymbol{s}'_2, u_{j_2})$$

where α is the learning rate, γ is the discount factor, and R is the reward that is given when one of desirable layout is achieved.

6.2 Update Rules for the Number of Cars to be Removed

In addition to learn the rearrange order and the removal destination by Q_1 and Q_2, Q_3 is defined as evaluation value for (s_3, u_{j_3}) when several freight cars can be moved in each rearrangement and removal. Here, $s_3 = [s, c_T, r_M]$. Then, $Q_1(s_1, u_{j_1})$, $Q_2(s_2, u_{j_2})$ and $Q_3(s_3, u_{j_3})$ are updated by following rules:

$$Q_1\ (s_1, c_T) \leftarrow \max_{u_{j2}} Q_2(s_2, u_{j_2}), \tag{13.3}$$

$$Q_2\ (s_2, r_M) \leftarrow \max_{u_{j3}} Q_3(s_3, u_{j_3}), \tag{13.4}$$

$$Q_3\ (s_3, p_M) \leftarrow \begin{cases} (1-\alpha)Q_3(s_3, p_M) + \alpha[R + \gamma^{q+1} V_1] \\ \quad \text{(next action is rearrangement)} \\ (1-\alpha)Q_3(s_3, p_M) + \alpha[R + \gamma V_2] \\ \quad \text{(next action is removal)} \end{cases} \tag{13.5}$$

6.3 Calculation of γ

A discount is conducted in each movement of freight cars. When the movement counts of freight car is evaluated, γ is set as constant. In addition, since the number of rearrangements is k for every initial layout, there is no necessary to discount for rearrangements, and thus, $q = 0$.

When the transfer distance of the locomotive is evaluated, γ is used to reflect the transfer distance of the locomotive and calculated by the following equation:

$$\gamma = \delta \frac{D_{\max} - \beta D}{D_{\max}}, \quad 0 < \beta < 1, 0 < \delta < 1 \tag{13.6}$$

Propagating Q-values by using (13.3)–(13.6), Q-values are discounted in accordance with the number of movements of cars, and the shorter transfer distance of locomotive obtain the better evaluation. In other words, by selecting the car to be rearranged, the removal destination, and the number of cars to be moved that have the largest Q-value, the transfer distance of the locomotive can be reduced.

In the learning stages, each u_j $(1 \leq j \leq 2m + \min\{p_S, p_d\} - 1)$ is selected by the soft-max action selection method [9]. Probability P for selection of each candidate is calculated by

$$\tilde{Q}_i(s_i, u_{j_i}) = \frac{Q_i(s_i, u_{j_i}) - \min_u Q_i(s_i, u)}{\max_u Q_i(s_i, u) - \min_u Q_i(s_i, u)} \tag{13.7}$$

$$P(s_i, u_{j_i}) = \frac{\exp(\tilde{Q}_i(s_i, u_{j_i})/T)}{\sum\limits_{u_j \in u_{j_i}} \exp(\tilde{Q}_i(s_i, u_j)/T)},$$
$$(i = 1, 2, 3). \tag{13.8}$$

In the addressed problem, Q_1, Q_2, Q_3 become smaller when the number of discounts becomes larger. Then, for complex problems, the difference between probabilities in candidate selection remain small at the initial state and large at final state before achieving desired layout, even after repetitive learning. In this case, obtained evaluation does not contribute to selections in initial stage of marshaling process, and search movements to reduce the transfer distance of locomotive is spoiled in final stage. To conquer this drawback, Q_1, Q_2, Q_3 are normalized by (13.7), and the thermo constant T is switched from T_1 to T_2 ($T_1 > T_2$) when the following condition is satisfied:

$$\begin{aligned} [\text{The count of } Q_i(s_i, u_{j_i})] > \eta, \\ \text{s.t. } Q_i(s_i, u_{j_i}) > 0, \\ 0 < \eta \leq [\text{the number of candidates for } u_{j_i}] \end{aligned} \tag{13.9}$$

where η is the threshold to judge the progress of learning.

The proposed learning algorithm can be summarized as follows:

1. Initialize all the Q-values as 0
2. (a) When no cars are placed on candidates of c_T, all of them are rearranged
   (b) Update corresponding $Q_3(s_3, p_M)$ by (13.5)
   (c) Store s_1, c_T
3. If no cars are in sub tracks, go to 9, otherwise go to 4
4. (a) Determine c_T among the candidates by roulette selection (probabilities are calculated by (13.8))
   (b) Put reward as R = 0
   (c) Update the corresponding $Q_3(s_3, p_M)$ by (13.5)
   (d) Store s_1, c_T
5. (a) Determine r_M(probability for the selection is calculated by (13.8))
   (b) Update corresponding $Q_2(s_2, r_M)$ by (13.4)
   (c) store s_2, r_M
6. (a) Determine p_M(probability for the selection is calculated by (13.8))
   (b) Update corresponding $Q_3(s_3, p_M)$ by (13.5)
   (c) Store s_3, p_M
7. Remove p_M cars and place at r_M
8. Go to 2
9. Receive the reward R, update $Q_1(s_1, c_T)$ by (13.3)

7 Computer Simulations

Computer simulations are conducted for $m = 12, n = 6, k = 36$ and learning performances of following five methods are compared:

(A) Proposed method that evaluates the transfer distance of the locomotive, considers the number of cars to be moved, and uses two thermo constants T_1, T_2 with normalized evaluation values.
(B) A method that the number of cars to be moved is 1, and uses two thermo constants with normalized evaluation values.
(C) A method that evaluates the number of movements of freight cars, considers the number of cars to be moved, and uses two thermo constants with normalized evaluation values.
(D) A method that evaluates the transfer distance of the locomotive, considers the number of cars to be moved, and uses one thermo constant T_1.
(E) The same method as (D) that the thermo constant is T_2.

The desirable layout of groups in the main track and the initial arrangement of cars in sub tracks are described in Fig. 13.6. In this case, the rearrangement order of groups is group_1, group_2, group_3, group_4. Cars $c_1, \cdots, c_9$ are in group_1, $c_{10}, \cdots, c_{18}$ are in group_2, $c_{19}, \cdots, c_{27}$ are in group_3, and $c_{28}, \cdots, c_{36}$ are in group_4. Other parameters are set as $\alpha = 0.9, \beta = 0.2, \delta = 0.9, R = 1.0, \eta = 0.95, T_1 = 0.1, T_2 = 0.05$. In method (C), the discount factor γ is assumed to be constant, and set as $\gamma = 0.9$ instead of calculating by (13.6).

Figures 13.7 and 13.8 show the results. In Figs. 13.7 and 13.8, horizontal axis expresses the number of trials and the vertical axis expresses the minimum transfer distance of locomotive to achieve a desirable layout found in the past trials. Vertical lines in Fig. 13.7 indicate dispersions at the corresponding data points. Each result is averaged over 20 independent simulations. In Fig. 13.7, as the number of trials increases, the transfer distance of locomotive reduces, and method (A) derives solutions that require smaller distance of movements of locomotive as compared to method (B). The total transfer distance can be reduced by method (A), because method (A) learns the number of cars to be moved, in addition to the solutions derived by method (B). In Fig. 13.8, the learning performance of method (A) is better than that of methods (D) and (E), because normalized evaluation and switching thermo constants in method (A) is effective for reducing the transfer distance of the locomotive. In method (C), the learning algorithm evaluates the number of movements of freight cars, and is not effective to reduce the total transfer distance of locomotive. Total transfer distances of the locomotive at 1×10^6th trial are described in Table 13.1 for each method. In the table, since method (C) evaluates the movement counts of freight cars, the number of movements is better than that of other methods. Methods (A), (D) and (E) obtain plans that consists of 31 movements of freight cars, then, method (A) progressively improve the solution so that the transfer distance of locomotive is reduced as compared to methods (D) and (E).

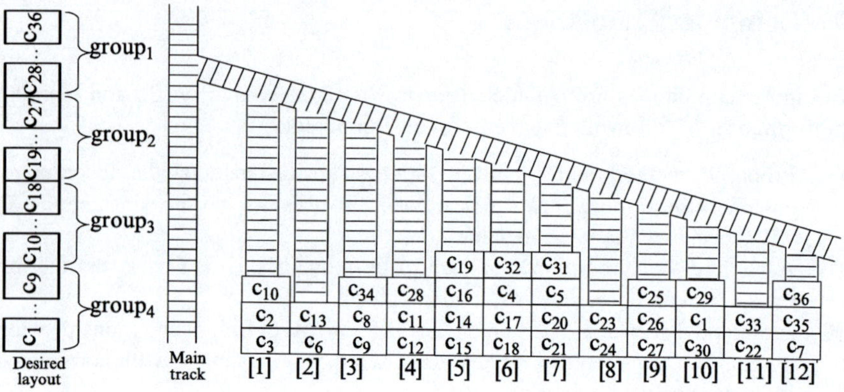

Fig. 13.6 Initial yard setting

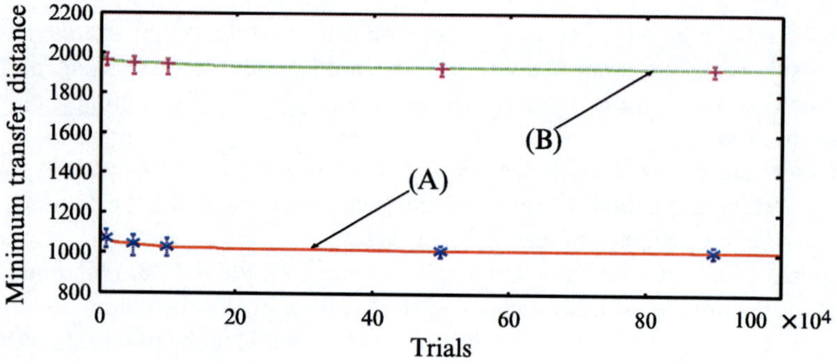

Fig. 13.7 Minimum transfer distances

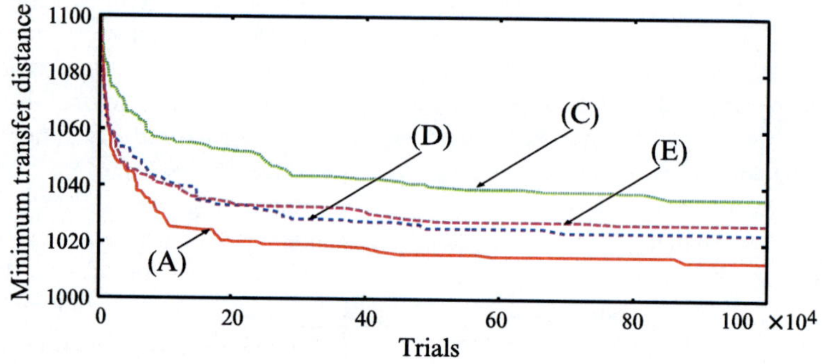

Fig. 13.8 Comparison of learning performances

Table 13.1 Total transfer distances of the locomotive

	Transfer distances			Movements at
Methods	Best	Average	Worst	the best result
Method (A)	981	1,013.60	1,040	31
Method (B)	1,892	1,929.0	1,954	60
Method (C)	1,002	1,035.75	1,078	22
Method (D)	999	1,026.75	1,051	31
Method (E)	984	1,023.35	1,049	31

8 Conclusions

A new scheduling method has been proposed in order to rearrange and line cars in the desirable order onto the main track. The unified learning algorithm of the proposed method is derived based on the reinforcement learning, considering the total transfer distance of locomotive as well as the number of movements of freight cars. In order to reduce the movement of locomotive, the proposed method learns the number of cars to be moved, as well as the layout of main track, the rearrangement order of cars, and the removal destination of cars, simultaneously. In computer simulations, learning performance of the proposed method has been improved by using normalized evaluation and switching thermo constants in accordance with the progress of learning.

References

1. Blasum, U., Bussieck, M.R., Hochstättler, W., Moll, C., Scheel, H.-H., Winter, T. (2000). Scheduling trams in the morning. Math. Meth. Oper. Res. 49(1):137–148.
2. Dahlhaus, E., Manne, F., Miller, M., Ryan, J. (2000). Algorithms for combinatorial problems related to train marshalling. In: Proceedings of the 11th Australasian workshop on combinatorial algorithms, pp 7–16.
3. Eggermont, C., Hurkens, C.A.J., Modelski, M., Woeginger, G.J. (2009). The hardness of train rearrangements. Oper. Res. Lett. 37:80–82.
4. He, S., Song, R., Chaudhry, S.S. (2000). Fuzzy dispatching model and genetic algorithms for railyards operations. Eur. J. Oper. Res. 124(2):307–331.
5. Hirashima, Y. (2011). A new rearrangement plan for freight cars in a train: Q-learning for minimizing the movement counts of freight cars. Lecture notes in electrical engineering, 70 LNEE, pp 107–118.
6. Hirashima, Y. (2011). A new reinforcement learning method for train marshaling based on the transfer distance of locomotive. Lecture notes in engineering and computer science: proceedings of the international multiconference of engineers and computer scientists 2011, IMECS 2011, Hong Kong, March 2011, 16–18, pp 80–85.
7. Jacob, R., Marton, P., Maue, J., Nunkesser, M. (2007). Multistage methods for freight train classification. In: Proceedings of 7th workshop on algorithmic approaches for transportation modeling, optimization, and systems, pp 158–174.
8. Kroon, L.G., Lentink, R.M., Schrijver, A. (2008). Shunting of passenger train units: an integrated approach. Transport. Sci. 42:436–449.

9. Sutton, R.S., Barto, A.G. (1999). Reinforcement learning. MIT Press, MA.
10. Tomii, N., Jian, Z.L. (2000). Depot shunting scheduling with combining genetic algorithm and pert. In: Proceedings of 7th international conference on computer aided design, manufacture and operation in the railway and other advanced mass transit systems, pp 437–446.
11. Watkins, C.J.C.H., Dayan, P. (1992). Q-learning. Mach. Learn. 8:279–292.

Chapter 14
Cost Adjusted MIQ: A New Tool for Measuring Intelligence Value of Systems

Tarik Ozkul and Ismail H. Genc

Abstract Most systems that require the operator control can be considered as man–machine cooperative systems in whose functioning, humans, machines and other unintelligent parts play specific roles. Each role has a value. The recently developed machine intelligence quotient (MIQ) measures the contribution provided by the machines to a system. However, for a more practical decision making process, one needs to also consider the cost of improvements. We propose a simple measure of the cost-benefit criterion which enhances the aforementioned concept by adjusting it for the cost, the cost adjusted MIQ (CAMIQ). The method can be especially useful when trying to determine the best solution among several contenders which are similarly valued, but costwise different.

Keywords Machine intelligence quotient • MIQ • Man–machine cooperative system • Human–machine cooperative systems • Intelligent systems • Cost of intelligence

1 Introduction

The cost-benefit analysis is an economics related concept typically used by decision makers to evaluate the desirability of a given intervention. This technique is ultimately used as a way of analyzing the cost effectiveness of different alternatives to see whether the benefits outweigh the costs. The purpose is to measure the

T. Ozkul (✉)
Computer Science and Engineering Department, American University of Sharjah, PO Box 26666, Sharjah, UAE
e-mail: tozkul@aus.edu

I.H. Genc
Economics Department, American University of Sharjah, PO Box 26666, Sharjah, UAE
e-mail: igenc@aus.edu

S.I. Ao et al. (eds.), *Intelligent Control and Innovative Computing*, Lecture Notes in Electrical Engineering 110, DOI 10.1007/978-1-4614-1695-1_14,

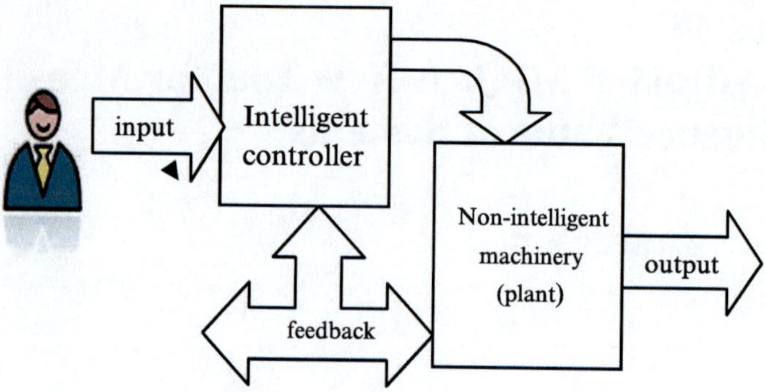

Fig. 14.1 Typical man–machine cooperative system

efficiency of the added value with respect to the status quo. By evaluating the cost-benefit ratio of different alternative approaches, one can determine which alternative solution provides the most benefit for the least cost [1–3]. Cost-benefit ratio of better solutions typically yields a smaller ratio than competing alternatives. In economics related projects, costs and benefits of proposed additions are measured and evaluated in terms of monetary values. However, measurement of benefits in monetary terms is a difficult task which requires estimation of many of the involved parameters. Therefore, in real life it is not surprising to see the actual values deviate from the estimated ones by wide margins [4, 5].

Similar assessment difficulties may occur while adding value added features to man–machine cooperative systems. Man–machine cooperative systems typically are comprised of a human controller, a computer controlled system and some additional non-intelligent parts.

A typical man–machine cooperative system is shown in Fig. 14.1. The system functions as a whole with all parts working together. Human controller makes strategic choices; intelligent system executes the orders and generates a subset of orders for non intelligent parts of the plant. Addition of a unit to the man–machine cooperative system involves the cost of the added unit as well as installation costs. Benefits of the added system appear as an increase in performance, or reduction in complexity of the whole man–machine cooperative system which may be translated into some monetary terms. By determining the ratio of cost to benefit, one may assess the value of the addition.

While adding value-added-features to a man–machine cooperative system, assessment of benefit can be a troublesome metric to measure. Measurement of this, in some cases, requires extensive estimation of parameters [4]. In such cases the recently defined term, namely the machine intelligence quotient, MIQ, can be used as an alternative metric for measuring the "benefit" part of the addition. Thus MIQ can serve as a practical way of determining the intelligence level of a man–machine cooperative system setup. However, lack of attention paid to the cost side of this operation neglects a crucial economic measurement component in terms of the cost-benefit analysis. MIQ, being an objective measurement parameter, can also be utilized to obtain an alternative path to determining the cost-benefit ratio of the set-up. In this paper, we intend to enhance the MIQ calculations with the cost adjustment to obtain a both technically and economically efficient system adjustment decision rule.

The rest of the paper is organized as follows: in the next section we describe the man–machine system, which is followed by a section devoted to the process of calculating the machine intelligence quotient. Then in Sect. 4 we provide the definition of an adjusted MIQ to incorporate cost consideration, as well. In Sect. 5, as a case study, we attempt to determine the intelligence level of a robot controller setup and the cost of MIQ for it which is designed to do an adhesive dispensing operation. Later we add a vision guidance system to the setup and find the MIQ for the modified version and the cost of MIQ thereof. Section 6 provides a comparative discussion on the two versions of the system. And Sect. 7 concludes the paper.

2 Basic Components of a Man–machine Cooperative System

Most general form of man–machine cooperative systems includes a human controller, an intelligent machine/controller, and many non-intelligent machine parts. In case of nuclear power plant, human controller is the person in charge of the control room monitoring and activating operation of the power plant, intelligent machine is a controller computer which monitors sensors, actuators and controls the process control loops. The plant contains many simple machinery parts like turbines, coolers, and heaters which are considered the non-intelligent part of the man–machine cooperative systems. In case of a robot work cell, one can consider the human operator as a decision maker making selection about the type of the operation to execute, robot itself as the intelligent machine carrying out the processing and the feeder or conveyor belt as the non-intelligent part of the plant that works under the command of the intelligent parts of the system which is robot controller in this case. Examples can be increased further to extreme cases where human operator may or may not exist or cases where the non-intelligent machinery may or may not exist. In this paper, it is assumed that intelligent system is in the most general form which contains all three elements stated below:

- Human operator.
- Intelligent machine.
- Non-intelligent plant components.

With the operation of these three distinct parts, the system works and fulfills its purpose. The interaction between these three parts and the sequence of operations is described by intelligent task graph.

3 What is Machine Intelligence Quotient?

Artificial intelligence and machine intelligence are two terms which are often used interchangeably. Artificial intelligence is widely regarded as the "science and engineering of designing intelligent machines". Machine intelligence quotient, on the other hand is regarded by many as the parameter that reflects the "intelligence level of autonomous systems". Machine intelligence has been a difficult concept to deal with and many researchers interpreted MIQ in different ways. Historically, Bien [6] and Kim [7] were the first researchers who have defined MIQ as an indicator of machine intelligence. Lately, there were other researchers who have used MIQ concepts designed by Bien and Kim to determine MIQ of different set ups. Park et al. [8, 9] used MIQ measurement to determine machine intelligence level of a nuclear power plant set up which is considered as a typical man–machine cooperative system which contained a human controller, intelligent automated controllers and dumb machines. Ulinwa [10] introduced some new interpretations to the MIQ concept. Finally, Anthony and Jannett [11] used MIQ methodology to determine the intelligence level of a distributed network setup.

As far as the context of this paper goes, we will be adopting the definition of MIQ used by Park et al. [8, 9]. According to this interpretation, systems or machines are not intelligent devices themselves. They are ultimately designed to assist a human operator who uses the system. MIQ is a parameter which reflects the ease of use of the system by the operator. Systems which are easy to use by the operator have high MIQ which reflects this ease of use. Systems which are found difficult to use by the operator have low MIQ which reflects the difficulty. In this context, all existing systems which are designed to be used by human operators have an MIQ of some sort.

Consider the following example of a household heating system to describe the concept. Imagine an old heating unit where the operator needs to light the pilot lamp, turn the gas valve, open and close several water valves and intervene frequently to keep the temperature at a comfortable level. Now picture a new unit where the operator sets the temperature level and simply pushes a button to activate the system. In both cases the systems are actually doing exactly the same thing, but in the case of the latter, the unit, i.e., the machine, is doing many of the steps like turning valves, lighting the pilot, etc. on its own. It is assumed that keeping a house warm at a comfortable level requires certain amount of work (intelligence). This work is partly being done by the operator and partly by the heating system itself. In the case of former unit which requires much intervention from the operator, operator spends considerable effort to keep the house warm at a specific temperature level. In the case of the second system, the operator effort is much less to achieve the desired heating quality. In terms of MIQ, we say that first system has low MIQ because many actions need to be done by the operator. In contrast, the second system has high MIQ since operator spends very little effort to achieve the desirable heating performance. The rest of the actions are being done by the heating system itself.

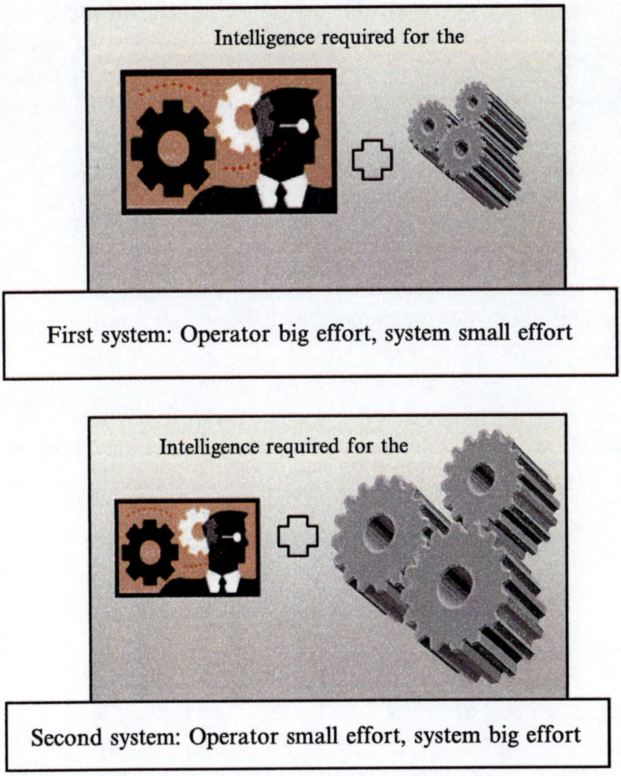

Fig. 14.2 Two identical systems doing the same job with different MIQ's

As one would notice, according to this version of understanding, the concept of MIQ does not really require the system to have a processor or built in intelligence inside. It assesses MIQ by the end value it provides toward achieving a certain goal. Accordingly any system, regardless whether there is a processor inside or not, has some MIQ which can be quantitatively measured. The concept is illustrated in Fig. 14.2.

3.1 Intelligence Task Graph

An intelligence task graph is a state diagram of operation which describes all the stages of the process the system has to go through from the beginning until the completion of the task. The intelligence task graph (ITG) also indicates which tasks are being executed by which part of the intelligent system. The tasks which are done by the human operator, the tasks which are done by the intelligent controller, and the tasks which are performed by non-intelligent plant components are all displayed by the ITG graph.

ITG graph specifies processes by the use of bubbles and directed arrows. Bubbles show processes, and arrows identify the sequence and flow of information from one process to another. Following the terminology developed by Park et al. [8], T indicates the set of tasks or processes to be completed where T is a vector with n components.

The task set is defined as:

$$\mathbf{T} = \{\mathrm{T1}, \mathrm{T2}, \mathrm{T3}, \ldots.\mathrm{Tn}\} \tag{14.1}$$

The task intelligence cost is a subjective parameter that indicates the difficulty of a specific task. This parameter is estimated by the user and it is denoted by τ.

$$\tau = \{\tau 1, \tau 2, \tau 3, \ldots.\tau \mathrm{n}\} \tag{14.2}$$

The data transfer matrix F defines the amount of information transferred from one task Ti to another task Tj.

$$\mathrm{F} = \begin{bmatrix} 0 & f12 & f13 & f14 & . & . & f1n \\ f21 & 0 & f23 & f24 & . & . & f2n \\ f31 & f32 & 0 & f34 & . & . & f3n \\ f41 & f42 & f43 & 0 & . & . & f4n \\ f51 & f52 & f53 & f54 & 0 & . & f5n \\ f61 & f62 & f63 & f64 & f65 & 0 & f6n \\ fn1 & fn2 & fn3 & fn4 & fn5 & . & 0 \end{bmatrix} \tag{14.3}$$

Another matrix used in the intelligence task graph is the task allocation matrix which defines which tasks are executed by which part of the system. The task allocation matrix is represented by A. By definition, the number of columns in matrix A is restricted to 3, where the first column indicates the tasks assigned to the intelligent machine controller, the second column indicates the tasks which are assigned to human controller, and the third column indicates the tasks which are executed by the non-intelligent part of the plant. Matrix A is a binary matrix with values of either 1 or 0.

$$\mathrm{A} = \begin{bmatrix} a11 & a12 & a13 \\ a21 & a22 & a23 \\ a31 & a32 & a33 \\ . & . & . \\ . & . & . \\ an1 & an2 & an3 \end{bmatrix} \tag{14.4}$$

In the above matrix, if task 1 is executed by the intelligent controller part of the system, a11 = 1 and a12 = a13 = 0. If task 2 is executed by the human controller, than a22 = 1 and a21 = a23 = 0. With this consideration matrix has binary entries which consist of 1's and 0's.

3.2 Control Intelligence Quotient

The control intelligence quotient (CIQ) is defined as the total intelligence required for carrying out all the tasks in the man–machine cooperative system. The formula that describes CIQ is as follows;

$$\mathrm{CIQ} = \sum\nolimits_{i=1}^{n} ai1.\tau i + \sum\nolimits_{i=1}^{n} ai2.\tau i \tag{14.5}$$

which is the combination of the intelligence of the intelligent controller and the intelligence of the human controller together.

Putting it in a different way, control intelligence quotient CIQ of the system is a combination of the intelligence contributed by the human controller and the intelligence provided by the machine.

$$\mathrm{CIQ} = \mathrm{HIQ} + \mathrm{MIQ} \tag{14.6}$$

Using the aforementioned household heater example, the job of heating the house and keeping it at a comfortable set temperature is a job which requires certain amount of intelligence, which is indicated as CIQ in the formula. This task is being done by the cooperation of human controller and the machine. The contribution of human controller to the completion of the task CIQ is labeled as HIQ, and contribution of the machine controller toward this goal is represented by MIQ. In other terms, addition of HIQ and MIQ is a set number for a certain given task. Thus, for a given specific task, if the human operator involvement increases (high HIQ), the machine part involvement (low MIQ) decreases. Decreasing human involvement means (less HIQ), which ultimately points to a system with high machine intelligence (high MIQ).

3.3 Human Intelligence Quotient

Human intelligence quotient (HIQ) is defined as the intelligence quantity needed from the human controller for controlling the system. The formula for HIQ is given as follows:

$$\begin{aligned}\mathrm{HIQ} = {} & \sum\nolimits_{i=1}^{n} ai2.\tau i + Cmh \sum\nolimits_{i=1}^{n} \sum\nolimits_{i=1}^{n} ai1.aj2.fij \\ & + Chm \sum\nolimits_{i=1}^{n} \sum\nolimits_{i=1}^{n} ai2.aj1.fij\end{aligned} \tag{14.7}$$

The first summation term indicates the combination of all tasks done by the human controller multiplied by difficulty levels of tasks. Second term in the formula includes a multiplicand parameter C_{mh}, which is defined as the interface complexity, which describes the difficulty of transferring data from machine to human. Through this second term we include all the intelligence demanded from human controller for interpreting information that comes from machine to human. The third summation term uses a parameter labeled as C_{hm} which is defined as the interface complexity while transferring data from the human controller to the machine controller. C_{hm} and C_{mh} are two indexes which are decided subjectively to indicate the difficulty level. These parameters vary between 0 and 1, 0 being the easiest and 1 being the hardest in assessing the difficulty of entering or interpreting data. Typically difficulty levels of these parameters are expected to be around 0.05 for well designed systems which more or less corresponds to the difficulty of pressing a single button to initiate a process.

3.4 MIQ, HIQ and CIQ Relationship

MIQ is defined as the difference between CIQ and HIQ.

$$\mathrm{MIQ} = \mathrm{CIQ} - \mathrm{HIQ} \tag{14.8}$$

The rationale behind this definition is as follows; CIQ is the total amount of intelligence required to run the system which is partly contributed by human controller and partly contributed by intelligent machine controller. MIQ, which is the true intelligence due to machine intelligence is supposed to be found by taking away the contribution done by the intelligence of human controller (HIQ).

4 Derivation of the Cost Adjusted MIQ

In this section, we would like to enhance the definition of MIQ to also incorporate a consideration of the cost in system adjustments. MIQ can be used as a factor in cost-benefit analyses when determining the value of man–machine cooperative systems [12, 13]. That would make MIQ more appealing to decision makers in business world as it reflects both the cost and benefit of undertaking a system adjustment. We assume that the decision maker assigns a subjective pecuniary value to each unit of MIQ, yielding $\mathrm{MIQ_m}$, which is the monetary value of the intelligence of the machine as perceived by the decision maker. Then, when $\mathrm{MIQ_m}$ is used for benefit measurement, the division of the cost involved by the $\mathrm{MIQ_m}$ of the system reveals a parameter which determines what could be construed as the cost-per-MIQ. Naturally, the perception of intelligence is a subjective matter, and its identification in financial terms varies from person to person. Therefore, $\mathrm{MIQ_m}$ reflects that personal view of decision makers in system MIQ assessments. Thus, the cost-per-MIQ, K, is calculated according to the following formula.

$$\mathrm{K} = \mathrm{C}/\mathrm{MIQ_m} \tag{14.9}$$

where C stands for the cost of the system. Among the two competing solutions, if alternative #1 reveals K1 and alternative #2 reveals K2; and, if among the alternative solutions $K1 < K2$, it can be said that alternative #1 is a better solution which provides better MIQ per capital. One might call K the cost adjusted MIQ, CAMIQ. Obviously, CAMIQ may be utilized in comparing a number of systems and/or upgrades to an existent system.

Through CAMIQ, that is, by measuring the cost-per-MIQ of different alternatives, one may decide which alternative solution may present the most MIQ per capital spending. The advantage of using MIQ as an objective benefit measurement tool is such that, it provides a unique way of assessing the impact of addition (upgrade) in terms of increase in intelligence. Since MIQ_m reveals the decision maker's perception of this addition, CAMIQ provides a further tool in consideration of system upgrades and/or comparisons. Therefore, gauging MIQ against the cost in the form of CAMIQ enhances the usability of MIQ in decision making processes in practice. Nevertheless, a researcher should always keep in mind that measurement of benefits in monetary terms is a difficult task which requires estimation of many of the involved parameters (see [4, 5] for a discussion on the issue in empirical projects).

One of the most fundamental questions for decision makers, however, is to determine the amount/level of cost to tolerate in conjunction with the improvements in the system where the latter is measured by MIQ_m. In other words, what is the optimum decision making rule for the cost-benefit issue discussed in this paper? The answer to this question is easily derived by taking the total differentiation of (14.9), and equating to zero for an optimization problem. That is,

$$d(K) = d(C) - d(MIQ_m) = 0 \tag{14.10}$$

Hence, it is easy to observe that

$$d(C) = d(MIQ_m) \tag{14.11a}$$

and by approximation

$$D(C) = D(MIQ_m) \tag{14.11b}$$

which has a familiar economic interpretation. The left hand side of the equation above represents the marginal cost of an economic activity while the right hand side characterizes the marginal benefit thereof. By following the economic logic, one can say that a decision maker will continue upgrading a particular system for as long as the additional benefit of this operation is no less than the additional cost of this operation involved. As the economic theory goes, then, the decisions are made on an incremental basis comparing the cost and benefit of every action, which is what our cost adjusted MIQ, viz. CAMIQ definition finds.

5 Various Interpretations and an Example for the Cost Adjusted MIQ

5.1 *Comments on the Application of the Process to Real Life Cases*

Because of the subjectivity in the identification of certain costs or benefits, the process could be potentially challenging. Especially, compounding the identification problem is the interlaced nature of some costs and benefits. For example, operating costs may have to be deducted from the benefits of a project. This is known as the netting out of operating costs.

An additional interpretation of CAMIQ, the cost-per-MIQ, K, can be provided by noting that K resembles the so-called break-even point of a project with costs and benefits. With properly identified costs and benefits, decision makers can determine the time it takes, i.e. the payback time, for a project to break even. Obviously, projects with longer payback durations are less preferred to those with shorter time spans.

Yet, a further concern with our formula for K is the possibility of the benefits and costs accruing over time. As any student of economics knows, the "value of money" depends on time. In that both the numerator and the denominator of K have to be discounted per the time span of realization of costs and benefits. In its most simplistic form, a future amount is brought to today's terms by dividing it by $(1+r)N$ where r is the discount factor, which is the real interest rate for many economic considerations; and N stands for the time span of the realization of costs and/or benefits. It goes without saying that this modification of our formula is needed if costs and benefits have divergent realization time spans.

The formula can also be enriched with the consideration of the uncertainty by attaching probabilities to contingencies of costs and benefits. As these issues are well-known in practice, we do not attempt to provide a revised version of K.

Finally, one should note that not all benefits may be unambiguously net gains, especially when certain government services are switched to e-environment. That is why; rigorous analyses are prerequisites to justify transfer of resources from the current system to an alternative one.

5.2 *Example Case*

Briefly speaking, developing an analysis of K involves three steps: in what we call Step 0: the problem is indentified, its scope (pervasiveness) is determined. This is to justify the attempt to adopt a new system instead of the current, and supposedly inefficient, system. In Step 1, we identify alternative system(s). And finally, in Step 2, for each alternative, we determine the costs and benefits (monetary as well as non-monetary).

In all countries, preparing the proposal for the annual budget at the ministry of finance level is a cumbersome task for budget experts as they individually need to collect data, enter them into their own computers, and carry out analyses, to allocate public funds. This is Step 0, the problem identification stage. An alternative could be the centrally administered computerized data entry system which is accessible to all experts at all times. This is Step 1, identification of alternative. In the last step, we determine costs (C), and benefits (MIQ) of both systems. For the sake brevity we outline here C and MIQ for the centrally administered system.

On the cost side, we may consider:

1. Equipment.
2. Operation and maintenance.
3. Training.
4. More coordination in data entry.
5. More centralized data storage.

On the benefits side, we can list:

6. Wider and easier access to data.
7. More reliable data.
8. Printing costs (negative).
9. Storage costs of printed material (negative).
10. Customer (budget expert) satisfaction.

The researcher needs to put a price tag on each item and calculate K as a result.

A few explanations are in order: Some of the costs are one-off while others are continuous. For example, initial costs are usually one-off whereas operating costs are ongoing. Continuous development of the software for efficiency and security reasons make up a portion of operating costs. Similar considerations may be given to hardware related issues.

Some of the benefits could easily be classified as cost reductions. This is where the netting out of operating costs comes handy. Also, as alluded to above, attaching a price tag on customer satisfaction could be challenging.

6 Conclusion

An alternative way of assessing value of man machine cooperative systems is presented. The method is based on the concept of machine intelligence quotient (MIQ) calculation of the system and evaluating the MIQ gained for the money spent for various alternatives. The alternative which gives the best MIQ for money is the choice recommended by this method.

References

1. Belli P, Anderson JR, Barnum HN, Dixon JA, Tan J-P (2001) Economic analysis of investment operations. Analytical tools and practical applications. WBI World Bank Report, Washington DC
2. Brent RJ (1996) Applied cost-benefit analysis. Edward Elgar, Cheltenham
3. Florio M, Vignette S (2003) Cost benefit analysis of infrastructure projects in enlarged EU: an incentive-oriented approach. Paper presented at the fifth European conference on evaluation of the structural funds. "Challenges for Evaluation in an Enlarged Europe", Budapest, 26/27 June
4. Flyvbjerg B, Holm MS, Buhl SL (2005) How (In)accurate are demand forecasts in public works projects? J Am Planning Assoc 71(2):131–146
5. Flyvbjerg B, Holm MS, Buhl SL (2002) Underestimating costs in public works projects Error or Lie? J Am Planning Assoc 68(3):279–295
6. Bien Z (1998) How to measure the machine intelligence quotient (MIQ): two methods and applications. Proceedings of world automation congress, Anchorage, May 1998
7. Kim S-W, Kim BK (1998) MIQ (Machine Intelligence Quotient) for process control system. Proceedings of world automation congress, Anchorage, May 1998
8. Park H, Kim B, Lim K (2001) Measuring the machine intelligence quotient (MIQ) of human–machine cooperative systems. IEEE Trans Syst man Cybernet 31:89–96
9. Park H, Kim B, Lim G (1999) Measuring machine intelligence for human–machine cooperative systems using intelligence task graph. Proceedings of the 1999 IEEE/RSJ International Conference on Intelligent Robots and Systems, pp 689–694. http://dspace.kaist.ac.kr/bitstream/10203/1353/1/IC_1999hjpark_1.pdf
10. Ulinwa IC MIQ understanding a machine through multiple perspectives analysis. www.isd.mel.nist.gov/ research_areas/ research_engineering/ Performance_Metrics/ PerMIS_2003/ Proceedings/Ulinwa.pdf
11. Anthony A, Jannett TC (2006) Measuring machine intelligence of an agent-based distributed sensor network system. Proceedings of international joint conferences on computer, information, and systems sciences and engineering (CISSE), Dec 2006
12. Ozkul T (2009) Cost-benefit analyses of man–machine cooperative systems by assessment of machine intelligence quotient (MIQ) gain. Proceedings of mechatronics and its applications, 23–26 March 2009
13. Ozkul T, Genc IH (2011) Development of cost adjusted MIQ concept for measuring intelligence value of systems. Lecture notes in engineering and computer science: proceedings of the international multiconference of engineers and computer scientists 2011, IMECS 2011, Hong Kong, 16–18 March 2011, pp 33–37

Chapter 15
Harmony Search Algorithm with Various Evolutionary Elements for Fuzzy Aggregate Production Planning

Pasura Aungkulanon, Busaba Phruksaphanrat, and Pongchanun Luangpaiboon

Abstract This chapter presents an application of a fuzzy programming approach for multiple objectives to the aggregate production planning (APP). The APP parameter levels have been applied via a case study from the SMEs company in Thailand. The proposed model attempts to minimise total production cost and minimise the subcontracting units. Conventional harmony search algorithm (HSA) with its hybridisations of the novel global best harmony search (NGHSA) and the variable neighbourhood search of the HSA (VHSA) including the hunting search (HuS) element on the pitch adjustment (HuSHSA). Based on the experimental results, it can be concluded that each algorithm is suitable for different types of situations. However, for all situations VHSA and NGHSA can obtain good solutions. Furthermore, the proposed VHSA is more effective than other approaches in terms of superiority of solution and required CPU time.

Keywords Fuzzy aggregate production planning • Particle swarm optimisation • Hunting search algorithm and variable neighbourhood search algorithm

1 Introduction

Most manufacturing companies in Thailand [1] do not perform appropriate production planning even though it plays an important role for the companies. This may be the result from the following issues. Firstly, production planning system is not available. Secondly, manufacturing companies in Thailand are not interested in the mathematical model of production systems including related complex approaches

P. Aungkulanon (✉) • B. Phruksaphanrat • P. Luangpaiboon
The Industrial Statistics and Operational Research Unit (ISO-RU), Department of Industrial Engineering, Faculty of Engineering, Thammasat University, Pathumtani, 12120, Thailand
e-mail: pasurachacha@hotmail.com; lbusaba@engr.tu.ac.th; lpongch@engr.tu.ac.th

S.I. Ao et al. (eds.), *Intelligent Control and Innovative Computing*, Lecture Notes in Electrical Engineering 110, DOI 10.1007/978-1-4614-1695-1_15,

to solve such models. Thirdly, manufacturing companies require an approach that is easy to understand and verify in order to easily convince their management team to agree with its solution. Finally, an approach should not require the additional investment on any expensive software due to the ongoing economic crisis in Thai industries. Based on those problems this chapter intends to provide another solution for developing aggregate production plans to decrease the production cost.

Aggregate production planning [2, 3] is concerned with the determination of production, inventory, and workforce levels to meet fluctuating demand requirements over a planning horizon that approximately ranges from six months to one year [4, 5]. Typically, the planning horizon incorporates the next seasonal peak in demand. The planning horizon is often divided into periods. Given demand forecasts for each period of a finite planning horizon, the APP specifies production levels, workforce, inventory levels, subcontracting rates and other controllable variables in each period that satisfy anticipated demand requirements while minimising the relevant cost over that production planning horizon. The concern of the APP is to select the strategy to deal with uncertainty with the least cost to the manufacturing companies. This problem has been under an extensive discussion and several alternative methods for finding an optimal solution have been suggested in the literature [6, 7].

2 Aggregate Production Planning (APP)

2.1 Multi-Objective Linear Programming (MOLP) Model to APP

Objective functions of the APP model are to minimise the total production cost (Z_1) and to minimise subcontracting units (Z_2). The total production cost composes of permanent worker salary, temporary worker wage, overtime cost of permanent and temporary workers, hiring and laying off cost of temporary workers and inventory holding cost over a specific number of m monthly planning periods. In this model, the overtime costs of permanent and temporary workers during workdays and holidays are different. The decision variables are as follow.

W = the number of permanent workers
$TW(t)$ = the number of temporary workers in period t
$H(t)$ = the number of temporary workers to be hired at the beginning of period t
$L(t)$ = the number of temporary workers to be laid off at the end of period t
$OW_n(t)$ = overtime man-hours of permanent worker during normal workday in period t
$OTW_h(t)$ = overtime man-hours of temporary worker during holiday in period t
$P(t)$ = total production quantity in period t
$I(t)$ = inventory level in period t
$Sub(t)$ = amount of subcontracted unit in period t

Moreover, there are some related parameters in the model. $n(t)$ and $h(t)$ are the number of normal workdays and holidays that can apply OT in period t, respectively. R_H, OH_n and OH_h are the number of regular working hours and allowable overtime hours in each normal workday and holiday, respectively. *MIN W* and *MAX W* are the minimal and maximal number of permanent workers to operate the production line, respectively. K_W and K_{TW} are an average productivity rate per man-day of permanent and temporary worker, respectively. $D(t)$ and $SS(t)$ are forecasted demand and safety stock level in period t, respectively. $MAX\ O_n(t)$ and $MAX\ O_h(t)$ are the maximal overtime man-hours that can be applied during normal workday and holiday in period t, respectively.

MAX_TW is the maximal number of temporary workers to operate the production line. *MAX_I* is the maximal allowable inventory level. *MAX_Sub* is the maximal allowable subcontracting units. *CW* is an average salary per month of a permanent worker. *CTW* is an average wages per day of a temporary worker. *CH* is a hiring cost per person of temporary worker. *CL* is an average inventory holding cost per month per unit of product. COW_n and COW_h are overtime cost per man-hour of permanent worker during normal workday and holiday, respectively. $COTW_n$ and $COTW_h$ are overtime cost per man-hour of temporary worker during normal workday and holiday, respectively. *CSub* is a subcontracting cost per unit.

There are six APP constraints which consist of permanent worker, inventory, production, overtime, temporary worker and subcontracting constraints as shown in (15.3)–(15.15).

Thai labor law states that temporary workers could not be continuously hired longer than four months (The Labor Protection Act of Thailand, 1998). After four months they must become permanent workers. Hence, Thai industries will lay off temporary workers after four months if they do not want to transfer them to permanent ones. Constraints 16 and 17 are related to this law [8].

All decision variables are nonnegative and some decision variables representing number of workers, namely, W, $H(t)$, $L(t)$, and $TW(t)$ are integer. Thus, the objective function and constraints over the time period ($t = 1, 2, \ldots, 12$) can be formulated as the APP mathematical programming model below and all parameters and the initial values were taken from [8].

$$Min\ Z_1 = \sum_{t=1}^{m} \begin{bmatrix} CW(W) + COW_n(OW_n(t)) + COW_h(OWh(t)) + \\ CTW(TW(t))(n(t)) + COTW_n(OTW_n(t)) + \\ COTW_h(OTW_h(t)) + CH(H(t) + CL(L(t)) + \\ CI(I(t)) + Sub(t)(CSub) \end{bmatrix} \tag{15.1}$$

$$Min\ Z_2 = \sum_{t=1}^{m} Sub(t) \tag{15.2}$$

Subject to

$$MAX\ W \leq W \leq MIN\ W \tag{15.3}$$

$$I(t) = I(t-1) + P(t) - D(t) \tag{15.4}$$

$$I(t) \geq SS(t) \tag{15.5}$$

$$I(t) \leq MAX\_I \tag{15.6}$$

$$P(t) = W \bullet KW \bullet n(t) + KW \bullet [OW_n(t) + OW_h(t)]/RH + TW(t) \bullet K_{TW} \bullet n(t) + K_{TW} \bullet [OTW_n(t) + OTW_h(t)]/RH \tag{15.7}$$

$$OW_n(t) + OTW_n(t) \leq MAX\_O_n(t) \tag{15.8}$$

$$OW_h(t) + OTW_h(t) \leq MAX\_O_h(t) \tag{15.9}$$

$$MAX\ O_n(t) \geq OH_n \bullet n(t)(W + TW(t)) \tag{15.10}$$

$$MAX\ O_h(t) \geq OH_h \bullet n(t)(W + TW(t)) \tag{15.11}$$

$$OW_n(t)/W = OTW_n(t)/TW(t) \tag{15.12}$$

$$OW_h(t)/W = OTW_h(t)/TW(t) \tag{15.13}$$

$$TW(t) \leq MAX\_TW \tag{15.14}$$

$$Sub(t) \leq MAX\_Sub \tag{15.15}$$

$$TW(t) = H(t-3) + H(t-2) + H(t-1) + H(t) \tag{15.16}$$

$$L(t) = H(t-3) \tag{15.17}$$

2.2 Fuzzy Programming Approach

Fuzzy programming approach is one of the most effective methods for solving multiple objective decision making (MODM) problem. A fuzzy set can be characterised by a membership function, usually denoted μ which assigns to each objective [9]. In general, the non-increasing and non-decreasing linear membership functions are frequently applied for the inequalities with less than or equal to and greater than or equal to relationships, respectively. Since the solution procedure of the fuzzy mathematical programming is to satisfy the fuzzy objective, a decision in a fuzzy environment is thus defined as the intersection of those membership functions corresponding to fuzzy objectives [10–12]. Using the simplest type of membership function, it is assumed to be linearly increasing over the "tolerance interval" of p_k for the kth objective and $k = 1, 2, \ldots, K$.

$$\mu_k(x) = \begin{cases} 1 & \text{if} \quad c_k^T x \leq Z_k^{PIS} \\ 1 - \frac{Z_k^{PIS} - c_k^T x}{p_k} & \text{if} \quad Z_k^{PIS} < c_k^T x \leq Z_k^{NIS} \\ 0 & \text{if} \quad c_k^T x \geq Z_k^{NIS} \end{cases}, \tag{15.18}$$

where

$$p_k = \left| Z_k^{PIS} - Z_k^{NIS} \right|, k = 1, 2, \ldots, K. \tag{15.19}$$

The membership functions, $\mu_k(x)$ should increase monotonously from 0 to 1. p_k are subjectively chosen constants of admissible violations of the kth objective. It can be defined by the Positive-Ideal Solution (PIS) and the negative-ideal solution (NIS). The Z_k^{PIS} and Z_k^{NIS} are the best possible solution and the feasible worst value of kth objective. Under the concept of min-operator, the feasible solution set is defined by interaction of the fuzzy objective set. The DM makes a decision with a maximal value in the feasible decision set. That is

$$\max_{x\geq 0}\min_{k}\left(1-\frac{Z_k^{PIS}-c_k^T x}{p_k}\right) \tag{15.20}$$

or equivalently

$$\textit{Maximise}\ \lambda$$

subject to

$$\mu_k(x) > \lambda, k = 1, 2, .., K.$$
$$Ax \leq b, x \geq 0,$$
$$\lambda \in [0, 1]. \tag{15.21}$$

Various hybridisations of harmony search algorithm (HSA) in the following section for this fuzzy programming approach is applied to APP problem.

3 Harmony Search Algorithm

HSA is a new meta-heuristic optimisation method proposed by Geem et al. in 2001 [13]. It is considered a population based or socially-based inspiration algorithm with local search aspects. HSA is conceptually derived from the natural phenomena of musicians' behaviour when they play or improvise their musical instruments together. This comes up with a pleasing harmony or a perfect state of harmony as determined by an aesthetic quality via the pitch of each musical instrument. Similarly, the optimisation process seeks to find a global solution as determined by an objective function via the set of values assigned to each decision variable.

In the musical improvisation, aesthetic estimation is performed by the set of pitches played by each instrument. The harmony quality is enhanced practice after practice. Each type of music composes of specific instruments played by musicians. If all pitches bring a good harmony, that experience is stored in each player's memory, and the possibility to make a good harmony is increased for the next time. Assume there are a certain number of preferable pitches in each musician's memory. Each instrument provides various notes. In music improvisation, each player sounds any pitch in the possible range, together making one a harmony vector.

If all plays together with different notes there is a new musically harmony. If this leads to a better new harmony than the existing worst harmony in their memories, a new harmony is included in their memories. In contrast, the worst harmony is excluded from their memories. Three rules of musical improvisation consist of rules of playing any one pitch from his memory, playing an adjacent pitch of one pitch from his memory, or playing totally random pitch from the possible sound range. These procedures are repeated until a fantastic harmony is found.

Similarly in engineering optimisation, harmony of the notes or pitches generated by a musician is analogous to the fitness value of the solution vector. Each musician can be replaced with each decision variable. The musician's improvisations are analogous to local and global search schemes in optimisation techniques. During searching, if all decision variable values make a good solution, that experience is stored in each variable's memory, and the possibility to make a good solution is also increased for the next time. Similarly, when each decision variable chooses one value in the HSA, it follows three rules which are to choose any one value from the harmony memory (HM) or memory considerations, choose an adjacent value of one value from the HM defined as pitch adjustments, or totally choose a random value from the possible value range defined as randomisation. These three rules in the HSA are associated with two parameters of a harmony memory considering rate (P_{HMCR}) and a pitch adjusting rate (P_{PAR}).

HSA is very successful in a wide variety of optimisation problems. It also presents several advantages with respect to conventional optimisation techniques. HSA does not require initial values for the decision variables and it imposes fewer mathematical requirements. Furthermore, instead of a gradient search like conventional algorithms, the HSA provides a stochastic search with no derivative information which is based on the HM consideration rate or P_{HMCR} and the pitch adjustment rate or P_{PAR} so that it is not necessary to derive the associated function during the problem analysis. HSA generates a new vector, after considering all of the existing vectors, whereas other meta-heuristics, such as the genetic algorithm, only considers the two parent vectors. The pseudo code is used to briefly explain to all the procedures of the HSA shown in Fig. 15.1.

As concerned in the literature for the algorithm parameter levels, an HMS of 20–50, a P_{HMCR} of 0.7–0.95, and a P_{PAR} of 0.3–0.7 were frequently recommended in HSA applications. However, the *IM* and *BW* were determined based on the number of objective function and possible value ranges of decision variable evaluations from other competitive algorithms, respectively.

4 Evolutionary Elements on the HSA

4.1 The Hunting Search Element on the Pitch Adjustment Rate (HuSHSA)

The hunting search (HuS) algorithm [14] is applied in this section to replace the original pitch adjustment operator of the original HSA and it is called as the

```
Procedure HSA Meta-heuristic()
Begin;
  Initialise algorithm parameters:
      IM:       the preset number of improvisations
      HMS:      the size of the harmony memory
      BW:       the 'distance bandwidth' or the amount of maximal change for pitch adjustment
                between two neighbouring values in discrete candidate set
      P_HMCR:   the rate of considering from the harmony memory
      P_PAR:    the pitch adjustment rate
  Initialise the HMS harmony memories;
  Evaluate the fitness values for all HMS;
  For j - 1 to IM
        Randomly select a position of [1, 2, ..., HMS] to improvise;
        Generate a random number in the range [0, 1] or RN1;
        Check RN1 with P_HMCR;
          If    RN1 < P_HMCR better, then pick the component from memory;
            Generate a random number in the range [0, 1] or RN2;
            If RN2 < P_PAR better, then adjust the harmony by a small amount BW;
              Generate a random number in the range [0, 1] or RN3;
                If RN3 > 0.5
                  Pitch Adjustment Harmony vector increase;
                Else
                  Pitch Adjustment Harmony vector decrease;
                End if;
            Else
                  Do nothing;
            End if;
          Else
            Pick a new random value in the allowed range;
          End if;
        Replace a new harmony if better;
  End for;
End;
End procedure;
```

Fig. 15.1 Pseudo code of the HSA meta-heuristic

HuSHSA. A prominent characteristic of the hunting group reorganisation is used as a chance for the harmony to be trapped in a local optimum. If this happens, the harmony will be reorganised themselves to get another opportunity to find the better outcome. Parameters of Ra_{MAX} and Ra_{MIN} are the maximal and minimal relative search radius of the hunter.

4.2 Variable Neighbourhood Search Element on the HSA (VHSA)

Variable neighbourhood search method (VNS), initially introduced by Mladenovic and Hansen in 1997 [15], is one among meta-heuristics designed for solving combinatorial and global optimisation problems. It exploits systematically the idea of neighbourhood change within a local search method to approach a better solution. Contrary to other local search methods, VNSM proceeds by a descent method to a local minimum exploring then, systematically or randomly, increasingly distant neighbourhoods of this incumbent solution.

This variant, called VHSA, uses maximal (or minimal) point in each iteration for the HMS improvement. This system needs improvement solutions of nearly optimal design point. Because HMS improves from an experience of solutions, these are then stored in groups of HMS. This system wants improvement in the best of the optimal points in each iteration.

4.3 Particle Swarm Optimisation Element or Novel Global Best Harmony Search Algorithm (NGHSA)

A prominent characteristic of particle swarm optimisation (PSO) [16] is that the individual is inclined to mimic its successful companion. Inspired by the swarm intelligence of particle swarm, a new variation of the HSA is proposed in this chapter. The new approach, called NGHSA, modifies the improvisation step of the HSA such that a new harmony can mimic the global best harmony in the HM. NGHSA and HSA are different in three aspects as follows.

Firstly, in the first step the HM considering and pitch adjustment rates are excluded from the NGHSA. The minimal and maximal bandwidth of BW_{MIN} and BW_{MAX}, the genetic mutation probability of P_m including the minimal and maximal pitch adjusting rate of $P_{\mathrm{PAR(MIN)}}$ and $P_{\mathrm{PAR(MAX)}}$ are introduced in the NGHSA. Secondly, in the third step, the NGHSA modifies the improvisation step of the HSA, and it works. The most important step of the HSA is Step 3, and it includes memory consideration, pitch adjustment and random selection. P_{PAR} and BW have a profound effect on the performance of the HSA. Mahdavi et al. [17] proposed a new variant of the HSA, called the improved harmony search (IHS). The IHS dynamically updates P_{PAR} and BW according to (15.22)–(15.24):

$$P_{\mathrm{PAR}}(gn) = P_{PAR(MIN)} + \frac{(P_{PAR(MAX)} - P_{PAR(MIN)})}{Iteration(MAX)} \times Iteration(current) \quad (15.22)$$

$$BW(gn) = BW_{MAX}\, exp(c \times Iteration(current)) \quad (15.23)$$

where

$$c = \frac{Ln\left(\frac{BW_{MIN}}{BW_{MAX}}\right)}{Iteration(MAX)} \tag{15.24}$$

Thirdly, the "best" and "worst" are the indexes of the global best and the worst harmony in the HM, respectively. The terms of r and *rand*() are all uniformly generated random number in the range of [0, 1]. A new harmony, as shown in Fig. 15.1, is used to illustrate the principle of position updating with the step i of $\left|x_i^{Best} - x_i^{Worst}\right|$; where i is defined as adaptive step of the ith decision variable. The region between the worst and best solution spaces is defined as a trust region for the ith decision variable. The trust region is actually a region near the global best harmony.

5 Computational Results and Analyse

In this work with the computational procedures previously described, a computer simulation program of the fuzzy programming approach of the APP model was implemented in a Visual C#2008 computer program. The suitable linear and continuous membership function has been determined for quantifying the fuzzy aspiration levels. The corresponding linear membership functions can be defined in accordance with an analytical definition of membership functions (PIS and NIS). From the conventional HSA, an interval of the membership (p_k) calculated from all responses in both scenarios of the average minimal total production cost and the average subcontracting units in the HM.

The HSA and VHSA parameters of HMS, P_{HMCR} and P_{PAR} are set at 30, 0.90 and 0.35, respectively. The NGHSA parameters of HMS, P_{HMCR}, P_{PARMAX}, P_{PARMIN}, BW_{MAX}, BW_{MIN} and P_m are set at 30, 0.90, 0.99, 0.01, 0.005, 0.0000005 and 0.01 respectively. The HuSHSA parameters of Ra_{MAX} and Ra_{MIN} are set at 1,000 and 1, respectively. Minimal total production cost and subcontracting unit scenarios are calculated from all previous data in the HM with 200 iterations at the nominal demand including the demand decrease and increase at 10% (Tables 15.1–15.3). The situations of the nominal demand, the demand increase and decrease seem different. The capacity of the factory is appropriate for nominal demand. For decrease demand capacity of the factory is surplus. But, for increase demand the capacity of the factory is not sufficient. A mathematical expression of the proposed model for this APP problem with the nominal demand can be then shown as followed with the corresponding (15.3)–(15.17).

$$\begin{aligned}
&\max = [\lambda], \\
&\lambda \le 1 - \left(\frac{z_1 - 79{,}432{,}529}{9{,}829{,}338}\right) \\
&\lambda \le 1 - \left(\frac{z_2 - 0}{15{,}715}\right)
\end{aligned} \tag{15.25}$$

Table 15.1 Initial results by optimising individual objective with the nominal demand

Scenarios	Total production cost	Subcontracting units
Feasible best so far with the minimal cost	66,343,158	–
Simple-HSA with the minimal cost	79,432,529	15,715
Simple-HSA with the minimal subcontracting	89,261,867	0
p_k	9,829,338	15,715

Table 15.2 Initial results by optimising individual objective on the 10% demand decrease

Scenarios	Total production cost	Subcontracting units
Simple-HSA with the minimal cost	80,159,075	16,934
Simple-HSA with the minimal subcontracting	89,471,737	0
p_k	9,312,662	16,934

Table 15.3 Initial results by optimising individual objective on the 10% demand increase

Scenario	Total production cost	Subcontracting units
Simple-HSA with the minimal cost	100,030,561	156,131
Simple-HSA with the minimal subcontracting	108,041,936	100,025
p_k	8,011,375	56,106

Table 15.4 Experimental results categorised by the algorithms on the nominal demand

Algorithm	Max λ	Total production cost	Subcontracting units
HSA	0.889	80,073,238	0
HuSHSA	0.752	80,530,984	0
VHSA	0.960	80,012,432	0
NGHSA	0.949	80,071,863	0

The comparisons are made for three different levels of demand in the APP. The demand deviation is taken to be independently and normally distributed with mean of the nominal level with the 10% increase and decrease. Using the fuzzy approach of the APP an aim is to simultaneously minimise total production costs and subcontracting units over a 12-month period. At the nominal demand (Table 15.4), total production costs and subcontracting units seemed to be better at 0.960 of the overall levels of decision making satisfaction via the VHSA. VHSA exploits

Table 15.5 Experimental results categorised by the algorithms on the 10% demand decrease

Algorithm	Max λ	Total production cost	Subcontracting units
HSA	0.675	83,164,217	0
HuSHSA	0.681	82,556,653	0
VHSA	0.679	82,859,019	0
NGHSA	0.725	81,632,893	0

Table 15.6 Experimental results categorised by the algorithms on the 10% demand increase

Algorithm	Max λ	Total production cost	Subcontracting units
HSA	0.880	100,956,577	105,103
HuSHSA	0.933	100,166,298	105,528
VHSA	0.915	100,536,630	104,038
NGHSA	0.910	100,616,218	102,163

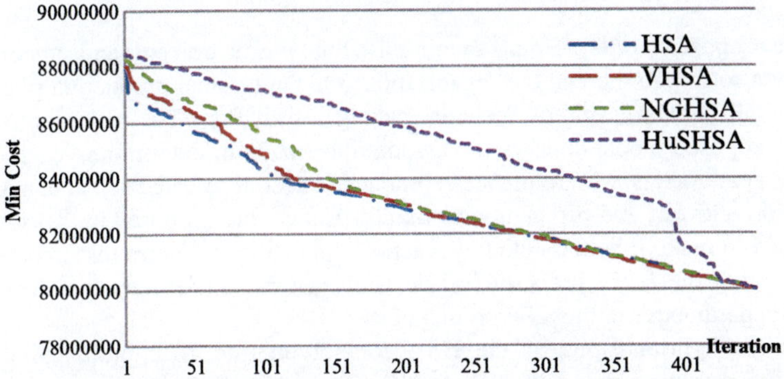

Fig. 15.2 Speed of convergence on the nominal demand

systematically the idea of neighbourhood change within a local search method to approach a better solution. It proceeds systematically or randomly. It is a appropriate for nominal demand with appropriate capacity situation that has medium set of solutions. The NGHSA seemed to provide the better satisfaction level when there is the demand decrease at 10% as shown in Table 15.5. It modifies the improvisation step of the HSA such that a new harmony can mimic the global best harmony in the HM. It is suitable for the application of APP with low demand but high capacity situation which has large set of solutions.

The HuSHSA seemed to obtain the better satisfaction level only when there is the demand increase at 10% and resources are limited. So, subcontracting units are necessary as shown in Table 15.6. HuSHSA performed the best for tight capacity factory because possible solutions are limited be resource constraints so it has a chance for the harmony to be trapped in a local optimum to find the better outcome. When the performance of the HSA variants of HSA, VHSA, NGHSA and HuSHSA was compared, the VHSA seems to be better in terms of speed of convergence (Fig. 15.2). The basic idea is the change of neighbourhoods during searching for a better solution. The hybridisations proceed by a descent method to a local minimum exploring then, systematically or at random, increasingly distance neighbourhoods of this local solution. Furthermore, in some additional experiments, small BW values with large P_{PAR} values usually cause the improvement of best solutions in final generations to converge to the optimal solution vector.

6 Conclusions

The APP is concerned with the determination of production, the inventory and the workforce levels of a company on a finite time horizon. Objectives are to reduce the total production cost to fulfill a non-constant demand and to reduce subcontracting units assuming fixed sale and production capacity. In this study we proposed an application of a fuzzy programming approach to the APP with various level of the demand.

The proposed model attempts to minimise total production cost and subcontracting units so that in the end the organisation gets the optimal production plan with the overall highest levels of decision making satisfaction. The major limitations of the proposed model concern the assumptions made in determining each of the decision parameters, with reference to production cost, forecasted demand, maximal work force levels, and production resources. Hence, the proposed model must be modified to make it better suited to practical applications. Future researchers may also explore the fuzzy properties of decision variables, coefficients and relevant decision parameters in the APP problems.

Various hybridisations of the HSA for solving the APP problems with a fuzzy programming approach. VHSA, NGHSA and HuSHSA employ the variable neighbourhood search, novel global best and HuS algorithms and for generating new solution vectors that enhances accuracy and convergence rate of HSA. The result of experiments can be concluded that each algorithm is suitable for different types of factory situations. VHSA is suitable for nominal demand. NGHSA is appropriate for low demand but high capacity and HuSHSA is proper for high demand but low capacity. However, for all situations VHSA and NGHSA can obtain good solutions. In this chapter the impacts of parameters, on various versions on the conventional HSA, such as BW seem to be evident. They need to be discussed and a strategy for tuning these parameters should be carried to enhance the performance measures of the algorithms. However, based on the preliminary experiments, VHSA is more effective than other approaches in terms of superiority of solution and required CPU time.

Acknowledgment This work was supported by the National Research University Project of Thailand Office of Higher Education Commission.

References

1. Techawiboonwong A (2003) Aggreate production planning and master production scheduling: problems and methods for labor-intensive Thai industries, PhD dissertation, SIIT Thammasart University
2. Belmokaddem M, Mekidiche M, Sahed A (2009) Application of a fuzzy goal programming approach with different importance and priorities to aggregate production planning. J Appl Quant Method 4(3):317–331
3. Fahimnia B, Luong L, Marian R (2005) Modeling and optimisation of aggregate production planning – a genetic algorithm approach. Int J Math Comput Sci 1: 1–6
4. Baykoç Ö, Sakalli Ü (2009) An aggregate production planning model for brass casting industry in fuzzy environment. Eng Technol 52:117–121
5. Chen Y (2010) A production planning problem solved by the particle swarm optimisation. Proceeding of the international multiconference of engineers and computer scientist 2010 Vol III, IMECS 2010, Hong Kong, 17–19 March 2010
6. Leung S, Wu Y (2004) A robust optimisation model for stochastic aggregate production planning. Product Plan Control 15(5):502–514
7. Aliev R, Fazlollahi B, Guirimov B, Aliev R (2007) Fuzzy-genetic approach to aggregate production–distribution planning in supply chain management. Inform Sci 177: 4241–4255
8. Aungkulanon P, Phruksaphanrat B, Luangpaiboon P (2011) Various hybridisations of harmony search algorithm for fuzzy programming approach to aggregate production planning, Lecture notes in engineering and computer science. Proceedings of the international multiconference of engineers and computer scientist 2011, vol. I, IMECS 2011, Hong Kong, 16–18 March 2011, pp.73–79
9. Zimmerman HJ (1992) Fuzzy set theory and its application. Kluwer Academic, pp 336–339
10. Zimmerman HJ (1985) Applications of fuzzy sets theory to mathematical programming. Inform Sci 35:29–58
11. Zimmerman HJ (1976) Description and optimisation of fuzzy systems. Int J Gen Syst 2:209–215
12. Zimmerman HJ (1978) Fuzzy programming and linear programming with several objective functions. Fuzzy Sets Syst 1:45–56
13. Lee KS, Geem ZW (2004) A new meta-heuristic algorithm for continues engineering optimisation: harmony search theory and practice. Comput Math Appl Mech Eng 194:3902–3933
14. Oftadeh R, Mahjoob MJ, Shariatpanahi M (2010) A novel meta-heuristic optimization algorithm inspired by group hunting of animals: hunting search. Comput Math Appl 60: 2087–2098
15. Mladenovic N, Hansen P (1997) Variable neighborhood search. Comput Oper Res 24: 1097–1100
16. Zou D, Gao L, Wu J, Li S, Li Y (2010) A novel global harmony search algorithm for reliability problems. Comput Ind Eng 58:307–316
17. Mahdavi M, Fesanghary M, Damangir E (2007) An improved harmony search algorithm for solving optimisation problems. Appl Math Comput 188:1567–1579

Chapter 16
A Compression Technique for XML Element Retrieval

Tanakorn Wichaiwong and Chuleerat Jaruskulchai

Abstract The main objective of compression technique has been changed, not only to reduce the storage but also to other efficiency. For instance, the large scale of XML collection, compression techniques is required for improving retrieval time. In this paper, we propose new XML compression algorithm that allows supporting absolute document XPath indexing and score sharing function by a top-down scheme approach. It has been discovered that these steps reduce the size of the data down by 82.29%, and reduce the length of retrieving time down by 51.38% when compare to GPX system. In addition, It has been reduced the length of score sharing processing time down by 44.18% when compared to before the compression.

Keywords XML retrieval • Compression strategies • Ranking strategies • Indexing unit

1 Introduction

The extensible markup language (XML) [1] is one of the most important formats for data interchange on the Internet. Recently, the initiative for the evaluation of XML retrieval (INEX) [2] has provided an excellent test corpus on XML information retrieval and queries. The corpus contains marked up with context, and queries included articles from IEEE journals and Wikipedia collections. There are two main performance issues in information retrieval; effectiveness and efficiency. In the past, much research was mainly aimed to improve only effectiveness. In recent years, research has been focused on the efficiency of the trend of retrieval large-scale collection. XML documents are used for digital libraries and to store the

T. Wichaiwong (✉) • C. Jaruskulchai
Department of Computer Science, Faculty of Science, Kasetsart University, Bangkok, Thailand
e-mail: g51840411@ku.ac.th; fscichj@ku.ac.th

S.I. Ao et al. (eds.), *Intelligent Control and Innovative Computing*, Lecture Notes in Electrical Engineering 110, DOI 10.1007/978-1-4614-1695-1_16,

large amount of data led to the development of information retrieval methods specifically designed for XML collections. Moreover, retrieval of XML elements involves not only the document level but also document's components, potentially consuming more disk space and retrieval time than the flat retrieval. Therefore, we have investigated the research question in this article on the efficiency issue, and then we present our approach toward improving this issue by using compression technique.

This paper is organized as follows: Section 2 reviews related works. Section 3 explains the new XML compression algorithm. Section 4 describes the implementation of our system, Section 5 show the experiment results, conclusions and further work are drawn in Section 6.

2 Related Work

In this section, we provide some historical perspectives on areas of XML research that have influenced to this article as follows.

2.1 XML Data Models

The basic data model of XML [1] is a labeled and ordered tree. Figure 16.1 shows the data tree of the XML document based on the node-labeled model. There are basically three types of nodes in a data tree:

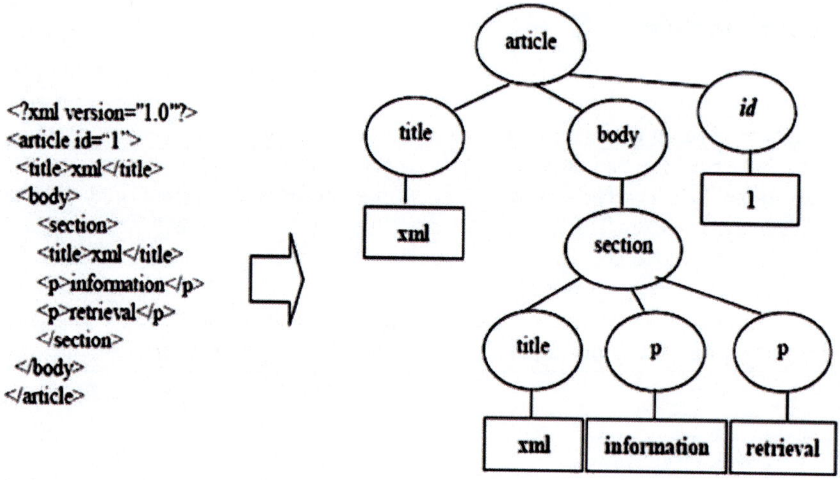

Fig. 16.1 The example of XML element tree

The element nodes, these correspond to tags in XML documents, for example, "body" and "section" node.

The attribute nodes, these correspond to attributes associated with tags in XML documents, for example, "id" node. In contrast to element nodes, attribute nodes are not nested (that is, an attribute cannot have any sub elements), are not repeatable (that is, two same-name attributes cannot occur under one element), and are unordered (that is, attributes of an element can freely interchange their occurrence locations under the element).

The leaf nodes (text nodes), these correspond to data values in XML documents, for example, "xml" node.

2.2 XML Indexing Schemes

Classical retrieval models have been adapted to XML retrieval. Several indexing strategies [3] have been developed in XML retrieval and more details will be described as follows:

Element-Base indexing [4] allows each element to be indexed based on both direct text and the text of descendants. This strategy has a main drawback, as it is highly redundant. Text occurring at the nth level of the XML logical structure is indexed n times and thus requires more index space. *Leaf-only indexing* [5] allows indexing of only leaves through element or elements directly related to text. This strategy addresses the issues of redundancy noted above. However, it requires a certain level of efficiency of the propagation algorithm for the retrieval of non-leaf elements. *Aggregation-based indexing* [6] uses concatenated text in an element to estimate a term statistic. This strategy has been used to aggregate term statistics based directly on the text and its descendants. *Selective indexing* [4] involves eliminating small elements and elements of a selected type. *Distributed indexing* [7] is separately created for each type of element in conjunction with the selective indexing strategy. The ranking model has to run against each index separately and retrieve ranked lists of elements. These lists are merged to provide a single rank across all element types. To merge lists, normalization is performed to take into account the variation in size of the elements in the different indices, so that scores across indices are comparable.

2.3 XML Compression Schemes

Recently, the main objective of compression technique has been changed, not only to the reduction of XML data storage but also to improve the efficiency. For instance, the large scale of XML collection, compression techniques are required for improving retrieval time. The XML data compression can be divided into two types respect to ability of supporting queries: (1) non-queriable XML compressors and (2) queriable XML compressors. Several compression strategies have been developed in XML as follows;

Non-Queriable XML Compressors

The non-queriable XML compressors are aim to achieve the highest of compression ratio as following.

XMill [8] is a technique which compresses both data and tag in order to reduce the size by starting with separating the tag, which is composed of elements and attributes, from the data, which is a character. After that, the data groups' relationships will be organized. The same data will be in the same group. The next step is the data compression by using gzip [9] so that the data will come out in the same file since grouping requires understanding of the data definitions which depend on the application type. XMill allows user check the data definition. The disadvantage of XMill that data cannot be search through the compressed data. However, XMill is the first research that made researchers realizes the importance of the problem and how to solve it in XML data compression. Data that has been compressed is not in the form of XML schema structure.

XPACK [10] is a way to compress XML data, which uses grammatical approaches in XML data compression and decompresses. The main component of XPACK is the grammar generator, which creates the grammar. The second component is the compressor which compresses the data. The last component is the decompressor which decompresses the compressed data by using the old structure of the data. However, XPACK cannot manage XML data that has mixed content element (which is an element composed of element and characters), limiting users to search for data in compressed XML.

The GPX [5] search engine is using a relational database to implement an inverted list data structure. It is a compromise solution provided by the convenience of a DBMS at the cost of somewhat reduced performance, which may otherwise be possible. For example, the XPath as following:

/article[1]/body[1]/section[1]/p[1]

This could be represented by two expressions, a Tag-set and an Index-set as below;

Tag-set:/article/body/section/p
Index-Set: 1/1/1/1

The original XPath can be reconstructed from the tag-set and the index-set. The GPX assigns to each tag set and each index-set with a hash code and creates auxiliary database tables and the hash codes to the corresponding tag-set and index-set entries. These hash tables are small enough to be held in memory and so decoding is efficient. The GPX takes 15 seconds to load all table data and takes an average of 7.2 seconds per topic. Sometimes, it takes longer than 30 seconds, depending on the type of query and these run is on a 3 GHz PC with 2 GB RAM. Unfortunately, this method has not been focused on the efficiency.

Queriable XML Compressors

The queriable XML compressors are aim to perform direct queries on compressed data as following.

XGrind [11] is a technique which compressed data and tag but the user can still search for data after the compression. This qualification results from the fact that the compressed data still maintain the structure of the old data. However, XGrind will compress only XML data that has DTD structure so some data set that does not have DTD will result in having the user waste time in creating DTD for XML data set that they wanted to compress.

XPRESS [12] uses the technique in compressing both the data and the tag. Its advantages are the same as XGrind: it can search for the data after the compression. Nevertheless, XPRESS does not use DTD. In addition, XPRESS presented a new idea which uses reverse arithmetic encoding, which is a method in organizing data so that the search for XPath expressions can be done effectively. Furthermore, XPRESS has developed the search of data type without having to use the information from users. However, the use of XPRESS is limited because XPRESS cannot understand documents that use ID and IDREF. It also does not have a way to decompress data back into normal XML.

XSchemaTag [13] is a technique that compresses only XML tag and that technique still enables to search and maintain documents because the data is already in the form of XML. The quality comes from compressed data, which has the old data structure. However, the XSchema Tag scheme does not take into account of the frequency of tag occurrences and the counter of tag position.

The representation of the absolute document XPath indexing (ADXPI) [14] is more problematic, because each unique XPath is repeated in the inverted list for each term in the same node, and the XPath repeated in many files. We find out the way to encode tags and the compression algorithm like XMill might be effective, but we considered this again to be unnecessary, particularly given the processing overheads. We have adopted the following simple compression scheme using dictionary mapping (DM) and easy to reconstruct the original XPath and more details will discuss in next section.

3 Our Propose

3.1 Compression Technique

In this proposed work, we are enhancing the existing compressors which use DM algorithm, which we call the compression of ADXPI (cADXPI) [15]. It is based on the principle of extracting XPath and position from the document, and grouping it based on the name of tag. The document is encoded as a sequence of integers, while the data grouping is based on XML tags. The main disadvantage of using XML documents are their large sizes caused by highly repetitive structures of those documents and often long tag names, for example "management_note [1]", "broadcasting_station[1]", and "system_of_measurement[1]" in INEX-Wikipedia 2009 collection. Therefore, a need to compress XML, both efficiently and easily to process on post-processing and filtering of the retrieval system. The re-organized data is now compressed by adaptive DM; it has extremely accurate compression as well as it eliminates the repetition of the dictionary based words in the database. Using DM algorithm, we derived probabilities, which dynamically changed with the frequency of tag name and position that allows us to directly retrieval the path in the compressed data. Since the data volumes reduced, such compressing of data path may be even faster than the original data.

Figure 16.2 depicts the details of a compression algorithm and in the following algorithm description; indentation is used to denote the details of algorithm processing:

- Fetch all leaf node entries from the collection list.
- For each list, create data structure to store tag name and frequency, we call Dictionary<tag,freq> data type.

```
Counter :=0;
For each List in LeafNodeList
    For each Path in List.Split('/')
        If NodeList.ContainKey(Path) Then
            NodeList[Path] ← NodeList[Path] +1;
        Else
            NodeList.Add(Path,1);
        End If
    End For
End For
NodeList ←NodeList.SortbyValue();
For each Path in NodeList
    FinalList.Add(Path,Counter);
    Counter ← Counter + 1;
End For
Return FinalList;
```

Fig. 16.2 The detail of the compression algorithm

- Split all tag and counter from the leaf and add to Dictionary<tag,freq>, for instance, the leaf node is:

  /article[1]/body[1]/section[1]/p[1].
  We can split them as follows;
  1st tag is "article[1]", frequency is 1.
  2nd tag is "body[1]", frequency is 1.
  3rd tag is "section[1]", frequency is 1.
  and "p[1]", frequency is 1.

- For each tag has to check in Dictionary<tag,freq> list as follows;

If Dictionary<tag,freq> has contain tag then freq is accumulate by freq = freq + 1
Otherwise add new tag and 1 to Dictionary<tag,freq> list.

- When already processed all of a list then create the Final Dictionary <tag,map> list by sorting freq from Dictionary<tag,freq>list, map a sequence of tag in Final list.
- Return the Final Dictionary<tag,map> list to store in DB.

Remind to our example, the compression algorithm processing is following;

article[1]/title[1]: xml
article[1]/body[1]/section[1]/title[1]:xml
article[1]/body[1]/section[1]/p[1]:information
article[1]/body[1]/section[1]/p[2]: retrieval

We can split all leaf-node and construct the dictionary list as follows;

1st tag is "article[1]", frequency is 4.
2nd tag is "title[1]", frequency is 2.
3rd tag is "body[1]", frequency is 3.
4th tag is "section[1]", frequency is 3.
5th tag is "p[1]", frequency is 1.
6th tag is "p[2]", frequency is 1.

Following the result list as above, we sort the dictionary list by frequency then the final dictionary with map as follows;

1st tag is "article[1]", frequency is 4.
2nd tag is "body[1]", frequency is 3.
3rd tag is "section[1]", frequency is 3.
4th tag is "title[1]", frequency is 2.
5th tag is "p[1]", frequency is 1.
6th tag is "p[2]", frequency is 1.

As a result, we have shown the XML element tree in Fig 16.3(a), and the compressed of XML element tree as shown in Fig 16.3(b). The data of leaf-node indices are stored in database and more details as shown in Table 16.1 as follows;

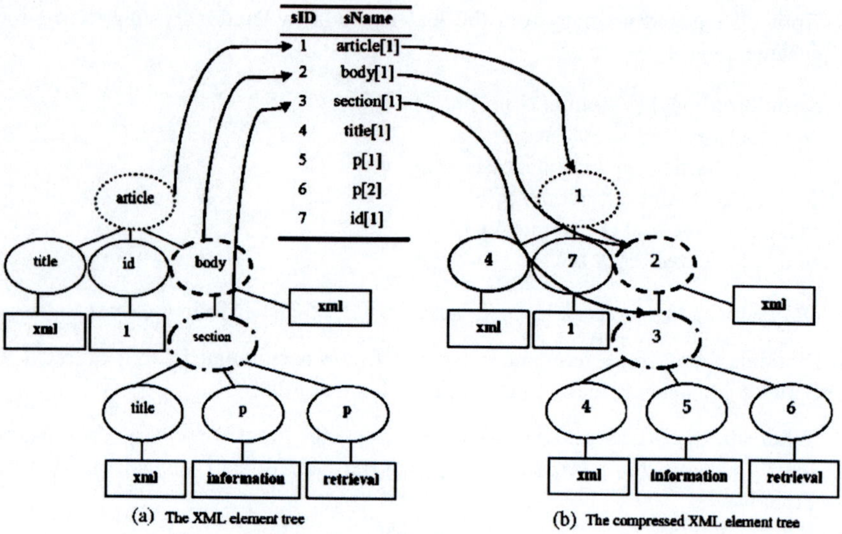

Fig. 16.3 Example of compressed XML element tree

Table 16.1 The details of leafnode table

ID	xPath	Details
1	x1/1/4	Xml
2	x1/1/2/3/4	Xml
3	x1/1/2/3/5	Information
4	x1/1/2/3/6	Retrieval
5	x1/1/7	1
6	x1/1/2	Xml

4 XML Retrieval Model

The more efficient XML information retrieval (MEXIR) [16] is a base on the leaf-node indexing scheme and uses a relational DBMS as a storage back-end. We discussed the schema setup using the MySQL [17] and the full-text engine Sphinx [18, 19] using the MySQL dumps function.

Sphinx has two types of weighting functions:

- Phrase rank: based on a length of the longest common subsequence (LCS) of search words between the document body and query phrases. If there's a perfect phrase match in some document then its phrase rank would be the highest possible, and equal to query words count.
- Statistical rank: based on classic BM25 function [20], which only takes word frequencies into account. If the word is rare in the whole database, it receives more weight.

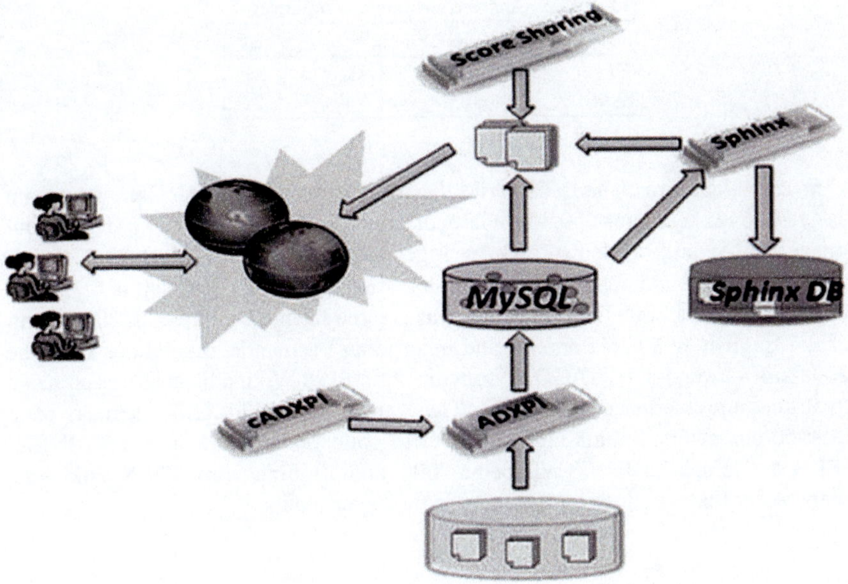

Fig. 16.4 MEXIR XML retrieval system overview

For the initial step, we consider a simplified XML data model but disregard any kind of Meta mark-up, including comments, links in the form of XLink or ID/IDRef and attributes. Fig. 16.4 depicts the overview of XML retrieval system. The main components of the MEXIR retrieval system are follows:

- When new documents are entered, the ADXPI Indexer parses and analyzes the tag and content data to build a list of leaf-nodes.
- The cADXPI compressor that analyzes the tag and position to build the structure index store in MySQL database.
- The Sphinx is used to analyze and build the indices.
- The score sharing function [21] is used to assign parent score by sharing score from leaf nodes to their parents using a top-down scheme approach.

5 Experiment Setup

5.1 INEX Collection Tests

The document collections are from the INEX-IEEE document collection, which contains a total of 16,819 articles from 24 IEEE Computer Society journals covering the period is 1995–2005 and totaling 764 megabytes in size and 11 million elements

Table 16.2 Compare environment resources

Resource	MEXIR	GPX
CPU	Dual-Core 1.87 GHz	3 GHz
RAM	1 GB	2 GB

in its canonical form. The INEX-Wiki06 XML Corpus for English Wikipedia from early 2006 [22] contains 659,338 Wikipedia articles; the total size is 4.6 GB without images and 52 million elements. On average, an article contains 161.35 XML nodes, whereas the average depth of a node in the XML tree of a document is 6.72. The INEX-Wikipedia 2009 [23] collection was created from the October 8, 2008 dump of the English Wikipedia articles and incorporates semantic annotations from the 2008-w40–2 version of YAGO. It contains 2,666,190 Wikipedia articles and has a total uncompressed size of 50.7 GB. There are 101,917,424 XML elements of at least 50 characters. Sphinx indexing of these collections took 1 minute for INEX-IEEE, 5 minutes for INEX-Wikipedia 2006, and 30 minutes for INEX-Wikipedia 2009. After that, our system uses them in the experiments

5.2 Evaluations

Our experiment targets in CO Task only as well as systems that accept CO queries. Note that CO queries are terms enclosed in the <title> tag. In the experiment of data compression, the effectiveness in data compression is the proportion of compression, which can be found by using;

$$\text{Size} = 1 - \left[\frac{\text{Compressed data size}}{\text{Actual data size}} \right] \tag{16.1}$$

And the effectiveness of response time is the proportion which can be found by using;

$$Time = 1 - \left[\frac{\textit{Processing time in compressed}}{\textit{Processing time in actual}} \right] \tag{16.2}$$

5.3 Experiment Results

In this section, we present the results of evaluation of the Score Sharing scheme with and without cADXPI technique. Although, in principle, any XML document part can be retrieved, some document parts tend to be more likely to be relevant.

Our system has shown the environment resources in Table 16.2 compare to GPX system. As such, we report the effectiveness of our system on INEX collections as follows:

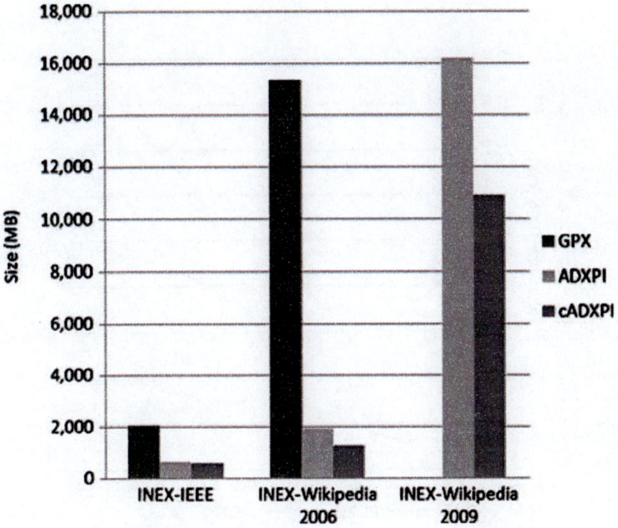

Fig. 16.5 Graph showing the size of data

Table 16.3 Compare the retrieving time (second)

Collections	MEXIR	GPX	%
INEX-IEEE	1.0	–	–
INEX-Wikipedia 2006	3.5	7.2	51.38
INEX-Wikipedia 2009	10.0	–	–

The performance of our purpose model is evaluated. It has been discovered that these steps reduces the size of the data down by 82.29% as shown in Fig. 16.5 and more details are following:

The cADXPI has shown 60.99% improvements over GPX on INEX-IEEE, and 82.29% improvements over GPX on INEX-Wikipedia 2006. However, GPX did not perform experiment on INEX-Wikipedia 2009 that we cannot report for this collection.

Another conclusion, in terms of retrieving time, our system required an average of one second per topic on INEX-IEEE, an average of up to four seconds per topic on INEX-Wikipedia 2006 better than GPX system, and an average of up to ten seconds per topic on INEX-Wikipedia 2009 as shown in Table 16.3. In addition, it also reduces the length of score sharing processing time down by 44.18% when compared to before the compression as shown in Fig. 16.6.

Thus, we have confirmed the compression technique was necessary in XML retrieval system.

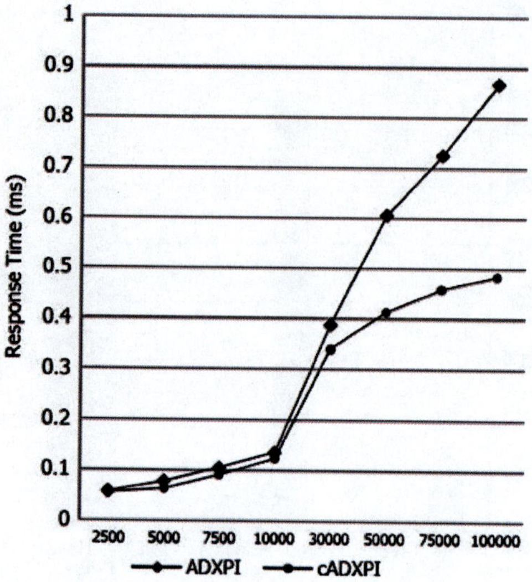

Fig. 16.6 Compare score sharing processing time with and without compression technique

6 Conclusion and Future Work

The main disadvantages of using XML documents are their large sizes caused by highly structures of those documents and often long tag and attribute names. Therefore, a need to compress XML, both efficiently and easier to processing in the retrieval system.

In this paper, we report experimental results of our approach for retrieval large-scale XML collection, to improve the efficiency of XML element retrieval. We propose new XML compression algorithm that allows supporting ADXPI indexing and score sharing function by a top-down scheme approach. The main focus of this technique is to avoid document decompression during query execution. In fact, the ability to perform direct queries on compressed data.

As our future work, we are going to study how to infer structural hints from Content and Structure (CAS) queries.

References

1. Extensible Markup Language (XML) 1.1 (Second Edition) [Online May, 2011]. Available: http://www.w3.org/TR/xml11/
2. INitiative for the Evaluation of XML Retrieval (INEX) [Online May, 2011]. Available: https://inex.mmci.uni-saarland.de/
3. Kamps J (2009) Indexing units. In: Liu L, Tamer Özsu M (ed) Encyclopedia of database systems (EDS). Springer, Heidelberg

4. Kamps J, de Rijke M, Sigurbjörnsson B (2005) The importance of length normalization for XML retrieval. Inform Retrieval 8(4):631–654
5. Geva S (2005) GPX – Gardens point XML information retrieval. INEX 2004. Advances in XML information retrieval. Third International Workshop of the Initiative for the Evaluation of XML, Springer Verlag, Berlin, Heidelberg, Vol. 3977, pp 211–223
6. Ogilvie P, Callan J (2005) Hierarchical language models for XML component retrieval. INEX 2004, Lecture Notes in Computer Science, Springer Verlag, Berlin, Heidelberg, Vol. 3493
7. Mass Y, Mandelbrod M (2005) Component ranking and automatic query refinement for XML retrieval. INEX 2004, Springer Verlag, Berlin, Heidelberg, Vol. 3493
8. Liefke H, Suciu D (2000) XMill: an efficient compressor for XML data. In Proceeding of the 2000 ACM SIGMOD International Conference on Management of Data, ACM, pp 153–164
9. Gailly JL, Adler M (2011) gzip: The compressor data. [Online May, 2011]. Available: http://www.gzip.org/
10. Maireang K, Pleurmpitiwiriyavach C (2003) XPACK: a grammar-based XML document compression. Proceeding of NCSEC2003 the 7th national computer science and engineering conference, 28–30 October 2003
11. Tolani PM, Haritsa JR (2002) XGRIND: A query-friendly XML compressor. In Proceedings of 18th International Conference on Databases Engineering, IEEE Computer Society Press, Los Alamitos, February 2002, pp 225–234
12. Min J-K, Park M-J, Chung C-W (2003) XPRESS: a queriable compression for XML data. Proceeding of the 2003 ACM SIGMOD international conference on management of data, 9–12 June 2003
13. Wichaiwong T, Jaruskulchai C (2007) Improve XML web services' performance by compressing XML schema tag. The 4th international technical conference on electrical engineering/electronics, computer, telecommunications and information technology, Chiang Rai, 9–12 May 2007
14. Wichaiwong T, Jaruskulchai C (2011) XML retrieval more efficient using adxpi indexing scheme. The 4th international symposium on mining and web, Biopolis, Singapore, 22–25 March 2011
15. Wichaiwong T, Jaruskulchai C (2011) XML retrieval more efficient using compression technique, lecture notes in engineering and computer science. Proceedings of the international multiconference of engineers and computer scientists 2011, IMECS 2011, Hong Kong, 16–18 March 2011
16. Wichaiwong T, Jaruskulchai C (2011) MEXIR: An implementation of high performance and high precision on XML retrieval. Computer Technology and Application, David Publishing Company, 2(4):301–310
17. MySQL Full-Text Search Functions [Online May, 2011]. Available: http://dev.mysql.com/doc/refman/5.1/en/fulltext-search.html
18. Sphinx Open Source Search Server [Online May, 2011]. Available: http://www.sphinxsearch.com/
19. Aksyonoff A (2011) Introduction to Search with Sphinx, O'Reilly Media
20. Robertson SE, Walker S, Jones S, Hancock Beaulieu MM, Gatford M (1995) Okapi at TREC-3. In: Harman DK (ed) Proceedings of the Third Text REtrieval Conference (TREC-3), NIST Special Publication, pp 500–225
21. Wichaiwong T, Jaruskulchai C (2010) A simple approach to optimize XML retrieval. The 6th International Conference on Next Generation Web Services Practices, Goa, India, November 23–25, pp 426–431
22. Denoyer L, Gallinari P (2006) The Wikipedia XML corpus. SIGIR Forum, pp 64–69
23. Schenkel R, Suchanek FM, Kasneci G (2007) YAWN: A semantically annotated Wikipedia XML corpus. In BTW, pp 277–291

Chapter 17
Words-of-Wisdom Search Based on User's Sentiment

Akiyo Nadamoto and Kouichi Takaoka

Abstract With the rapid advance of the Internet, everybody can get the information from it easily. There are, however, few system which extracts and presents the information suitable for user's sentiment. We propose the system that searches for the information based on user's sentiment. In this paper, we propose words-of-wisdom search system as a first step of the research. We propose two types of words-of-wisdom search based on user's sentiment. One is positive/negative (P/N) sentiment, the other is multi-dimensional sentiment. Both of two methods, we calculate sentiment value of words which consists of words-of-wisdom. After that we calculate sentiment value of words-of-wisdom by using sentiment value of words.

Keywords Sentiment search • Positive/negative • Multi-dimensional sentiment • Words-of-wisdom

1 Introduction

There are many search engines on the Internet, such as Google and Yahoo!. Almost search engines are keyword based search. There are also many recommendation services on the Internet, such as Amazon et al. Mr. Steve Johnson says, "To fully understand an individual's preferences, search engines must move beyond

A. Nadamoto (✉)
Department of Intelligence and Information, Konan University, Okamoto 8-9-1 Higashinada-ku Kobe, Japan
e-mail: nadamoto@konan-u.ac.jp

K. Takaoka
Graduate School of Natural Science Graduate Course, Konan University Graduate School, Okamoto 8-9-1 Higashinada-ku Kobe, Japan
e-mail: tmk_kouichi@yahoo.co.jp

S.I. Ao et al. (eds.), *Intelligent Control and Innovative Computing*, Lecture Notes in Electrical Engineering 110, DOI 10.1007/978-1-4614-1695-1_17,

the simple keyword and focus instead on capturing the consumer's intent at that moment." [1], we consider one of the prominent example of "the consumer's intent at that moment" is a sentiment.

However, recent search engine and recommendation services do not care about user's sentiment. For example, when we input "book, happy, feeling", search system replays "Q&A site" on the top lists of the search results. The answer of the Q&A sites are a respondent's subject view, and user may not feel same feeling of the answer. In this research, we propose the novel search system which searches for the content based on user's sentiment. We consider there are two types of user's sentiment. Sometimes user just wants to make positive or sometimes he/she wants to make more complex sentiment such as happy, relax, amazing and so on. Then we propose two types of methods to search the information based on user's sentiment. One is positive/negative (P/N) sentiment, the other is multi-dimensional sentiment.

In this paper, it is a first step of the research, we target on word-of wisdom and propose the search system which searches for it based on user's sentiment. People sometime focus on words-of-wisdom to become feeling positive or happy. In this way, words-of-wisdom make people to change the feeling, and it is strongly related to people's various sentiments.

In Japan, the book of "words of Nietzsche" [2] ranked 1 in the category of self-development books. In this way, the words-of-wisdom becomes popular. Furthermore, there are many words-of-wisdom sites on the Internet, for example there are 30,000 words-of-wisdom on the Meigen-Navi [3]. Everybody can read many kinds of words-of-wisdom. It, however, is difficult to search the words-of-wisdom which user's want to read. Because recent search engines do not care about sentiment but users sometime want to get words-of-wisdom based on his/her sentiment. In this paper, we quantify the words-of-wisdom's sentiment and user's sentiment. Then, we propose the search system which searches for the words-of-wisdom based on two types of user's sentiment which are P/N and multi-dimensional sentiment Both of two methods, we calculate sentiment value of words which consists of words-of-wisdom. After that we calculate sentiment value of words-of-wisdom by using sentiment value of words.

The flow of extracting words-of-wisdom can be represented as follows:

- A user selects sentiment type which is P/N or multidimensional.
- A user inputs information about the desired sentiment.
- We quantify the expression of words-of-wisdom based on sentiment value which is P/N value or multi-dimensional value, and calculate sentiment value of words which consists of words-of-wisdom. In this paper, we designate the value as "value of sentiment words".
- We calculate sentiment of words-of-wisdom by using the value of sentiment words.
- We extract the words-of-wisdom that is similar to the user's desired sentiment.

2 Related Work

Recently, studies of sentiment expression extraction are active. Various sentiment models have been proposed. Multi-dimensional sentiment vectors are proposed in various sentiment models. For example, the model of Plutchik [4, 5] is a typical sentiment model. The model of Plutchik includes clustering of four-dimensional sentiment vectors based on eight emotions: acceptance↔disgust, anticipation↔surprise, joy↔sadness, anger↔fear.

Tokuhisa et al. [6] propose a method for inferring the emotion of a speaker conversing with a dialog system from the semantic content of his/her utterance. For example, from the sentence "I was disappointed because the shop was closed and I'd traveled a long way to get there." pulled from the Web, they learn that the clause "the shop was closed and I'd traveled a long way to get there." is an example of an event that provokes disappointment. In this paper, they refer to such an example as an emotion-provoking event and a collection of event-provoking events as an emotion-provoking event corpus. Sentiment models is (happiness, pleasantness, relief, fear, sadness, disappointment, unpleasantness, loneliness, anxiety, anger) and (neutral), used eleven kinds of emotion.

Kawai et al. [7] classify news articles extracted from multiple news sites using the user profile. Their system is called MPV Plus, which presents a user's interests. In fact, MPV Plus extracts interest words and sentiment vector from the user's browsing history. They propose four-dimensional sentiment vectors of the impression of the article, presented for information: happy↔unhappy, acceptance↔rejection, relaxation↔strain, anger↔fear. As described above, many studies have extracted the emotion of a multi-dimensional vector.

Dave et al. [8] divide the reviews of web sites into positive reviews and negative reviews. Kobayashi et al. [9] propose a dictionary-based method for acquiring a P/N lexicon that specifies whether each entry means a positive or negative concept. Their method is based on a bootstrap method. Fujimura et al. [10] classify a Web into positive or negative and extract a reputation from the Web. They propose a scoring method to extract reputation. Takamura et al. [11] propose a method that calculates sentiment orientations of words. They regard sentiment orientations of words as spin of electron, and modeling the word network as a spin model.

Kumamoto et al. [12] specifically examine sentiments related to multi-dimensional emotions and then extract emotion from web news. The emotion model comprises two-dimensional sentiment vectors sad↔glad and angry↔pleased, which are between 0 and 1. Their hypothesis is "A news article that includes emotion word 'e' represents the emotion. Their system was designed using co-occurrence between the word in the news article and emotion words. Kumamoto et al. [12, 13] propose a method extracted from impression that people feel they see or hear the content. In this paper, propose a method extracted form not information originator's emotion but information receiver feeling emotion.

3 Search by P/N Value

When users search the words-of-wisdom based on their desired sentiment, they sometimes search by easy sentiment such as P/N. Then we propose the words-of-wisdom search by P/N value [14]. We calculate the word-of-wisdom P/N value using value of sentiment words which consists of P/N value. In this section, we call value of sentiment words which consists of P/N value. as "the word P/N value". We calculate the word-of-wisdom P/N value by using the word P/N value. Our method is based on the term frequency from each positive and negative words-of-wisdom.

3.1 Extracting Teaching Data

We extract teaching data to calculate the word P/N value from the Internet as follows:

1. We manually extracted 1,000 words-of-wisdom which consist of 500 positive words-of-wisdom and 500 negative words-of-wisdom from the Internet.
2. There are eight subjects into which the 1,000 words-of-wisdom were divided; they were further classified as positive or negative ones. We regard as positive or negative words-of-wisdom those for which five or more subjects have judged as positive or negative. Consequently, 377 words-of-wisdom were positive and 434 words-of-wisdom were negative.
3. We extracted each 350 words-of-wisdom from those 377 positive words-of-wisdom and 434 negative words-of-wisdom. They become teaching data when we calculated the word P/N value.

Table 17.1 shows extracted teaching data.

3.2 Calculating Word P/N Value

We calculate a word P/N value according to the following steps:

Table 17.1 Example of extracting words of wisdom in our experiment

Positive words of wisdom
A man is not finished when he's defeated; he's finished when he quits.
While there's life, there's hope.
Our greatest glory consists not in never falling, but in rising every time we fall.
Negative words of wisdom
I have generally found that a man who is good at excuses is usually good at nothing else.
The tragedy of life is what dies inside a man while he lives.
To mourn a mischief that is past and gone. Is the next way to draw new mischief on.

Table 17.2 Examples of the word P/N values

Word	Word P/N value	Word	Word P/N value
Success	0.75	Lost	−0.35
Happy	0.723	Loneliness	−0.51
Change	0.26	Die	−0.80
Laugh	0.25	Unfortunate	−0.85

1. We calculate each term frequency of nouns, adjectives, adverbs, and verbs in the teaching data of 700 words-of-wisdom.
2. We delete stop words for the results of step (1) because some words are not meant to be judged for P/N, such as nouns, verbs, and adjectives.
3. We calculate all words of words P/N value WPN_i using the following equation.

$$\mathrm{WPN}_i = \frac{X_{Pi}}{X_{P\max}} - \frac{X_{Ni}}{X_{N\max}} \tag{17.1}$$

Therein, X_{Pi} denotes the term frequency of word i in a positive words-of-wisdom, $X_{P\max}$ denotes a term frequency of word $P_{\max}$ which is the same part of speech as i and the maximum term frequency in a positive words-of-wisdom. Furthermore, X_{Ni} represents the term frequency of word i in a negative words-of-wisdom, $X_{N\max}$ is a term frequency of word $N_{\max}$, which is the same part of speech as i and the maximum term frequency in negative words-of-wisdom. We calculate the word P/N value in advance; then we save the word in the database. We calculate and save 1,933 words. Table 17.2 shows some word P/N values.

3.3 *Word-of-wisdom P/N Value*

A word-of-wisdom consists of one sentence or multiple sentences. We propose a word-of-wisdom based on one sentence or multiple sentences.

3.3.1 Consisting of One Sentence

When a words-of-wisdom consists of only one sentence, we propose a method using only our proposed word P/N value. We calculate the word-of-wisdom P/N value MPN_j using the following formula.

$$\mathrm{MPN}_j = \frac{\sum_{k=1}^{n} \mathrm{WPN}_k}{n} \tag{17.2}$$

Therein, n represents number of words that have a word P/N value in word-of-wisdom j. Also, WPN_k is the word P/N value of the word.

3.3.2 Consisting of Multiple Sentences

Many words-of-wisdom comprise multiple sentences. In this case, when the words-of-wisdom is positive, not all sentences are positive. For example, Einstein says "I never think of the future. It comes soon enough." In this words-of-wisdom, the first sentence is negative sentence, but the second sentence is a positive sentence. Therefore, this words-of-wisdom is a positive words-of-wisdom. When the system ignores sentence structure, the system might judge the word-of-wisdom as negative. Then we analyze words-of-wisdom that comprise multiple sentences. We propose a formula of method of word-of-wisdom P/N values. We conducted a preliminary experiment to analysis to determine which sentences most influence the word-of-wisdom P/N value. We first randomly extracted 700 words-of-wisdom from the database. Of the 700 words-of-wisdom, 252 words-of-wisdom are simple sentences, 292 words-of-wisdom are two sentences, and 156 words-of-wisdom are three sentences. Next the six subjects judged all sentences in 448 words-of-wisdom (292 + 156) as positive or negative according to their own viewpoint. They also judged the 700 words-of-wisdom as positive or negative by their own viewpoint. Then, we analyze the results to determine the number of sentences that are the same as the results obtained using the words-of-wisdom system. In the results, last sentence is 88%, middle sentence is 63%, and first sentence is 63%. It means that 88% of the final sentence are the same as the P/N word-of-wisdom.

We propose a formula of word-of-wisdom P/N value of multiple sentences using results of our preliminary experiment.

$$MPN_j = \frac{\sum_{K=1}^{n-1} MPN_k + \alpha \times MPN_n}{n} \tag{17.3}$$

In that equation, MPN_j signifies the word-of-wisdom P/N value of multiple sentences words-of-wisdom j. In addition, n denotes the number of sentence in j, MPN_k represents the word-of-wisdom P/N value of each sentence in j. MPN_n stands for the last sentence of word-of-wisdom P/N value. α denotes the weight of the last sentence. In our result, the last sentence effect the word-of-wisdom is 88%; the other sentence effect the word-of-wisdom is 63%. That is, the last sentence has a 1.4 times greater effect on the result than other sentences do. Therefore, we regard α as 1.4.

4 Search by Multi-dimensional Sentiment Value

A user sometimes wants to search more suitable his/her sentiment, because user's sentiment is complication. Then we also propose search for words-of-wisdom by using multi-dimensional sentiment value. In this paper, we propose multi-dimensional sentiment for words-of-wisdom based on Nakamura's proposed ten-categories sentiments [15].

Table 17.3 Ten type of sentiment categories

Sentiment category English(Japanese)	Word
Joy (ki/Yorokobu)	Fun, interesting, good, happy
Anger (do/okoru)	Crazy, irritating
Sorrow (ai/aware)	Dowdy, gammy, gaunt, sad
Fear (fu/kowagari)	Terrible, worried
Shame (haji)	Disgraceful, squirmy
Liking (kou/suki)	Lovely, honey, pleasantly
Dislike (iya)	Hateful, gloomy
Excitement (takaburu)	Irritate, vexation
Relief (an/yasuragi)	Relax, simple, sporty
Surprise (kyou/odoroki)	Amazement, consternation

4.1 The Ten-dimensional Sentiment Vector

There are some sentiment vector, we use Nakamura's [15] sentiment ten-dimensional vector. Nakamura call the name of element for ten-dimensional vector as sentiment categories. In this paper, we call his categories as sentiment categories. Table 17.3 shows the sentiment categories and example of the words which consists of sentiment categories.

First, we determine sentiment of words-of-wisdom by using ten-dimensional sentiment vector. In this time, when there is a sentiment words in a words-of-wisdom, we determine the sentiment category which is indicated sentiment words, as a sentiment category of the words-of-wisdom.

For example, "Happiness isn't something you experience: it's something you remember." includes word of "Happiness". The word of "Happiness" is indicate "Joy" in ten-dimensional vector. In this case, we regard the words-of-wisdom as "Joy". When there are multiple sentiment words in a words-of-wisdom, we regards the sentiment of it as each sentiment category. For example, "God send you joy, for sorrow will come fast enough." includes "joy" and "sorrow". Joy is in Joy category and sorrow in sorrow category. In this case, we regard the words-of-wisdom has two categories. In this way, we determine sentiment vector automatically. We call the sentiment of the words-of-wisdom as system's judgment.

We had preliminary experiment to determine best suited sentiment of words-of-wisdom by using the system's judgment. In our preliminary experiment, there are 15 subjects, and target data is 100 words-of-wisdom from the Internet randomly. When a user was his/her calm, they judged sentiment words-of-wisdom in their subject view. They could select sentiment categories up to two. We assumed the third high rank place to be a correct answer in user's judgment. They become correct answer. The average of the results are 55%, it means the results of ten-dimensional sentiment is not good. Then we consider we should reduce the dimension from ten-dimensional sentiment.

4.2 *Join of the Sentiment Vector*

4.2.1 Two Methods of Join the Dimensions

When we join the sentiment vector, we focus on the co-occurrence sentiment in a words-of-wisdom. There are multiple sentiments in a words-of-wisdom in the results of user's judgments. For example "To love is to suffer. To avoid suffering one must not love. But then one suffers from not loving." has three type of sentiments. They are "Sorrow", "Fear", and "Liking". In this case, we call these sentiment categories are co-occurrence sentiment. When we join the sentiment vector, we regard co-occurrence sentiment may similar sentiment. Then we focus on the co-occurrence sentiment category and co-occurrence probability of sentiment category. First, we count co-occurrence sentiment category from each 100 words-of-wisdom which we used in our before experiment. We call the number of it as co-occurrence frequency CW_{ij}. Next we calculate co-occurrence degree of sentiment category CA_{ij} as follows:

$$CA_{ij} = \frac{CW_{ij}}{AW_j} \tag{17.4}$$

i is a sentiment category, j is a co-occurrence category of i, and AW_j is a number of words-of-wisdom which j appears in 100 words-of-wisdom.

Table 17.4 shows example of it.

To reduce the dimensions of sentiment vector, we create directed graph whose nodes are sentiment category, based on co-occurrence frequency and co-occurrence ratio. We join the nodes in the directed graph. We propose two types of reducing the dimension of sentiment vector.

Join the Dimensions Based on Top Three of Co-occurrence Frequency

From the users' experiment, we create category graph which consists of top three of co-occurrence frequency in a sentiment category that means out degree of each nodes are 3. We call the category graph as sentiment graph. For example, when a category A has top three co-occurrence frequency B, C, and D, sentiment graph has A→B, A→C, and A→D. In this way, we create sentiment graph each sentiment category. When the two nodes are bi-directional link and other node is also bi-directional link with two nodes, we join three nodes.

For example in Fig. 17.1, we first compare n_1 and n_2. n_1 and n_2 are bi-directional link and other nodes which are n_3 and n_4 are also bi-directional link with n_1 and n_2. We can join n_1 and n_2. Next, we compare n_1 and n_3, they are bi-directional link However, n_1 connected with n_2 and n_4, but n_3 connected with n_2 and n_5 They are connected with different nodes. In this case, we do not join n_1 and n_3 Figure 17.2 shows the result of the join. In this case, category of "Joy" and "Liking" are joined, and "Sorrow" and "Anger" are joined. The dimensional of sentiment vector becomes 8.

Table 17.4 Co-occurrence frequency and co-occurrence degree based of joy category

	Joy	Anger	Sorrow	Fear	Shame	Liking	Dislike	Excitement	Relief	Surprise
Joy	0	0	5	1	4	32	2	30	21	6
All AW_j	47	6	47	24	24	36	41	39	28	19
Co-occur ratio			0.10	0.04	0.16	0.90	0.04	0.78	0.76	0.32

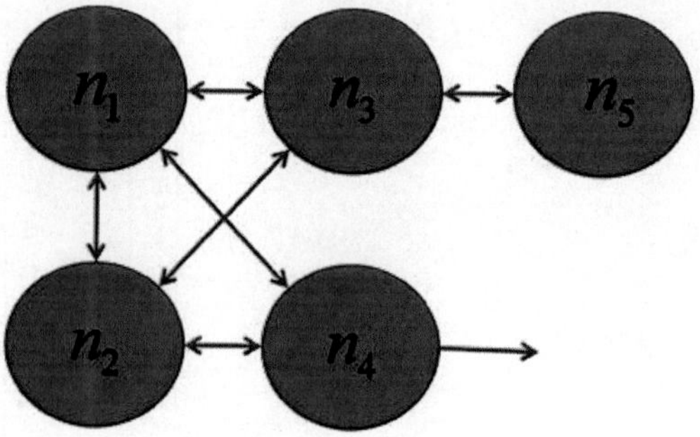

Fig. 17.1 Example of dimension binding

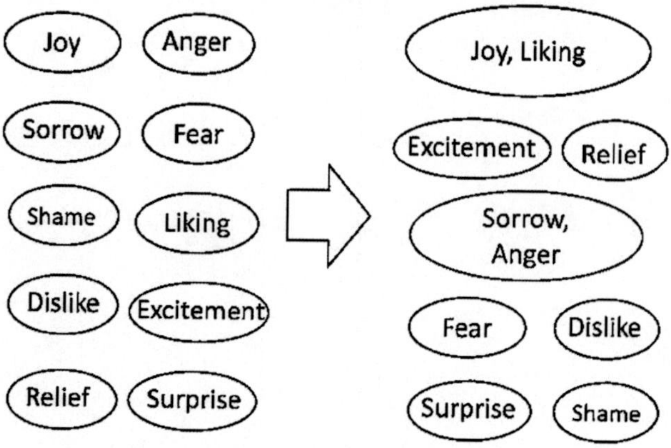

Fig. 17.2 Dimensions binding using threshold value

Join the Dimensions Based on Co-occurrence Ratio

We create sentiment graph by using co-occurrence ratio. The nodes of the sentiment graph are a sentiment categories, and we join the nodes which are bigger than threshold β of co-occurrence ratio. For example, in the Table 17.5, if the β is 0.35, "Joy" category becomes co-occur with "Liking", "Excitement", and "Relief" whose co-occurrence ratios are bigger than 0.35. We had three types of conditions. If the sentiment graph meets all conditions, we join the nodes.

- Condition 1: Bi-directional link

Both of categories are co-occurrence and their co-occurrence ration is bigger than threshold β. For example, co-occurrence ratio of "Joy" with "Liking" is bigger than threshold β, they become bi-directional link.

- Condition 2: Closed chain

We consider the relations of the nodes which consist of closed chain are strong. Then we extract closed chain from the sentiment graph.

Table 17.5 Result of comparison experiment of precision

	Ten-dimensional	Top 3	Threshold				
			0.2	0.3	0.35	0.4	0.5
The number of dimensions	10	8	4	4	6	7	8
Precision	55%	59%	68%	68%	63%	60%	59%

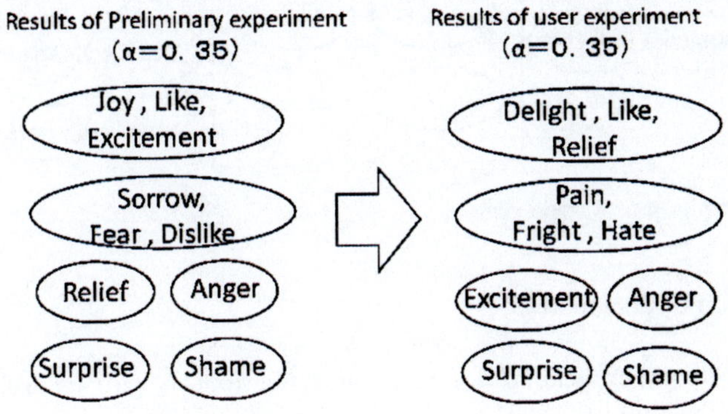

Fig. 17.3 Example 2 of dimensions binding

- Condition 3: When a node has multiple closed chain, we join nodes which has highest co-occurrence ratio.

In Fig. 17.1, n_1 has two closed chain. One is n_1, n_2 and n_3, the other is n_1, n_2, and n_4. When co-occurrence ratio n_1, n_2, n_3 higher than n_1, n_2, n_4, closed chain n_1, n_2, n_3 are joined.

4.2.2 Comparison of the Precisions

We had experiments to determine best method of reduce multi-dimensional vector. We compare the precision which are ten-dimentional vector, top three of co-occurrence and threshold of co-occurrence ratio. Table 17.5 shows the results of these experiments. Fig. 17.3 the results of join based on $\beta = 0.3, 0.35, 0.4$. In our experiments, threshold $= 0.3$ is best results. However, it is clear that the small number of dimension becomes higher results. We consider we have a threshold of the dimension. Wilhelm Max Wundt [16] clusters six sentiment such as pleasurable, unpleasurable, agreeable, disagreeable, excited and quiet. Descartes [17] also says humans emotion is only six which are surprise, love, hate, desire, delight, and sad. Then, we regard the threshold of the sentiment dimension is six. In our experiment, the best result of co-occurrence threshold is 0.35.

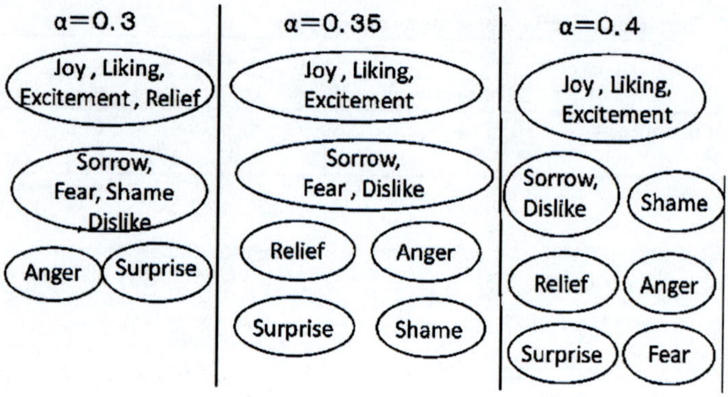

Fig. 17.4 Results of user experiment

4.2.3 Users Experiment

We had users experiment to decide multi-dimensional vector by using the co-occurrence threshold 0.35. The data is as same as preliminary experiment. We divide subjects age into 16–22 years old, 23–34 years old, 35–44 years old, 45–54 years old, 55–64 years old, and over 65 years old. Subjects are 10 each woman and man and each age. Total subjects are 120. The results of the declare randomly the dimension in Fig. 17.4.

4.3 Value of Sentiment Words

We calculate sentiment of words-of-wisdom by using six-dimensional vector in Fig. 17.4. When we calculate sentiment value of words-of-wisdom, first we focus on the words which consist of words-of-wisdom. We call the words as sentiment words. We calculate value of sentiment words as following methods.

- We randomly extract words-of-wisdom form the Internet, and divide into 6 dimensional sentiment vector by hands. In fact, we extract 17,446 words-of-wisdom from the Internet, and divide them.
- We divide the words-of-wisdom into words by using morphological analysis. We use Japanese morphological analysis Juman [18]. In this time, we use some rules which is proposed Kumamoto et al. [13].
- We calculate term frequency (*tf*) of nouns, adjective, and verb from each sentiment vector. We normalize *tf* of them based on each sentiment vector, it becomes value of sentiment of words.

Table 17.6 shows the example of value of sentiment of words. There are some words which are no-meaning for sentiment of words-of-wisdom, we regard them as stop words and delete them.

Table 17.6 Example of value of sentiment words

Word	Joy, liking, relief	Anger	Sorrow, fear, dislike	Shame	Excitement	Surprise
Happiness	0.81	0.16	0.39	0.2	0.2	0
Shame	0.02	0	0.11	1.0	0	0
Tear	0.23	0.07	0	0.07	1.0	0
Anger	0.03	1.0	0.06	0	0	0
Love	1.0	0.07	0.33	0	0.4	0
Surprise	0.01	0	0.03	0	0	1.0

4.4 Sentiment Value of Words-of-wisdom

After we calculate value of sentiment words WPN_i, we calculate sentiment value of words-of-wisdom by using WPN_i. In this time, we calculate sentiment value based on two types of words-of-wisdom. One is single sentence words-of-wisdom, the other is multiple sentences words-of-wisdom. They are as same as P/N value. Then we calculate multi-dimensional sentiment value of words-of-wisdom by using same formula in Sect. 3.3.

5 Experiments

5.1 Experiments Conditions

We conducted two experiments. One experiment was conducted to examine the extraction of the word-of-wisdom P/N value. We next examined the extraction of the word-of-wisdom by multi-dimensional sentiment. In the experiment, our purpose is measuring the benefits of our proposed methods. We did two experiments by using the condition as follows:

- Datasets: We extracted 250 words-of-wisdom from the Internet randomly.
- Subjects: nine people
- Condition of the subjects: When user was his/her calm, they judged sentiment words-of-wisdom in their subject view.
- Judge:
  1. Subjects could select one sentiment from "Positive", "Negative", and "Neutral".
  2. They could select sentiment categories up to two from six sentiment categories.

We compare the results of systems with subjects judgments.

Table 17.7 Results of multi-dimensional sentiment

	Joy, liking, relief	Anger	Sorrow, fear, dislike	Shame	Excitement	Surprise
Recall (%)	63.6	9.1	79.0	6.8	13.7	13.0
Precision (%)	70.5	3.5	53.1	23.8	56.3	23.8
F-Measure (%)	66.9	5	63.6	10.5	24.2	23.0

5.2 Results

- P/N value

When the results of system's P/N value is bigger than 0, it becomes "Positive". When the result is less than 0, it becomes "Negative". "Neutral" is 0. We assumed correct answer is biggest number of results which subjects selected. Precision is 46%, recall is 49%, and F-measure is 47.6%.

- Multi-dimensional sentiment

We assumed the two high rank place to be a correct answer in system's judgment. We also assumed correct answer is the two high rank place to be a correct answer in subjects' judgment. Precision is 52%, Recall is 41%, and F-measure is 46%. Table 17.7 show the results.

5.3 Discussion

F-Measure of P/N value is not so high. Because it is difficult to measure the word-of-wisdom by only P/N value and our sentiment is more complex.

In the multi-dimensional sentiment, the results of "Joy, Liking, Relief" and "Sorrow, Fear, Dislike" are good, but other results are not good. Because there are few sentiment words which have "Anger" and "Shame", and we should more consider the value of sentiment words.

6 Conclusion and Future Work

We proposed words-of-wisdom search system as a first step of to search for the information based on user's sentiment. We proposed two types of sentiment search, one is P/N sentiment and the other is multi-dimensional search. In the search by P/N value, we proposed a method of calculating the word P/N value for words-of-wisdom and the word-of-wisdom P/N value based on it. Then, we created a word-of-wisdom database based on our word-of-wisdom P/N value. In the search by multi-dimensional sentiment, we proposed six-dimensional sentiment vector for words-of-wisdom based on joining of ten-categories sentiments. Then, we calculate value of sentiment words, and calculate sentiment value of words-of-wisdom by using it.

Subjects for future work include the following.

- Considering sentiment distance.

In this paper, we consider only user's desired sentiment, however we must consider the sentiment distance which is between user's current sentiment and desired sentiment.

- Adaptation of other content.

In this paper, we target on words-of-wisdom, however our purpose of this research is to search for the information based on user's sentiment. Then we must to adapt to other content such as news articles, music, and so on.

References

1. International Association of Engineers [Online]. http://www.iaeng.org, 1 September, 2011
2. Friedrich Wilhelm Nietzsche (2010) Words of Nietzsche. Discover 21, Tokyo
3. Meigen-Navi (Words-of-Wisdom navigation). http://www.meigennavi.net/, 9 November, 2011
4. Plutchik R (2010) The multi factor-analytic theory of emotion. Psychology 50:153–171
5. Plutchik R (2001) The nature of emotions. Amer Sci 89:344–350
6. Tokuhisa R, Inui K, Matsumoto Y (2008) Emotion classification using massive examples extracted from the web. Proceedings of the 22nd international conference on computational linguistics—Volume 1, Manchester, pp 881–888
7. Kawai Y, Kumamoto T, Tanaka K (2007) Fair news reader: recommending news articles with different sentiments based on user preference. In international conference on knowledge-based intelligent information and engineering systems (KES '07), Vietri sul Mare, Italy, pp 612–622
8. Dave K, Lawrence S, Pennock DM (2003) Mining the peanut gallery: opinion extraction and semantic classification of product reviews. In Proceedings of the 12th international world wide web conference (WWW '03), Budapest, Hungary, pp 519–528
9. Kobayashi N, Inui T, Inui K (2001) Dictionary-based acquisition of the lexical knowledge for p/n analysis (in japanese). In Proceedings of Japanese society for artificial intelligence, Nara Japan, pp 40–45
10. Fujimura S, Toyoda M, Kitsuregawa M (2005) A reputation extracting method considering structure of sentence (in japanese), Institute of Electronics, Information and Communication Engineers, Data Engineering Workshop (DEWS2005), 6C-i8, Nagasaki, Japan
11. Takamura H, Inui T, Okumura M (2005) Extracting semantic orientations of words using spin model. In: Proceedings 43rd annual meeting of the association for computational linguistics (ACL'05), Ann Arbor, Michigan, pp 133–140
12. Kumamoto T, Tanaka K (2005) Proposal of impression mining from news articles, Lecture Notes in Computer Science. In international conference on knowledge-based intelligent information and engineering systems (KES '05), Melbourne, Australia, pp 901–910
13. Kumamoto T, Kawai Y, Tanaka K (2011) Improving a method for quantifying readers' impressions of news articles with a regression equation, In: Proc. of the 2nd workshop on computational approaches to subjectivity and sentiment analysis (WASSA2.011), Portland, Oregon, USA, pp 87–95

14. Takaoka K, Nadamoto A (2011) Words-of-wisdom search system based on positive negative degree, Lecture notes in engineering and computer science: Proceedings of the international multiconference of engineers and computer scientists 2011 (IMECS 2011), Hong Kong, pp 368–373
15. Nakamura A (1993) Sentiment dictionary. Tokyodo, Tokyo
16. WM Wundt (1896) Grundriss der Psychology, Verlag von Wilhelm Engelmann, Leipzig
17. Rene D (2008) Les passions de l'ame, iwanami library, Tokyo
18. Kurohashi S, Nakamura T, Matsumoto Y, Nagao M, Improvements of Japanese morphological analyzer JUMAN. In: Proc. of the Int. workshop on social networks and social media mining on sharable natural language resources, Nara Japan, pp 22–28

Chapter 18
Configurable Application Designed for Mining XML Document Collections

Mihaela Juganaru Mathieu, Cristal Karina Galindo-Durán, and Héctor Javier Vázquez

Abstract In this chapter we present a flexible and configurable application for mining large XML document collections. This work is centered on the process of extracting document features related to structure and content. From this process, an attribute frequency matrix is generated and, depending on the cluster algorithm, it is transformed and/or used to obtain similarity measures.

Keywords Clustering • Conceptual framework • Methodology • Modularity • UML • XML mining

1 Introduction

The eXtensible Markup Language XML [1, 2] is a self-describing language for data representation, exchange and storage of information in structured and/or semi-structured form [3], without the need to specify how information will be viewed. This signifies that content and presentation are independent. XML also provides the means to share information in a reliable and easy manner and offers compatibility between different systems and platforms and between different applications. These

M.J. Mathieu
Laboratoire en Sciences et Technologies de l'Information, Institut H. Fayol, Ecole Nationale Supérieure des Mines de Saint Etienne, 158, cours Fauriel, 42023, SaintEtienne Cedex 2, France
e-mail: mathieu@emse.fr

C.K. Galindo-Durán (✉)
Computer Science, MSE Program, Universidad Autónoma Metropolitana, Unidad Azcapotzalco
e-mail: cdgalindod@gmail.com

H.J. Vázquez (✉)
Departamento de Sistemas, Universidad Autónoma Metropolitana, Unidad Azcapotzalco, Avenida San Pablo 180, Col. Reynosa Tamaulipas, Mexico D.F. C.P. 02200, Mexico
e-mail: hjv@correo.azc.uam.mx; statfor@yahoo.com

S.I. Ao et al. (eds.), *Intelligent Control and Innovative Computing*, Lecture Notes in Electrical Engineering 110, DOI 10.1007/978-1-4614-1695-1_18,

reasons make XML a preferred language to encode text and non-text data, to build Web servers and Web applications and for creating collections of documents. Given that most non text data are stored in databases, XML is generally used to store collections of text and data containing links to other documents (within the same collection or in another collection) and references to images and other multimedia files.

The aim of the first part of this chapter is to present the methodological and design principles for building an application to mine XML documents, highlighting specific issues such as parsing and tokenization. This methodology is based on a general data mining process model and a specification design [4] that resulted in a modular and configurable XML mining application. It is modular in the sense that it is easy to change modules, functions or algorithms and is configurable in that it allows the user to:

- Mine various large collections.
- Choose all documents, or a random sample of them from a given collection.
- Choose, as items to mine, different attributes about structure, content and/or hyperlinks (within and/or outside documents).
- Count frequencies or count only the presence/absence of structural or content features.
- Produce different data transformations from the frequency matrix.
- Generate different metrics of similarity as well as similarity matrices, depending of the counting method used in the previous step.
- Choose from different mining algorithms.

The second part of this chapter discusses basic software design elements and gives details about the implementation of the application, in particular parsing and tokenization.

2 Methodology and Design for an XML Mining Application

For general data mining, two well-known process models [5] are classically presented: an eight-step model proposed by Anand and Buchner in 1998 [6] and a nine-step model proposed by Fayyad et al. in 1998. The latter, presented by Minei [7], might be summarized as follows:

1. To understand the problem in order to define the tasks to solve and goals to pursue, i.e., association, classification, cluster analysis, sequential pattern discovery, temporal modeling, regression or deviation detection.
2. To identify the data target and the variables to achieve the goal.
3. To pre-process and to clean data by removing noise, identifying outliers, identifying missing values and selecting methods for correcting and handling them.
4. To reduce data dimensionality or to transform it to find invariant representations of data.

5. To choose the data mining task depending on the goal, i.e. description or prediction.
6. To select methods and algorithms for searching and extraction of data patterns, including parameter setting.
7. To generate patterns of interest; the result might take the form of classification rules, decision trees, regression models or trends.
8. To evaluate and/or to interpret the mined patterns. This refers to reconfiguring procedures and then considering a possible return to one or more of the previous steps.
9. To evaluate the performance of the system and to resolve potential conflicts with previous or a priori beliefs or extracted knowledge.

Each step is discussed, in detail, in texts about data mining [8]. Most of the time, however, definitions of terms such as *data, information or knowledge* are not clarified. Although it is not the goal of this chapter to discuss issues about those terms, it is important to understand and differentiate them [9].

The application is specified and designed to mine collections of XML documents (the data target) oriented to clustering and not to classification [10]. Clustering is the process of partitioning data into a set of groups, clusters or subclasses [11]. This process is performed without any information about possible groups and without examples or learning sets. Once the process finished, all data in the cluster share common traits. Given that the user is not directly involved until the end of the process, clustering is considered to be unsupervised classification. Conversely, in supervised classification, which is used to predict group membership for data instances, the user intervenes and monitors the process.

In the case of information coded in XML, the general process model does not clarify how to take into account document's content (data variables) and its structure (attributes about structure) [12, 13].

An XML document is in fact a graph representing a collection of atomic and complex objects in the form of an ordered and labeled tree, where each node represents an XML element and each element might have one or more XML attributes. Attributes are also provided to annotate elements. An XML element is made up of a start tag, an end tag, and content in between. The start and end tags describe the content within the tags (the value of the element). The content may be a Character Data (CDATA) section or an ordered set of XML elements. Complexity increases because the same tag may exist at different levels of the tree. This gives a multidimensional structure to an XML document and creates the need to manage information in different ways [14].

In the pre-processing phase, in addition to the tasks listed in step 3 of the nine step model, an extra task should be included: that is, to consider tags, their nesting and data references to reveal their structure in a XML document. Most of the methods work on the trees representing documents [15, 16]. Some of them compare trees using metric distances based on: the edit distance, the minimum number of mutations required to change a tree into another one or the number of paths they share. Other methods focus on sub-trees inside the document.

To reveal structure, it is necessary to include a parsing step and eventually some particular strategies for XML retrieval and extraction of data patterns from XML documents, such as partition of XML documents into non-overlapping indexing units or mapping to a set of lexicalized sub-trees [17]. On the other hand redundancy caused by nested elements makes it necessary to establish restrictions, such as discarding small elements, removing or keeping some types defined by users or collapsing several nested elements in the same result.

Fortunately, if the objective is just to mine the data, as in the case of our application, some tasks such as identifying tokens (content words, hyperlinks, tags and references, as well as their nesting level), can be reduced.

A token in a XML document is the smallest data element that can be processed to revel a meaning. As part of a CDATA section, a word is a token. A CDATA section separates any kind of character (including XML markup) that should be ignored by the parser. In well formed XML documents, an open tag always has a closed tag, so we can just consider the open tags. If we need to indicate a token's position in the XML tree, it's sufficient to consider the path of the node. In the case of a token inside a CDATA section, the path of the CDATA is just considered.

The simplest form of tokenization (obtaining the tokens) is to consider all the tags and all the words as they are presented in the textual parts. However, another problem might emerge: As, number of tokens increases data dimensionality also increases. This can be solved, as in text mining [18], by taking the invariant form of the word in its lexical category. It's also possible to perform a supervised tokenization and to take into account some external information, such as a list of the type of tags, indications about the importance of a type of code or ontology, lists of synonyms or specialized vocabulary indicating the "important" words.

An option for reducing and managing data dimensionality is to remove irrelevant characteristics, such as elements containing references to images, sound files and other multimedia formats. In the case of both internal and external links, just their number, not their content, might be considered.

Moreover this, for mining purposes it might be sufficient to consider the frequency or just simply the presence of tokens. All or a subset of tokens can be selected and counted (frequency and presence and/or absence). These counts can be integrated in a single table or organized in different count or frequency matrices; for example, one frequency count matrix to register tokens about document-structure (tags, markers, token path, depth or any another feature related to structure) and another matrix to register counts of tokens about document-content (composed mainly of text). The final result of this is a matrix containing frequencies about tokens (frequency matrix). In this way, XML mining research has led to the development of various techniques and tools [19, 20] dealing with data frequency patterns as well as frequency patterns within parse trees.

The frequency matrix, in its original or transformed form, is the basis for different well-known statistical techniques oriented to data prediction or just to description. Transformations or normalizations may be applied, for example to standardize values or to stabilize variance. For the present application a hard clustering and soft clustering technique are proposed. In hard clustering, each document belongs only to one cluster while in soft clustering the same document may belong, at different degree, to different clusters. The expectation maximum (EM) algorithm (soft clustering) can be applied directly on the frequency matrix [21], while for algorithms such as hierarchical clustering (hard clustering) [11], a metric should be chosen to obtain similarity matrices.

Depending on the nature of the counts (i.e., frequencies or binary data), there are different kinds of similarity measures. For example, in the case of frequency counts, it is common to use the Euclidean, Mahalanobis, Minkowski and $\chi 2$ distances [22]. In the case of the presence or absence of particular features, it is possible to assign more or less importance to presences than to absences: the Jaccard and Dice–Sorensen indexes are most often used for this purpose. However, if both presences and absences are considered to have the same importance, then the Sokal and Michener, Sokal and Sneath, Hamming or Roux indexes may be used [22].

We could also consider another type of frequency measure, a frequency function inspired by information retrieval and text mining: Tf-idf (Term frequency–inverse document frequency) is a measure that evaluates the importance of a word in a document or a set of documents [18]. This measure takes into account the direct frequency of a word in the document as well as the inverse frequency giving less weight to words that are present in many documents. For this type of frequency, we use similarity measures such as product cosine or Jaccard. Figure 18.1 provides a view of the sequence of steps for pre- processing document structure and content. This forms part of the conceptual model proposed to design this application [4].

Once the clusters have been obtained, they can be parsed to find data structures and explore their characteristics, which can then be interpreted and evaluated, and the mining process eventually repeated.

3 Software Description

User-oriented, flexibility and modularity are the basic principles that guide the design of this application. The application should be user-oriented and flexible in the sense that it gives the user the ability to choose among different types of tokens, different frequencies, different data transformations, different types of similarity distances and even different types of clustering algorithms. All these options can be applied on a large collection or on a random sample of the collection. At the end of the process and during user interpretation of results, the application would "feed" the user's knowledge into the XML mining process. Modularity means that the application can be partitioned to allow easy implementation and maintenance. This would allow a user to modify a part of the application or to integrate a new procedure without redesigning the entire application. In a first level, the design process was separated into two packages, as follows:

- Package one, to obtain documents, to tokenize them and store them, and to obtain the frequency matrix.
- Package two, to deal with the mining process itself.

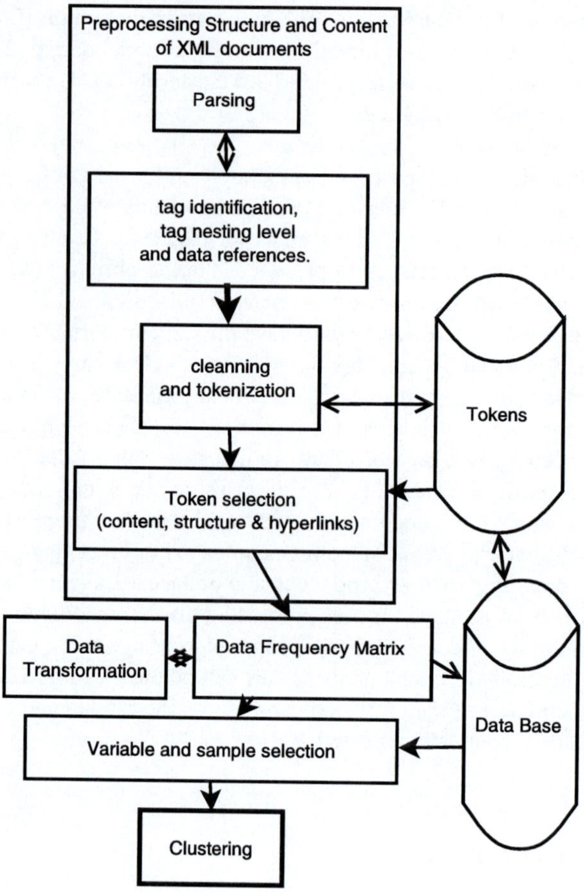

Fig. 18.1 Conceptual model

The unified modeling language (UML[1]) is used to present different elements of the proposed design such as use case and class diagrams. In the following section, class diagrams and implementation of an entity relational model to store information are presented. Details of the use case diagrams are presented in [4].

3.1 Class Diagram

In the class diagram, classes are first defined for the XML collection and its documents. These two classes are directly associated with a multiplicity [1, *],

[1]http://www.uml.org [Accessed November 8, 2011].

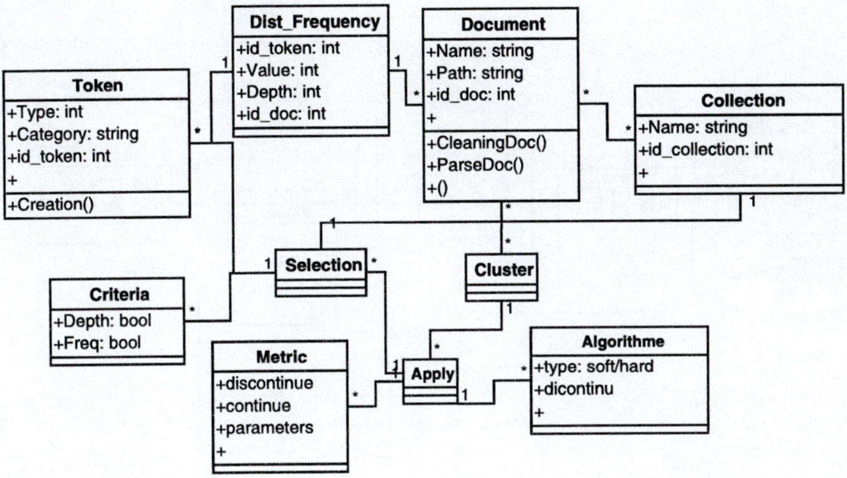

Fig. 18.2 Class diagram

given that one collection has several documents. An important class is "Distribution_frequency", which is associated with several documents and several tokens. Tokens are the objects identified in the document and may have several categories (for example, structural or grammatical). The remaining classes are related to the mining process, which applies cluster algorithms, based on a given metric to the frequency matrix (revealed by classes: Token and Distribution_frequency). The classes proposed are: **Collection_C, Document_C, Distribution_Frequency_C, Token_C, Criteria_C, Choice_C, Metrics_C, Apply_C, Cluster_C and Algorithm_C**.

Figure 18.2 shows a static view of the class diagram, with the ten classes and their associations proposed for this application.

3.2 Implementation

The INEX collection[2] is used as a data target to implement this application. It currently consists of two collections: a large one that has about 2.7 million documents and a smaller one that contains approximately 463,000 XML documents. However, for testing purposes only 60,000 documents are considered. This collection also includes bags of words that are commonly found in documents and frequencies of various structures in the form of XML tags, trees, links and names of entities. One feature is that documents are well formed and there is no need for a document type

[2]This collection contains an image of Wikipedia articles, and was extracted from a 2007 copy.

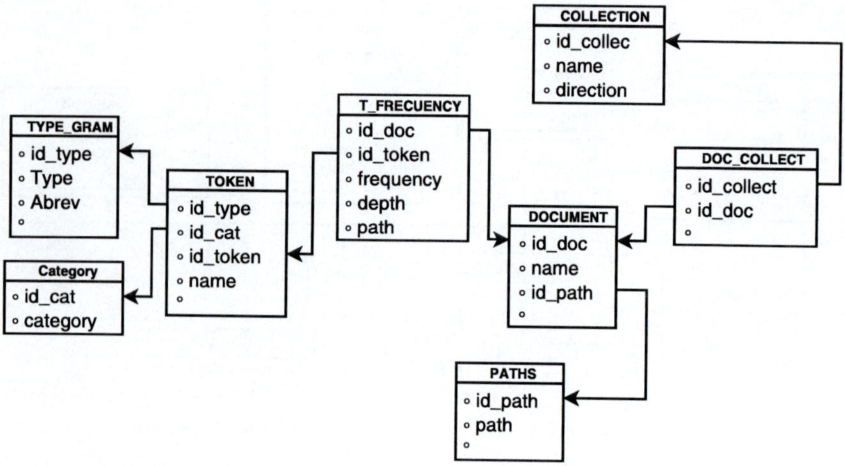

Fig. 18.3 Relational model

definition (DTD). The collection is grouped in 22 packages, with a mean number of five branches in each document. With these characteristics, work is greatly reduced because each document has a unique node on the first level, and an open tag is always closed in a coherent order. This also means that it is known where to begin parsing and that it is not necessary to check whether the tag is closed.

For package one, our major decision was to store in a DBMS (database management system) all the information about documents, tokens and frequencies in a database. The purpose is to avoid calculating all feature frequencies (structure, content and hyperlinks) each time the user analyses different sets of attributes. Frequency matrices and presence/absence matrices are directly extracted with aggregated SQL queries from the tables of the database. PostgreSQL[3] is used for the database implementation of the DBMS. The eight tables of the relational model are presented in Fig. 18.3, as follows:

1. COLLECTION table with a tuple for each XML document collection.
2. DOCUMENT table references PATH table and contains a tuple for each XML document.
3. DOC_COLLECTION table indicates the composition of collections.
4. PATHS table to indicate where the XML files are physically stored.
5. TOKEN table stores a unique tuple for each possible token.
6. CATEGORY table contains a tuple for each category of a token: word, tag, link or other.
7. TYPE_GRAM table indicates the grammatical type for a token word.
8. T_FREQUENCY table indicates frequency of tokens in documents.

[3]http://www.postgresql.org/ [Accessed November 8, 2011].

A first step is to obtain the XML collection, assign a number and identify the path where each document is stored. This information is stored in the DOCUMENT and PATH tables. There are several mechanisms for tracking and accessing information stored in XML documents. Basic mechanisms are implemented in parsers such as SAX (simple API for XML[4]) and DOM (document object model[5]). SAX tracks the document sequentially in depth while DOM randomly accesses the document and builds, in memory, a tree of the XML document, permitting the document to be parsed, in depth or in breadth, or its elements to be modified. Because DOM is less complex, it is used to obtain tags, hyperlinks and paths tags. The pseudocode is the following:

```
/* pseudocode for Parsing and Tokenization */
for each document D of a collection
      if D was tokenized then break;
      insert into DOCUMENT table;
      count[...] < −0$;
      let id_doc be the key into DOCUMENT table;
      depth < −0;
      obtain with DOM the XML tree
      /* treat all the nodes starting with the root */
      for each node n of XML tree
         if n is <outsidelink> or <collectionlink> node n is a link
         else
            n is a node to treat
            let <x> be the tag of the node;
            if <x> not exists in TOKEN table then
               insert <x> into TOKEN;
               indicate if is a hard or a soft tag;
            end if
            let textual be the CDATA of n;
call TreeTagger for textual: the output is: list of (word,
                                lexical_category, invariable_form)
     for each (word, lexical_category, invariable_form)
         if (word, lexical_category, invariable_form) ∉ TOKEN table
then insert into TOKEN table
               end if
               let be id_token the key value in TOKEN table
               count[id_token, depth] ++
            end for
            depth ++;
            recursive call to treat the sons of n, if any;
         end if
```

[4] http://www.saxproject.org/ [Accessed November 8, 2011].

[5] http://www.w3.org/DOM/ [Accessed November 8, 2011].

```
        end for
        for each count[id_t, depth] <> 0 do
            insert (id_doc, id_t, depth, count[id_t, depth]) into
          FREQUENCY
                table
        end for
end for
```

Another decision was to use a grammatical invariable generator such as Tree Tagger[6], a part of speech tagger that obtains the grammatical category and invariant form of every word [23]. After parsing, Tree Tagger is used to obtain content tokens.

From this, a token frequency matrix is created, which forms the basis of the content and structure frequency matrices. Token frequency matrix generation requires extra storage, but, as explained previously, it was decided to implement it, given that it is simple to obtain whole tokens counts, and because it later provides the user a choice of generating the frequency or the presence/absence matrix.

In this implementation with INEX, 23 soft tags (such as emph, normal list and item) and 7 hard tags (such as article, name and body) were considered, as well as, 2 types of links (collection link and outside link) and 36 tokens for content (such as cardinal number, determiner, existential and foreign word). This provides at least 66 variables for each document.

This architecture of the DBMS offers enough flexibility to center on a subset of the variables. For example, if the user wants to cluster documents based only on the words without tags and links, a filter is introduced in the CATEGORY table. Also, if the user wants to consider only the words having a meaning (excepting for example the stop words[7]), it is easy to filter in the TYPE_GRAM table to exclude determinants, conjunctions and prepositions. It is also possible in the TOKEN table to flag the stop word linking to external information.

Another characteristic of this application is its ability to work on the whole collection or just a random sample of it. As an example for a small collection with 545 XML documents, the most important tables have 8,336 rows for the TOKEN table and 156,793 for the T_FREQUENCY table. The smallest document (6 kb) has about 16 tokens, and the larger document (30 kb) has around 1,846.

Java[8], an object oriented language, was used to implement this application. Using Java allows us to introduce and consider new distance measures or mining algorithms. Java also offers the possibility of working with various APIs for DBMS or XML, to interface with Tree Tagger or to call mining procedures or libraries written in GNU applications such as R[9] or other languages.

---

[6] http://www.ims.uni-stuttgart.de/projekte/corplex/TreeTagger/DecisionTreeTagger.html [Accessed November 8, 2011].

[7] A stop word is a word frequently appearing in a text, such as "the", "and", "a".

[8] http://www.oracle.com/technetwork/java/javase/downloads/index.html [Accessed November 8, 2011].

[9] http://www.r-project.org/ [Accessed November 8, 2011].

4 Conclusion and Future Work

In this chapter, we presented the methodology and implementation for a modular and configurable application designed for XML mining and clustering in a large collection of documents. The specified design was based on UML, and the particularities of XML documents were fundamental to understand the mining process.

Concerning the generation of a token frequency matrix, the choice of a DBMS saves space and allows, if required, the creation of one or two separate frequency matrices one for structure and other for content, given that each token category can be easily identified.

This work highlights the importance of developing a configurable and modular application based on a specified design; other research has excluded the possibility of modifying parts of the pre-processing and mining application or changing its configuration.

The first stage (package one) of development of the software application has been completed and uses Java and PostgreSQL for implementing the database management system. The information saved in the database allows the user to mine: structure or content, hyperlinks (within and/or outside documents) and also to choose the type of the frequency (direct, presence/absence, standardized, transformed or Tf-idf) and, depending of the clustering method, to apply a similarity function.

This implementation is being tested with a representative sample of 537 from 60,000 documents to obtain the token frequencies and then the frequency matrix.

The frequency matrix is the basis for the second package, which focuses on the numerical part of the data mining process. Hierarchical clustering (hard clustering) [11] and expectation maximization (EM) algorithm (soft clustering) [21] are proposed for cluster analysis. They can be implemented in Java or linked with subroutines programmed in other applications such as R or in other mining tools [5].

Acknowledgments The authors acknowledge financial support given to the Promep group "Analysis and Information Management" and to the Universidad Autónoma Metropolitana, Unidad Azcapotzalco (UAM-A) for the financial support given to the project, "Study and Systems Modelling," Number: 2270217. They acknowledge Dr. Nicolas Dominguez Vergara (former head of the Systems Department) and Dr. Silvia Gonzalez Brambila (former Coordinator of the Master's Computer Science Program at the UAM Azcapotzalco) for their support and encouragement. They also acknowledge Peggy Currid for her advice in copyediting.

References

1. Bray T, Paoli J, Sperberg-McQueen CM, Maler E, Yergeau F (2008) eXtensibleMarkup Language (XML) 1.0 (5th Edition), X3C recommendation. http://www.w3.org/TR/REC-xml/ [Accessed November 8, 2011]
2. Harold ER, Scott Means W (2004) XML in a nutshell. O'Reilly Media, Sebastopol, CA, USA

3. Abiteboul S, Buneman P, Suciu D (2001) Data on the web: from relations to semi structured data and XML. Morgan Kaufmann, Series in data management systems, San Francisco, USA
4. Duran CKG, Juganaru-Mathieu M, Vazquez HJ (2011) Specification design for an xml mining configurable application, lecture notes in engineering and computer science. In: Proceedings of the international multiconference of engineers and computer scientists 2011, IMECS 2011, Hong Kong, 16–18 March 2011. (best paper award). http://www.iaeng.org/publication/IMECS2011/IMECS2011_pp378-381.pdf [Accessed November 8, 2011]
5. Mikut R, Reischl M (2011) Data mining tools, WIREs data mining knowledge discovery. 1(5):431–443. http://onlinelibrary.wiley.com/doi/10.1002/widm.24/pdf [Accessed November 8, 2011]
6. Büchner AG, Mulvenna MD, Anand SS, Hughes JG (1999) An internet-enabled knowledge discovery process (best paper award). In: Proceedings of the 9th international database conference, Hong Kong, July 1999. http://mcbuchner.com/HTML/Research/PDF/IDC99.pdf [Accessed November 8, 2011]
7. Minaei-Bidgoli B (2004) Data mining for a web-based educational system, PhD thesis, Michigan State University. http://www.lon-capa.org/papers/BehrouzThesisRevised.pdf [Accessed November 8, 2011]
8. Cios KJ, Pedrycz W, Swiniarski RW, Kurgan LA (2007) Data mining: a knowledge discovery approach, Springer Science+Business Media, New York, USA
9. Ackoff RL (1989) From data to wisdom. J Appl Syst Anal 16:3–9
10. Oracle text application developer's guide release 1 (10.1) (2010). http://www.stanford.edu/dept/itss/docs/oracle/10g/text.101/b10729/classify.htm [Accessed November 8, 2011]
11. Grabmeier J, Rudolph A (2002) Techniques of cluster algorithms in data mining, in data mining and knowledge discovery. Kluwer Academic Publishers, 6(4):303–360, Netherlands. http://www.springerlink.com/content/d6ekxxcu0d2ngamj/fulltext.pdf [Accessed November 8, 2011]
12. Candillier L, Denoyer L, Gallinari P, Rousset MC, Termier A, Vercoustre AM (2007) Mining XML documents. In: Poncelet P, Masseglia F, Teisseire M (ed) Data mining patterns: new methods and applications. pp 198–219
13. Büchner AG, Baumgarten M, Mulvenna MD, Böhm R, Anand SS (2000) Data mining and XML: current and future issues. In: Proceedings of the 1st international conference on web information systems engineering, Hong Kong. http://mcbuchner.com/HTML/Research/PDF/WISE00.pdf [Accessed November 8, 2011]
14. Tran T, Nayak R, Bruza P (2008) Document clustering using incremental and pairwise approaches. In: Focused access to XML documents 6th international workshop of the initiative for the evaluation of XML retrieval, INEX-2007. pp. 222–233. Springer-Verlag, Berlin Heidelberg
15. Candillier L, Tellier I, Torre F (2005) Transforming XML trees for efficient classification and clustering. Workshop on mining XML documents, INEX-2005, Schloss Dagstuhl, 28–30 November 2005. http://www.grappa.univ-lille3.fr/_candillier/publis/INEX05.pdf [Accessed November 8, 2011]
16. Dalamagas T, Cheng T, Winkel K-J, Sellis Timos (2003) A methodology for clustering XML documents by structure. Inform Syst 31:187–228
17. Manning CD, Raghavan P, Schtze H (2009) An introduction to information retrieval. Cambridge University Press, Cambridge, pp 201–210. http://nlp.stanford.edu/IR-book/pdf/irbookonlinereading.pdf [Accessed November 8, 2011]
18. Feldman R, Sanger J (2007) The text mining handbook: advanced approaches in analyzing unstructured data. Cambridge University Press, Cambridge
19. Tran T, Nayak R, Bruza P (2008) Combining structure and content similarities for XML document clustering. In: Proceedings 7th Australasian data mining conference, Glenelg, South Australia. CRPIT, 87. Roddick JF, Li J, Christen P, Kennedy PJ, (ed). http://crpit.com/confpapers/CRPITV87TranT.pdf [Accessed November 8, 2011]
20. Lee JW and Park SS (2004) Computing similarity between XML documents for XML mining. In Engineering Knowledge in the Age of the Semantic Web, 14th International Conference, EKAW, pp 492–493. Springer Verlag

21. Dempster AP, Laird NM, Rubin DB (1977) Maximum likelihood from incomplete data via the EM algorithm. J R Stat Soc B 39(1):1–38. Royal Statistical Society, London, http://web.mit.edu/6.435/www/Dempster77.pdf [Accessed November 8, 2011]
22. Roux M (1985) Algorithmes de Classification. Masson, Paris
23. Schmid H (1997) Probabilistic part-of-speech tagging using decision trees. In: Jones DB, Somers H (ed) New methods in language processing. Routledge, London

Chapter 19
Integrated Approach for Privacy Preserving Itemset Mining

Barış Yıldız and Belgin Ergenç

Abstract In this work, we propose an integrated itemset hiding algorithm that eliminates the need of pre-mining and post-mining and uses a simple heuristic in selecting the itemset and the item in itemset for distortion. Base algorithm (matrix-apriori) works without candidate generation so efficiency is increased. Performance evaluation demonstrates (1) the side effect (lost itemsets) and time while increasing the number of sensitive itemsets and support of itemset and (2) speed up by integrating the post mining.

Keywords Matrix-apriori • Privacy preserving data mining • Sensitive itemset hiding

1 Introduction

Data mining is simply defined as finding hidden information from large data sources. It became popular in last decades by the help of increase in abilities of computers and collection of large amount of data [1]. Although it is successfully applied in many fields such as marketing, forecasting, diagnosis and security, it is a challenge to extract knowledge without violating data owner's privacy [1–6]. Privacy preserving data mining (PPDM) come up with the idea of protecting sensitive data or knowledge to conserve privacy while data mining techniques can still be applied efficiently.

B. Yıldız (✉)
Department of Computer Engineering, Dokuz Eylul University, Buca-35160, Izmir, Turkey
e-mail: barisyildiz@computer.org

B. Ergenç
Department of Computer Engineering, Izmir Institute of Technology, Urla-35430, Izmir, Turkey
e-mail: belginergenc@iyte.edu.tr

S.I. Ao et al. (eds.), *Intelligent Control and Innovative Computing*, Lecture Notes in Electrical Engineering 110, DOI 10.1007/978-1-4614-1695-1_19,

PPDM can be categorized as data hiding and rule hiding. In data hiding the database is modified in order to protect sensitive data of individuals. In rule hiding this modification is done to protect sensitive knowledge which can be mined from the database. In other words data hiding is related to input privacy while rule hiding is related to output privacy where frequent itemsets, association rules or classification rules are considered as outputs. Association rule or frequent itemset hiding is most popular method to provide output privacy.

There may be some situations where knowledge extracted by rule mining algorithms includes rules or itemsets that should stay unrevealed. These itemsets are called sensitive itemsets. Itemset hiding intends to modify database in such a way that sensitive itemsets are hidden with minimum side effects on non-sensitive ones. Sanitization of the database by placing false or unknown values is NP-Hard problem so heuristic approaches are needed where the idea is to reduce the support and confidence of sensitive itemsets [7–10].

Most of the itemset or rule hiding approaches is based on Apriori algorithm which needs multiple database scans and pre-mining of association rules. On the other hand FP-Growth algorithm, which has a better performance compared to Apriori, makes two database scans for finding frequent itemsets [11]. The work presented in [12] uses hiding algorithm based on P-tree [13] similar to FP-tree of FP-Growth algorithm. They sanitize informative rules and eliminate need for pre-mining of association rules. Another, frequent itemset mining algorithm with two database scans is Matrix-Apriori [14]. It is simpler than FP-Growth in terms of maintenance of the compact data structure and performs better in finding the rules or itemsets in that data structure [15].

Our initial idea came with the idea of using Matrix-Apriori algorithm in itemset hiding in order to benefit from its advantages in terms of limited database scan and easily maintained compact matrix structure. In our previous work presented in [16], four versions of the same itemset hiding algorithm are proposed where the Matrix-Apriori algorithm is modified to have itemset hiding capabilities. Each version uses different heuristics in selecting the transaction and the item in itemset to distort. Our algorithm (1) inputs sensitive itemsets, no matter whether they are frequent or not, which prevents privacy breech caused by pre-mining, (2) finds supports during hiding process and at the end returns sanitized database and frequent itemsets of this database as outputs eliminating the need of post-mining on sanitized database and (3) uses simple heuristics in itemset hiding avoiding heavy optimization cost. In this work we extended our study and analyzed the impact of integrating post-mining step in sanitization process.

In [16] a group of case study had been done to show the performance of four versions of our itemset hiding algorithm in terms of side effect, hiding time and distortion on initial database while changing the size of the original database, the number of sensitive itemsets and support of sensitive itemsets. Results showed that (1) spmaxFI (select shortest pattern and maximum of frequent items) has better overall performance both as side effect and runtime, (2) side effect is related to given sensitive itemset, (3) support count or database size is not directly related to the number of lost itemsets and (4) time to hide sensitive itemset is a function of distortion and database size and (5) distortion is related to support count. Now, we carried out another group of case study with the most effective version, spmaxFI, to see the impact of integrating post-mining in sanitization process on the performance and discovered that speed up can reach up to 10%.

The structure of the paper is as follows. Next section gives a short survey about association rule hiding and itemset hiding. In Sect. 3, in order to be self-contained itemset hiding process presented in [16] is revisited. Performance evaluation is given with discussion on results in Sect. 4. Then the paper is concluded with final remarks on the study and foreseen future work plan in Sect. 5.

2 Related Work

PPDM has been proposed as a solution to the problem of violating privacy while sharing data for knowledge extraction. The aim of PPDM is to develop algorithms to modify original data or mining techniques in such a way that useful knowledge can still be extracted while private data or knowledge is hidden. PPDM is mainly categorized as input and output privacy [17]. Input privacy is known as data hiding and output privacy is mostly known as rule or itemset hiding.

In data hiding, sensitive data is modified or trimmed out from the original database so that individual's private data will not be revealed by the data mining algorithm. Wide range of data hiding techniques can be grouped as perturbation based techniques [18–20], cryptographic techniques [21, 22] and anonymization based techniques [23–25].

In rule or itemset hiding, sensitive knowledge which can be mined from the database is hidden while non-sensitive knowledge can still be mined [26]. It refers to the process of modifying the original database in such a way that certain sensitive association rules or frequent itemsets disappear without seriously affecting the data and non-sensitive rules or itemsets. The detailed survey about association rule hiding given in [27] where the association rule hiding methods are classified as heuristic, border based and exact approaches. Exact approaches provide completely sanitized database with no side effect but their computational cost is high. In [28, 29] exact techniques which formulate sanitization as constraint satisfaction problem and solve these by integer programming are given. Border based approaches uses border theory [30]. In [31–33] border based techniques for association rule hiding are proposed. The idea behind these approaches is that the elements on the border are boundary to the infrequent itemsets. During hiding process, instead of considering non-sensitive frequent itemsets, they are focused on preserving the quality of the border. Heuristic approaches uses heuristics for modifications in the database. These techniques are efficient, scalable and fast algorithms however they do not give optimal solution and may have side effects. These techniques based on support and confidence decreasing [7,9].

Another group of algorithms for hiding rules or itemsets can be labeled as fake transactions [34–36]. The idea behind this is to anonymize real transactions by insertion of fake transaction. The approach is practical since any rule or itemset hiding algorithm can be applied on the dataset including fake transactions while maintaining the high theoretical privacy [36]. The work presented in [35] tries to balance the high memory requirement and privacy. A combination of fake transaction randomization method and a new per-transaction randomization method is proposed in [34].

Most of the association rule hiding algorithms are Apriori [37] based and needs multiple database scans to find support of sensitive itemsets because these techniques require data mining done prior to the hiding process. In [12] a tree structure which is similar to FP tree [11] is used to store information about database. This algorithm gets predictive item and sanitize informative rule set which is the smallest set of association rules that makes the same prediction as the entire rule set. The algorithm does not need data mining to be done before hiding process and does not scan database many times.

In the framework presented in [8], algorithms require two database scans; at first scan the inverted file index is created and at second scan items are deleted from selected transactions. In [10] blocking is used instead of distortion of items in the database. The idea behind this approach is that replacing false values may increase side effects on non-sensitive rules so the algorithms use unknown values to hide given sensitive rules.

Our survey on rule or itemset hiding research motivated us to propose a hiding approach which eliminates the need of pre-mining, avoids multiple scans of database and heavy computational cost during hiding process. We used matrix-apriori algorithm [14] as the basis of our sanitization framework. It works without candidate generation and scans database only twice. It has a simpler data structure and performs better compared to FP-growth in [15]. The sanitization process is embedded into the itemset mining process and at the end frequent itemsets of sanitized database is available. In the following section our algorithm is introduced.

3 Integrated Sanitization Framework for Itemset Hiding

As displayed in Fig. 19.1, our privacy preserving frequent itemset mining approach gets database D, sensitive itemsets Ls and minimum support minsup as input and returns sanitized database Ds with frequent itemsets which can be found from Ds as FIs. Sensitive itemsets are given without any knowledge about their frequency. If any itemset given as sensitive is frequent in original database then it is hidden through itemset hiding process. Most hiding approaches first do mining and calculate support of all frequent itemsets then start hiding process. This has two disadvantages (1) it

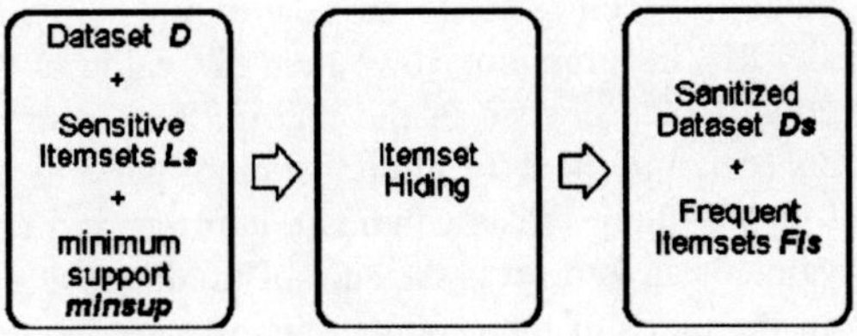

Fig. 19.1 Sanitization framework

```
INPUT: Original Database D, minimum support minsup, List of sensitive itemsets Ls

OUTPUT: Sanitized Database Ds, Frequent itemsets FIs of Ds

BEGIN
P1   Read D and find frequent items                     // first scan of database
P2   Read D and build MFI, STE and TList                // second scan of database
P3   Modify MFI

   FOR every itemset in Ls
       Calculate support of the sensitive itemset Is
       Number of iterations:=(Support of Is –minsup) * number of transactions in TList +1
       FOR 1 TO Number of iterations
           Select shortest pattern from MFI
           Select transaction from TList
           Select most frequent item of sensitive itemset in transaction
           Distort item in D
           Update MFI
           Update STE
           Update TList
       END
   END
   Find frequent itemsets FIs using up to date MFI
   Return Ds, FIs
END
```

Fig. 19.2 Itemset hiding algorithm

might cause a privacy breech if the one performing hiding process is not trusted because all frequent itemsets are required to be known before the hiding process and (2) it requires pre-mining causing decrease in efficiency. Our approach ensures that user does not know whether given sensitive itemset was frequent in original database because supports of sensitive itemsets are found during hiding process and eliminates the need for pre-mining process.

The overall process of itemset hiding algorithm is shown in Fig. 19.2. At first scan (P1), for the specified minimum support, frequent items are found. At second scan (P2), matrix MFI holding the frequent itemsets, vector STE holding the corresponding support counts of the itemsets in MFI and the TList holding the transaction ids of the rows of database D containing the itemsets in MFI is build. Columns of the matrix MFI show the frequent items; each row shows a different itemset. If the corresponding item is present in the itemset corresponding cell value is set to "1", "0" otherwise. After that step MFI is modified to speed up frequent pattern search (P3). For each column of MFI, beginning from the first row, the value of a cell is set to the row number in which the item is “1”. If there is not any “1” in remaining rows then the value of the cell is left as “1” which means down to the bottom of the matrix, there is no row that contains this item. After constructing the MFI matrix, finding patterns is simple. Beginning from the least frequent item, create candidate itemsets and count its support value. The support value of an itemset is the sum of the items at STE of which index are rows where all the items of the candidate itemset are included in MFI’s related row. Frequent itemset mining is done on this compact data structure which eliminates the need for database scan for itemset support counting. This part of matrix-apriori algorithm is modified to have itemset hiding capabilities (lines 1 to 15).

As explained above while building MFI and STE, we also construct a transaction list as TList which keeps the transaction ids of transactions containing the itemset in each row of MFI. In proposed approach, transaction selection for modifying is done on MFI and database scan in order to find transaction is eliminated.

Between lines 1 and 15 for every itemset in sensitive itemsets list Ls, hiding process is run. Support value for sensitive itemset is calculated using MFI and STE. If the support of the itemset is above minsup then the number of iterations to hide itemset is calculated (line 3). This number indicates number of distortions to be done on the dataset to reduce the support of the sensitive itemset Is below minsup. Following this, at each iteration transaction to modify is selected (lines 5 and 6). In [16] we used different heuristics in selecting the itemset and the item in itemset to distort and we discovered that selecting the shortest itemset and the most frequent item in that itemset causes minimum side effect. So in this work, we based our performance evaluation on spmaxMFI version. Selected item is distorted in transaction (line 8), the distortion technique is replacing “1” with “0” in related cell. Matrix structure MFI is updated after distortion (line 9). We decrease the value of related row in STE (line 10) and delete transaction modified in that row of TList (line 11). By this way it is ensured that we have compact mirror of semi-sanitized dataset in MFI, STE and TList throughout the hiding process.

The selection and distortion process is repeated until the support of sensitive itemset Is is below minsupport. After sanitization of a Is the next itemset is read from Ls and sanitized. At final step (line 15) frequent itemsets FIs of sanitized dataset Ds are found using up-to-date MFI and STE.

Now, let us explain an itemset hiding process using an example. Shortest pattern and most frequent item (spmaxFI) stragety is applied and itemset of BA is assumed to be sensitive (Is). In Fig. 19.3 sample database, MFI, STE and TList before hiding process is given. For minsupport value 3 (50%) 4 frequent itemsets (length 1 itemsets are not included) can be found. These are CB, CA, CBA and BA. But remember that our approach does not need frequent itemset mining to be performed before hiding process.

As in line 2 of the hiding algorithm, using MFI and STE support of BA is calculated to be 4 (66%). Since the minsupport value is 3 (50%), number of iterations to sanitize BA can be calculated as 2 (line 2). At first iteration shortest pattern that holds BA is found as third row of MFI and related transaction is T4 from TList. Most frequent item of sensitive itemset BA is A so it will be deleted from

Fig. 19.3 Database D, MFI, STE and TList before hiding process

TID	Items	MFI			STE	TIDs
		A	B	C		
T1	ABC	2	2	2		
T2	ABC	3	3	5	3	T1,T2,T3
T3	ABC	4	1	0	1	T4
T4	AB	1	0	0	1	T5
T5	AD	0	0	1	1	T6
T6	CD					

Fig. 19.4 Sanitized database Ds, MFI, STE and TList after hiding process

TID	Items
T1	ABC
T2	ABC
T3	BC
T4	B
T5	AD
T6	CD

MFI			STE	TIDs
A	B	C		
2	2	2		
3	3	5	2	T1,T2
4	6	0	0	
1	0	0	1	T5
0	0	7	1	T6
0	7	0	1	T4
0	1	1	1	T3

selected transaction (Fig. 19.4). Meanwhile STE value of selected row is decreased and modified transaction id is deleted from the list. After deletion the new pattern B is added to matrix and T4 is added to transaction list which is now the sixth row of the matrix. At second iteration second row is selected as shortest and T3 is selected for modification. In Fig. 19.4 sanitized database Ds, MFI, STE and TList after sanitization process are shown.

After sanitization process we are able to find frequent itemsets for sanitized database using up-to-date matrix structure. Support values of itemsets are calculated as CB(50%), CA(33%), CBA(33%) and BA(33%). Support of itemset BA is now under minsupport and it is hidden. CBA is also hidden because it is a superset of BA. However, CA is now under minimum support and cannot be find as frequent although it was not sensitive. This is the side effect and CA is called lost itemset.

4 Performance Evaluation

In this section, performance evaluation of our itemset hiding algorithm is given. Firstly, the performance of algorithm in two different cases is given. Secondly, impact of integrating post-mining on runtime is observed. Two synthetic databases are used to see effect of different database size. The algorithm is executed on databases (1) to see effect of increasing number of sensitive itemsets, (2) to see effect of increasing support of sensitive itemset. The effects observed are number of lost itemsets as side effect, runtime for hiding process and number of items distorted for hiding itemsets.

Table 19.1 Sensitive itemsets for case 1

Itemset No	Itemsets for 5 k database	Support (%)	Itemsets for 10 k database	Support (%)
1	37 31 32	2.96	36 20 6	3.00
2	7 47 41	3.06	50 13 10	3.01
3	5 6 4	2.92	33 49 42	2.93
4	24 13 46	3.08	29 14 11	3.07
5	45 34 20	2.94	39 41 18	2.95

4.1 *Simulation Environment*

Test runs are performed on a computer with 2.4 GHz dual core processor and 3 GB memory. During evaluations, it is ensured that the system state is similar for all test runs and results are checked for consistency.

When hiding is applied inputs are original database and sensitive itemsets where the outputs are sanitized database and frequent itemsets which can be mined from this sanitized database. Two synthetic databases generated by ARtool [38] are used in evaluations. One database has 5,000 transactions while number of items is 50 and average length of transactions is 5. Other database has 10,000 transactions while number of items is 50 and average length of transactions is 5. Minimum support is defined as 2.5% for all evaluations and if no hiding is applied then 2,714 frequent itemsets from 5 k database and 5,527 frequent itemsets from 10 k database can be found.

4.2 *Increasing Number of Sensitive Itemsets*

For both databases five of length three itemsets which are closest to 3.0% support are selected as sensitive itemsets. These itemsets are given in Table 19.1. Selected itemsets are mutual exclusive to ensure that one is not affected by hiding process of previous itemsets. The aim of this study is to understand the effect of increasing the number of sensitive itemsets on itemset hiding. For each run next itemset in the table is added to the sensitive itemsets given to program. At first run itemset no 1 is given as sensitive, at second run itemset no 1 and itemset no 2 are given as sensitive and so on.

The side effect and time to hide is given in Fig. 19.5. In both databases number of lost itemsets is increased while number of sensitive itemsets is increased. What more can be inferred from these figure is that side effect is related to the characteristics of sensitive itemsets, not to the database size. For instance, we come across higher number of lost itemsets for 5 k database(29 itemsets) compared to 10 k database(22 itemsets) at 5 itemset hiding point.

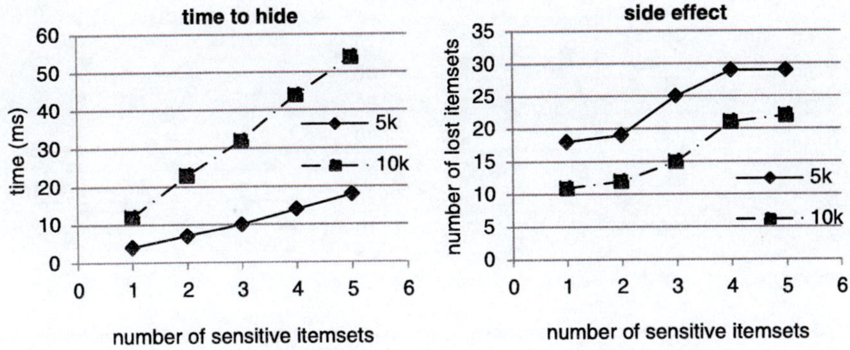

Fig. 19.5 Side effect and time to hide while increasing number of sensitive itemsets

Table 19.2 Sensitive itemsets for case 2

Itemset No	Itemsets for 5 k database	Support (%)	Itemsets for 10 k database	Support (%)
1	37 31 32	2.96%	36 20 6	3.00%
2	18 28 47	3.50%	4 49 42	3.54%
3	14 17 24	4.00%	9 8 3	4.23%
4	28 47 4	4.50%	7 33 18	4.47%
5	46 20 4	5.00%	24 39 13	5.03%

When we came to time cost of hiding we give pure hiding cost excluding the reading database and building up matrix to make observation simple. These are identical for the same databases. It is clear from the figure that database size effects time to hide itemsets for same cases. While the database size increases, time needed for hiding itemsets increases. The reason behind this is the cost of travelling on matrix to select pattern. It is clear that matrix size is bigger for 10 k database compared to 5 k database.

4.3 *Increasing Support of Sensitive Itemset*

For both databases five of length three itemsets which have increasing support values between 3.0% and 5.0% are selected as sensitive itemsets. These itemsets are given in Table 19.2. The aim of this study is to understand the effect of increasing the support value of sensitive itemsets on itemset hiding. For each run next itemset in the table is selected as the sensitive itemsets given to program. At first run itemset no 1 is given as sensitive, at second run itemset no 2 is given as sensitive and so on.

The side effect and time cost of increasing support value for sensitive itemset is given in Fig. 19.6. The statement "side effect is related to characteristics of selected itemsets" which was written in the first case study is approved in this study. For

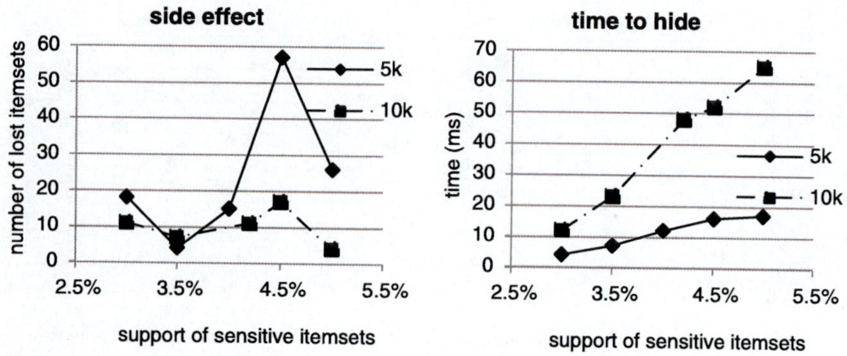

Fig. 19.6 Side effect and time to hide while increasing support of sensitive itemset

example, in the 5 k database for itemset no 1 the number of lost itemsets is 18 however, for itemset no 1 the number of lost itemsets is 4. Although the support is increased number of lost itemsets is decreased.

It is clear from the figure that database size effects time to hide itemsets for same cases. Like it was in first case study only pure hiding process is considered.

The number of distortions is related to support count of sensitive itemsets as it was stated in previous part and so we will have increasing number of distorted items with increasing support.

4.4 Integrating Post-mining into Itemset Hiding

The time efficiency of integrated post-mining is observed for both two case studies above. Firstly, the program is executed without post-mining. This includes reading database, building matrix and pure hiding process. Secondly, the sanitized database is given to the pure matrix-apriori for frequent itemset mining. The running time of these two steps are added and time cost without integration is calculated. Afterwards, the program is executed with integrated post-mining. The difference between first calculation and integrated approach gives the gain. In Fig. 19.7 the speed up by integrating post-mining into itemset hiding is given for two cases mentioned in previous parts.

The evaluations reveal that integrating frequent itemset mining into hiding is always faster choice compared to do hiding and mining separated. For both cases on 5 k database the speed up of integrated approach is at least 2.1%, at most 6.9% and in average 5.4%. The gain increases when database size is increased. For 10 k database integrated approach is 9.2% faster than separated approach in average. The speed up is at least 7.8% and at most 10.3%.

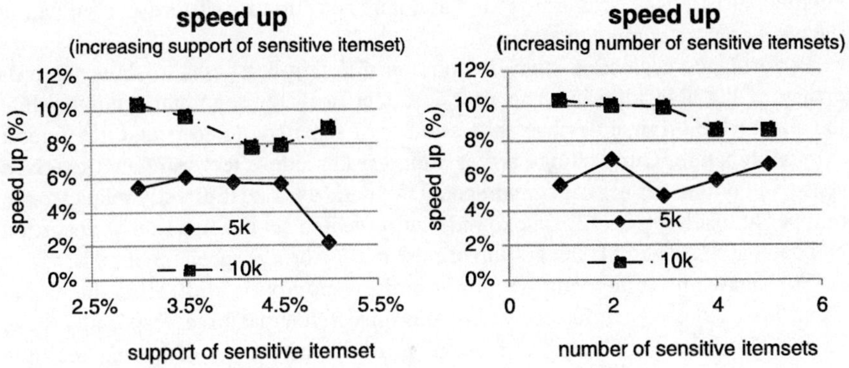

Fig. 19.7 The speed up of integrated post-mining

4.5 Discussion on Results

In this section, we analyzed effects of itemset hiding algorithm on number of lost itemsets, time for hiding process and number of distortions needed for hiding itemsets and time efficiency on post-mining. We used two different databases to understand the effect of database size and two different set of sensitive itemsets to understand the effects of number and support of sensitive itemsets. One of the results from these studies is that side effect is related to characteristics of selected sensitive itemsets because subsets or supersets of that itemset are affected too. Other important result is that integrating post-mining approach always performs faster than separated approach. The difference between 5 k and 10 k databases (5.4% and 9.2%) show that with bigger databases efficiency increases. Larger databases are likely to have more variety of patterns so larger matrix structures. Time for reading database and building up matrix are increased when database size is increased. Building this structure takes more time. The main aim of integrated approach is to eliminate time loss of reading database and rebuilding matrix structure. Our evaluations proved that the integrated approach is always faster than separated approach.

5 Conclusion and Future Work

In this paper we introduced an integrated approach for frequent itemset hiding. Main strengths of the algorithm are (1) it works without pre-mining so privacy breech caused by the knowledge obtained by finding frequent itemsets in advance is prevented, (2) efficiency is increased since no pre-mining is required, (3) supports are found during hiding process and at the end sanitized database and frequent itemsets of this database are given as outputs so no post-mining is required, (4) simple heuristic is used in transaction and item selection for distortion eliminating the need of extra computational cost.

Performance evaluation study is done on different databases to show the efficiency of the algorithm in terms of side effect (lost itemsets) and runtime while the size of the original database, the number of sensitive itemsets and the itemset supports change. Our findings are as follows: (1) side effect is related to given sensitive itemset, (2) neither support count nor database size is directly related to the number of lost itemsets, (3) time to hide sensitive itemset is a function of distortion and database size and (4) integration of post-mining brings speed up of 10%.

Our next aim is to compare our promising approach with different hiding algorithms and carry out further evaluations on different databases, especially those having bigger average transaction lengths, to see the impact of having multiple sensitive itemsets in a single transaction on distortion. Finally, we want to adapt this practical itemset hiding algorithm in dynamic environments allowing incremental itemset hiding.

References

1. Dunham M (2002) Data mining: introductory and advanced topics. Prentice Hall PTR Upper Saddle River, NJ, USA
2. Zhang N, Zhao W (2007) Privacy-preserving data mining systems. Computer 40(2):52–58
3. Grossman R, Kasif S, Moore R, Rocke D, Ullman J Data mining research: opportunities and challenges [Online]. http://pubs.rgrossman.com/dl/misc-001.pdf. Accessed on May 31, 2010
4. Yang Q, Wu X (2006) 10 challenging problems in data mining research. Int J Inform Technol Decision Making 5(4):597–604
5. Kantardzic M (2002) Data mining: concepts, models, methods, and algorithms. John Wiley & Sons, Inc. New York, NY, USA
6. Han J, Kamber M (2005) Data mining: concepts and techniques. Morgan Kaufman, Publishers Inc. San Francisco, CA, USA
7. Atallah M, Bertino E, Elmagarmid A, Ibrahim M, Verykios V (1999) Disclosure limitation of sensitive rules. In: Proceedings of 1999 workshop on knowledge and data engineering exchange, KDEX '99, Chicago, 7 November 1999
8. Oliveira S, Zaiane O (2002) Privacy preserving frequent itemset mining, Proceedings of 2nd IEEE international conference on data mining, ICDM'02, Maebashi City, 9–12 December 2002
9. Verykios V, Elmagarmid A, Bertino E, Saygin Y, Dasseni E (2004) Association rule hiding. IEEE Trans Knowledge Data Eng 16(4):434–447
10. Saygin Y, Verykios V, Clifton C (2001) Using unknowns to prevent discovery of association rules. ACM SIGMOD Records 30(4):45–54
11. Han J, Pei J, Yin Y (2000) Mining frequent patterns without candidate generation. ACM SIGMOD Records 29(2):1–12
12. Wang S, Maskey R, Jafari A, Hong T (2008) Efficient sanitization of informative association rules. Exp Syst Appl 35(1–2):442–450
13. Huang H, Wu X, Relue R (2002) Association analysis with one scan of databases. Proceedings second IEEE international conference on data mining, ICDM'02, Maebashi City, 9–12 December 2002, pp 629–632
14. Pavon J, Viana S, Gomez S (2006) Matrix apriori: speeding up the search for frequent patterns. Proceedings 24th IASTED international conference on databases and applications, DBA 2006, Innsbruck, 14–16 February 2006, pp 75–82

15. Yıldız B, Ergenç B (2010) Comparison of two association rule mining algorithms without candidate generation. In: Proceedings 10th IASTED international conference on artificial intelligence and applications, AIA 2010, Innsbruck, 15–17 February 2010, pp 450–457
16. Yıldız B, Ergenç B (2011) Hiding sensitive predictive frequent itemsets, lecture notes in engineering and computer science. In: Proceedings of the international multiconference of engineers and computer scientists 2011, IMECS 2011, Hong Kong, 16–18 March 2011, pp 339–345
17. Ahluwalia M, Gangopadhyay A (2008) Privacy preserving data mining: taxonomy of existing techniques. In: Subramanian R (ed) Computer security, privacy and politics: current issues, challenges and solutions. IRM, New York, pp 70–93
18. Agrawal D, Aggarwal C (2001) On the design and quantification of privacy preserving data mining algorithms. Proceedings 20th ACM SIGMOD SIGACT-SIGART symposium on principles of database systems, PODS'01, CA, 21–24 May 2001, pp 247–255
19. Agrawal R, Srikant R (2000) Privacy-preserving data mining. ACM SIGMOD Records 29(2):439–450
20. Liu L, Kantarcioglu M, Thuraisingham B (2008) The applicability of the perturbation based privacy preserving data mining for real-world data. Data Knowledge Eng 65(1):5–21
21. Lindell Y, Pinkas B (2002) Privacy preserving data mining. J Crytol 15(3):177–206
22. Pinkas B (2006) Cryptographic techniques for privacy-preserving data mining. ACM SIGKDD Explorations Newslett 4(2):12–19
23. Bayardo R, Agrawal R (2005) Data privacy through optimal k-anonymization. Proceedings of 21st international conference on data engineering, ICDE'05, Tokyo, 5–8 April 2005, pp 217–228
24. Sweeney L (2002) Achieving k-anonymity privacy protection using generalization and suppression. Int J Uncertain, Fuzziness Knowledge-Based Syst 10(5):571–588
25. Brickell J, Shmatikov V (2008) The cost of privacy: destriction of data-mining utility in anonymized data publishing. Proceedings of 14th ACM SIGKDD international conference on knowledge discovery and data mining, KDD'08, Las Vegas, 24–27 August 2008, pp 70–78
26. Verykios V, Bertino E, Fovino I, Provenza L, Saygin Y, Theodoridis Y (2004) State-of-the-art in privacy preserving data mining. ACM SIGMOD Records 33(1):50–57
27. Verykios V, Gkoulalas-Divanis A (2008) A survey of association rule hiding methods for privacy. In: Aggarwal C, Yu P (ed) Privacy-preserving data mining: models and algorithms. Springer, New York, pp 267–289
28. Gkoulalas-Divanis A, Verykios V (2006) An integer programming approach for frequent itemset hiding. Proceedings of 15th ACM international conference on Information and knowledge management, CIKM'06, Virginia, 5–11 November 2006, pp 748–757
29. Gkolalas-Divanis A, Verykios V (2008) Exact knowledge hiding through database extension. IEEE Trans Knowledge Data Eng 21(5):699–713
30. Mannila H, Toivonen H (1997) Levelwise search and borders of theories in knowledge discovery. Data Mining Knowledge Discov 1(3):241–258
31. Sun X, Yu P (2005) A border-based approach for hiding sensitive frequent itemsets. Proceedings of 5th IEEE international conference on data mining, ICDM'05, Houston, 27–30 November 2005, pp 426–433
32. Sun X, Yu P (2007) Hiding sensitive frequent itemsets by a border-based approach. J Comput Sci Eng 1(1):74–94
33. Mousakides G, Verykios V (2008) A max min approach for hiding frequent itemsets. Data Knowledge Eng 65(1):75–89
34. Boora RK, Shukla R, Misra AK (2009) An improved approach to high level privacy preserving itemset mining. Int J Comput Sci Inform Security 6(3):216–223
35. Mohaisen A, Jho N, Hong D, Nyang D (2010) Privacy preserving association rule mining revisited: privacy enchancement and resource efficiency. IEICE Trans Inform Syst E93(2): 315–325

36. Lin JL, Liu JYC (2007) Privacy preserving itemset mining through fake transactions. Proceedings of 22nd ACM symposium on applied computing, SAC 2007, Seoul, 11–15 March 2007, pp 375–379
37. Agrawal R, Srikant R (1994) Fast algorithms for mining association rules in large databases. Proceedings of 20th international conference on very large data bases, VLDB'94, Santiago de Chile, 12–15 September 1994, pp 487–499
38. Cristofor L artool project [Online]. http://www.cs.umb.edu/~laur/ARtool/. Accessed on May 13, 2010

Chapter 20
Mining General Fuzzy Sequences Based on Fuzzy Ontology

Mehdi Gholizadeh, Mir Mohsen Pedram, and Jamshid Shanbezadeh

Abstract Sequence mining, a branch of data mining, is recently an important research area, which recognizes subsequences repeated in a temporal database. Fuzzy sequence mining can express the problem as quality form that leads to more desirable results. Sequence mining algorithms focus on the items with support higher than a specified threshold. Considering items with similar mental concepts lead to general and more compact sequences in database which might be indistinguishable in situations where the support of individual items are less than threshold. This paper proposes an algorithm to find sequences with more general concepts by considering mental similarity between items by the use of fuzzy ontology.

Keywords Sequence mining • Subsequence • Similarity • Mental concept • Fuzzy ontology

1 Introduction

Sequential data is an important type of data with a wide range of applications in science, medicine, security and commercial activities. The sequential data is a set of sequences or sub-structures in a data set that repeats more than or equal to a minimum support. DNA sequence is an example that encodes the generic

---

M. Gholizadeh (✉)
Computer Engineering Department, Islamic Azad University Science and Research Branch, Tehran, Iran
e-mail: mhdgholizadeh@gmail.com

M.M. Pedram • J. Shanbezadeh
Computer Engineering Department, Faculty of Engineering, Tarbiat Moallem University, Karaj/Tehran, Iran
e-mail: pedram@tmu.ac.ir; shanbehzadeh@gmail.com

S.I. Ao et al. (eds.), *Intelligent Control and Innovative Computing*, Lecture Notes in Electrical Engineering 110, DOI 10.1007/978-1-4614-1695-1_20,

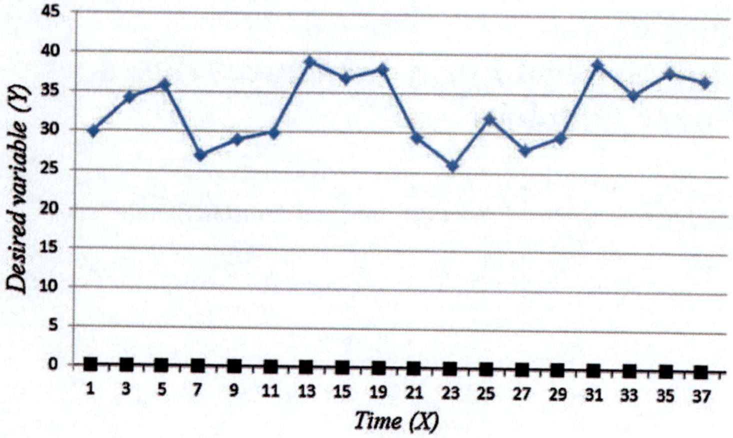

Fig. 20.1 A diagram of time series

makeup of humans and all other species; and protein sequence that expresses the information and functions of proteins. Besides, the sequential data is able to describe the individual human behavior such as the history of customers' purchases in a store. There are various procedures to extract data and patterns out of data sets such as time series analyzing, association rules mining, and, sequence mining. Time series is defined as a set of stochastic data gathered within a regular fixed time intervals and, time series analyzing refers to stochastic methods that operate on such data [1–3]. The diagram of time series could be figured by setting the horizontal and vertical axis representing the time and the desired variable respectively. Figure 20.1 demonstrates the general form of a time series diagram.

The association rule based mining algorithms try to find the dependencies and relations between various data in a database. These algorithms consist of two stages. The first one finds a set of highly repeated items and, the second one extracts some suitable rules from the highly repeated collections. The highly repeated items are collected by methods like the apriori algorithm based on the number of repetitions [4–7]. Then, the algorithm generates the data and patterns by using the collected items. Sequence mining identifies the repeated sub-sequences in a set of sequential data. The input data in sequence mining is comprised of a list of transactions and their occurrence time. Moreover, each transaction includes a set of items. Sequential patterns are also a set of sequentially happened items. The main purpose of sequence mining is to search and find all the sequential patterns with support values greater than or equal to a minimum support threshold [8–11]. Figure 20.2 shows the classification of frequent pattern mining studies.

There are several efforts to enhance sequence mining algorithm. Zaki introduced an efficient approach called SPADE. The algorithm is able to find repeated sequences effectively [13]. Yu and Chen proposed a method to mine sequences with multi-dimensional information [14]. Moreover, Zhao and Bhowmick suggested a

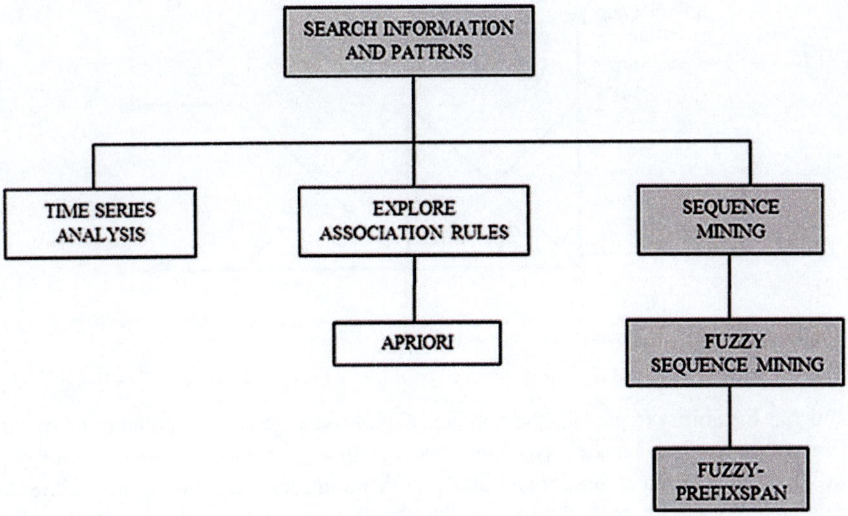

Fig. 20.2 Frequent pattern mining studies [12]

survey about sequential pattern mining [15]. In addition, Nancy and others offered a new method for finding fuzzy sequential pattern mining in negative form [16]. This paper presents a novel sequence mining algorithm that utilizes a common item set to describe the similar mental concepts. Therefore, we can find more effect and general sequences with higher support values. The organization of the rest of this paper is as follows. Section 2 defines the sequence mining by employing an example and introduces two useful sequence mining algorithms. Section 3 presents fuzzy prefixspan sequence mining algorithm. Section 4 introduces the algorithm of sequence mining based on similar mental concepts and investigates the proposed algorithm. Section 5 shows the numerical experiments and discussion and, Sect. 6 is the conclusion.

2 Fuzzy Sequence Mining

Sequence mining tries to find the sequential patterns with support values greater than or equal to a minimum support threshold (declared by the user). The following sentence is an example of sequential patterns: "Customers who have purchased a printer, are reasonably probable to purchase printer ink, too". In this example, the purchase of printer and printer ink can present a sequence.

Classic sequence mining algorithms show sequences like <printer, printer ink>. We can see there is no information about the amount of purchased items in this sequence and, essential information is ignored. There are two solutions; crisp or fuzzy sequence mining algorithms. The first one can mine the repeated sequences

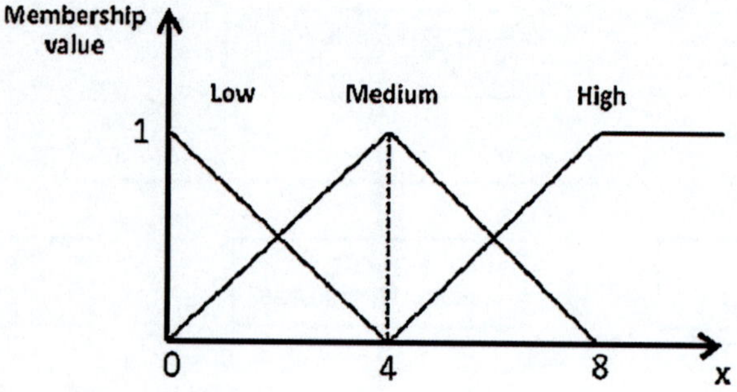

Fig. 20.3 The fuzzy membership functions [12]

and, has the ability to provide the number of items occurred in the sequences. So, the form of output will be such as <Item_1 : value_1 Item_2 : value_2 ... Item_n : value_n>, in which each item is mentioned along with its amount, for example <printer:2 printer ink:5>. The second case has the ability to provide the fuzzy term of the number/amount of items occurred in the sequences. Therefore, the algorithm has an output like <printer: *low*, printer ink: *medium*>.

The first method shows the number of each item but, the major problem is the severe decrease in the sequences support values compared to the classic sequence mining. In fact, to find the support value, these algorithms must consider both the number of items' occurrence and their values. This will decrease the support value. For example, to find sequences with support threshold equal to 2, in the classic sequence mining method it is just sufficient to consider the item at least two times; but in crisp sequence mining, the item must occur at least two times with the value of 2 for printer and 5 for printer ink.

Fuzzy sequence mining expresses items' repetition in Fuzzy linguistic terms. This method introduces a criterion to determine the number of each items' occurrence and moreover, somewhat moderates the problem of the first method because, in this case the supported value of the sequences increase, due to be fuzzy terms. Figure 20.2 shows that we need fuzzy perfixspan to discover the desired sequences. The perfixspan algorithm [17], the fastest one among sequence mining algorithms [18], has been used as the base algorithm in this paper. This paper utilizes the fuzzy membership functions shown in Fig. 20.3 to fuzzify the crisp values.

3 Fuzzy PrefixSpan Algorithm [12]

At first we have to introduce the concepts of *prefix, suffix* and *projected database* that are the basis and essential terms in fuzzy prefixspan algorithm. Then fuzzy prefixspan algorithm will be presented.

3.1 Prefix

Suppose that all the items within an element are listed alphabetically. For a given sequence α, where $\alpha = < \mathrm{p}_1 : \mathrm{k}_1\ \mathrm{p}_2 : \mathrm{k}_2 \ldots \mathrm{p}_n : \mathrm{k}_n >$, each $\mathrm{p}_i : \mathrm{k}_i (1 \leq i \leq n)$ is an element. A sequence $\beta = < \mathrm{p}'_1 : \mathrm{k}'_1\ \mathrm{p}'_2 : \mathrm{k}'_2 \ldots \mathrm{p}'_m : \mathrm{k}'_m > (m \leq n)$ is called a prefix of α if (1) $\mathrm{p}'_i : \mathrm{k}'_i = \mathrm{p}_i : \mathrm{k}_i$ for $i \leq m-1$; (2) $\mathrm{p}'_m : \mathrm{k}'_m \subseteq \mathrm{p}_m : \mathrm{k}_m$; and (3) all items in $(\mathrm{p}_m : \mathrm{k}_m - \mathrm{p}'_m : \mathrm{k}'_m)$ are alphabetically after those in $\mathrm{p}'_m : \mathrm{k}'_m$.

For example, consider sequence s =<(a:*low*)(a:*low* b:*medium* c:*medium*)(a:*high* c:*high* d:*low*)>. Either sequences <(a:*low*)(a:*low* b:*medium*)> or <(a:*low* b:*medium* c:*medium*)> are prefixes of s, but neither <(b:*medium* a:*high*)> nor <(a:low a:*low*)> is a prefix.

3.2 Suffix

Consider a sequence $\alpha = < \mathrm{p}_1 : \mathrm{k}_1\ \mathrm{p}_2 : \mathrm{k}_2 \ldots \mathrm{p}_n : \mathrm{k}_n >$ and each $\mathrm{p}_i : \mathrm{k}_i (1 \leq i \leq n)$ is an element. Let $\beta = < \mathrm{p}'_1 : \mathrm{k}'_1\ \mathrm{p}'_2 : \mathrm{k}'_2 \ldots \mathrm{p}'_m : \mathrm{k}'_m > (m \leq \mathrm{n})$ be a subsequence of α. Sequence $\gamma = < \mathrm{p}''_1 : \mathrm{k}''_1\ \mathrm{p}_{1+1} : \mathrm{k}_{1+1} \ldots \mathrm{p}_n : \mathrm{k}_n >$ is the suffix of α with respect to prefix β, denoted as $\gamma = \alpha / \beta$, if is the suffix of α with respect to prefix β, denoted as $\gamma = \alpha / \beta$, if

1. $1 = i_m$ such that there exist $1 \leq i_1 \leq \ldots \leq i_m$ such that there exist $\mathrm{p}'_j : \mathrm{k}'_j \subseteq \mathrm{p}_{ij} : \mathrm{k}_{ij} (1 \leq j \leq m)$ and i_m is minimized. In other words, $\mathrm{p}_1 : \mathrm{k}_1 \ldots \mathrm{p}_1 : \mathrm{k}_1$ is the shortest prefix of α which contains $\mathrm{p}'_1 : \mathrm{k}'_1\ \mathrm{p}'_2 : \mathrm{k}'_2 \ldots \mathrm{p}'_{m-1} : \mathrm{k}'_{m-1}\ \mathrm{p}'_m : \mathrm{k}'_m$ as a subsequence; and
2. $\mathrm{P}''_1 : \mathrm{k}''_1$ is the set of items in $\mathrm{p}_1 : \mathrm{k}_1 - \mathrm{p}'_m : \mathrm{k}'_m$ that are alphabetically after all items in $\mathrm{p}'_m : \mathrm{k}'_m$.

If $\mathrm{P}''_1 : \mathrm{k}''_1$ is not empty, the suffix is also denoted as (-items in $\mathrm{P}''_1 : \mathrm{k}''_1$) $\mathrm{p}_{1+1} : \mathrm{k}_{1+1} \ldots \mathrm{p}_n : \mathrm{k}_n$. Note that if β is not a subsequence of α, the suffix of α with respect to β is empty.

For example, consider sequence s=<(a:*low*)(a:*low* b:*medium* c:*medium*) (a:*medium* c:*high*)(d:*high*)(c:*low* f:*low*)> and <(a:*low* b:*medium* c:*medium*) (a:*medium* c:*high*) (d:*high*)(c:*low* f:*low*)> is the suffix with respect to (a:*low*), and <(c:*medium*) (a:*medium* c:*high*)(d:*high*)(c:*low* f:*low*)> is the suffix with respect to (a:*low*)(b:*medium*) and <(a:*medium* c:*high*)(d:*high*)(c:*low* f:*low*)> is the suffix with respect to (a:*low*)(a:*low* c:*medium*).

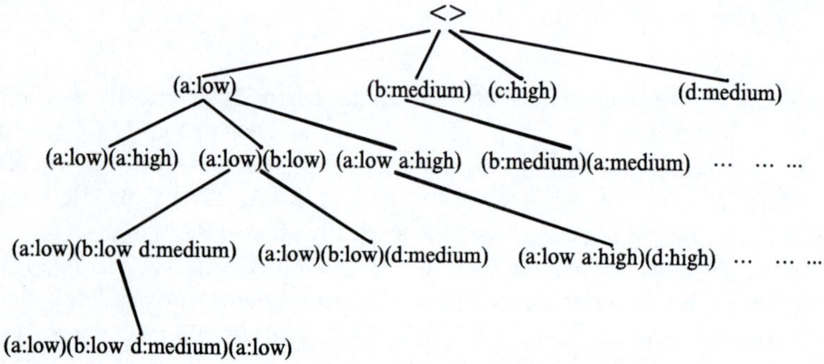

Fig. 20.4 The fuzzy sequence enumeration tree on the set of items {a, b, c, d, d} [12]

3.3 Projected Database

Let α be a fuzzy sequential pattern in a fuzzy sequence database S. The fuzzy α-projected database, denoted as $S|\alpha$, is the collection of suffixes of sequences in S with respect to prefix α.

So, an enumeration tree will emerge such as you see in Fig. 20.4. Then, the depth of the first search method is applied to the enumeration tree. In Fig. 20.4, sub-trees related to each node indicate all sequence patterns which are prefixes of the node. This tree is called as sequence enumeration tree.

So, based on the above discussion, we present the algorithm of fuzzy prefixspan as follows.

Input: A sequence database S, and the minimum support threshold min_support.
Output: The complete set of fuzzy sequential patterns.
Method: Call fuzzy prefixspan(Ø, 0, S).
Subroutine fuzzy prefixspan(α, l)

The parameters are:

- **α is a fuzzy sequential pattern;**
- l is the i-length of α; and
- Is the fuzzy α-projected database if $\alpha \neq$ Ø, otherwise, it is the sequence database S.

Method:
Scan once, find each fuzzy frequent item as (b:k) that leads to face with two states:

- b can assembled to the last element of α as at different times to form a sequential pattern like (α)(b:k);
- b can be append to α as simultaneous to form a sequential pattern like (α b:k).

For each fuzzy frequent item (b:k), append it to α to form a sequential pattern α', and output α';
For each α', construct fuzzy α'-projected database $_{\alpha'}$, and call the fuzzy prefixspan(α', $l + 1$, $s|\alpha'$).

4 Fuzzy Sequence Mining for Similar Mental Concept

Sequence mining algorithms often work in binary form. In other words, an item is in a desired sequence if its repetition is more than a minimum support. This definition ignores the inter-items' mental similarity. If we use these similarities, we can achieve more general sequences. In other words, we have to consider the items' repetition and, their mental similarity to gather them in one group and put them under a general concept. This case, rather than studying the items one by one, we calculate the repetition of the general concepts. This results in sequences with upper support and besides, more general sequences. For this purpose, in addition to the data collection that shows the transactions, there should be another collection which represents the items similarities.

Ontology can be used to show the similar mental concepts. Ontology is a method to represent knowledge in an understandable format for both human and machine and provides the ability to share the information between different programs. All the concepts in the desired range, associated with their hierarchical structure and the existing relations between concepts are defined in ontology. In fuzzy ontology, we can also model and represent the uncertainty of the real world [19].

The proposed algorithm receives two sets as inputs. The first one is a collection including identification number, time, number of items and the items' repetition. The second data set describes the similarity between each item and each general concept by a membership function, i.e., the fuzzy ontology database. The first dataset is transformed into a new dataset in which items are substituted with general concepts described by the fuzzy ontology; then, the fuzzy prefixspan algorithm is employed on the new dataset and the final results are sequences with more general concepts.

4.1 *Nomenclature*

A_i: ith general concept,
a_j: jth item which has mental similarity with the ith general concept,
C_k: Identification number,
t_m: Transaction date,
$n_{aj}(t_m)$: Number of item a_j in the transaction with date t_m,
$Similarity(A_i, a_j)$: The measure describing the similarity of item a_j and the general concept A_i,

Count(A_i, t_m, C_k): Number of times that concept A_i occurred by the identification number C_k at the time t_m,

Fuzzified(n): The fuzzified term for n,

Fuzzy-Count(A_i, t_m, C_k): Fuzzy value of times that concept A_i occurred by the identification number C_k at the time t_m.

4.2 Algorithm

a. Inputs

I. The dataset including identification number, time, items and the number of items happening.

II. The dataset containing a list of similar mental concepts by which their similarity is determined via fuzzy ontology.

b. Outputs

General sequences that indicate the items regularity and priority.

c. Steps

I. Receive the first and second datasets and build the new one as follows:

1. The identification number (C_k) and the transaction date (t_m) get no change,
2. The items a_j are replaced with the concepts A_i,
3. The number of occurrence of the concept A_i in fuzzy form is calculated as:

$$\text{Count}(A_i, t_m, C_k) = \text{Count}(A_i, t_m, C_k) + \text{Similarity}(A_i, a_j) \times n_{aj}(t_m) \tag{20.1}$$

$$\text{Fuzzy} - \text{Count}(A_i, t_m, C_k) = \text{Fuzzified}(\text{Count}(A_i, t_m, C_k)) \tag{20.2}$$

II. Use the fuzzy prefixspan algorithm for the new dataset.

III. Return the mined general sequences in step 2.

IV. End.

5 Illustrated Example

As an example, consider the transactional dataset shown in Table 20.1. The table includes transactions of two customers, purchased items, date and number of each purchased item. Table 20.2 shows the fuzzy ontology, in which general concepts as well as items are shown. In fact, the similarity degree for item a_j and general

Table 20.1 Transactions of some customers [12]

Customer identification number	Purchase time	Product	Number
100100002	95/07/22	Tea	1
100100002	95/07/22	Cream	3
100100003	95/07/23	Butter	5
100100003	95/07/23	Coffee	1
100100002	95/07/27	Fruit juice	6
100100003	95/07/29	Fruit juice	2

Table 20.2 An instant of items with similar mental concept [12]

Product	Hot drink	Fat dairy
Tea	1	0
Coffee	1	0
Cream	0	0.9
Butter	0	0.9
Fruit juice	0.1	0

Table 20.3 Transactions of the customers with similar mental concept [12]

Customer identification number	Purchase time	Product	Number	Fuzzy values		
				Low	Medium	High
100100002	95/07/22	Hot drink	1	0.75	0.25	–
100100002	95/07/22	Fat dairy	2.7	0.32	0.68	–
100100003	95/07/23	Fat dairy	4.5	–	0.78	0.13
100100003	95/07/23	Hot drink	1	0.75	0.25	–
100100002	95/07/27	Hot drink	0.6	0.85	0.15	–
100100003	95/07/29	Hot drink	0.2	0.95	0.05	–

concept A_i is shown by the table. As you see there are two general concepts named *Hot drink* and *Fat dairy*. For example similar mental concept between *Fat dairy* and *Cream* is 0.9.

The original data set (Table 20.2) is transformed into Table 20.3 by (20.2), in which general concepts are used. Table 20.3 shows that more general transactions can be mined. This table, unlike Table 20.3, uses more general concepts such as hot drink and fat dairy.

Making Table 20.3 is the most important phase in our proposed algorithm. So, we will describe it in more details. The table has comprised 5 major columns. Based on the steps of the propose algorithm, the customer identification number and time have no changed and are same as Table 20.1. Product is replaced by the mental concepts. Also number is calculated by formula 1. At last fuzzy values are calculated by formula 2 and fuzzy membership function shown in Fig. 20.3.

Fuzzy prefixspan algorithm has been applied on the dataset in Table 20.3 without considering special minimum support (it means, all the sequences with no threshold in repetition will have been produced). Table 20.4 lists the results and shows each item with its fuzzy values. Indeed the table made by applying Fuzzy prefixspan

Table 20.4 Output sequences found by proposed fuzzy method [12]

Sequences	Support	Sequences	Support
<(Hot drink: *Low*)>	1.8	<(Fat dairy: *Low*) (Hot drink: *Medium*)>	0.15
<(Hot drink: *Medium*)>	0.5	<(Fat dairy: *Medium*) (Hot drink: *Low*)>	0.68
<(Fat dairy: *Low*)>	1	<(Fat dairy: *Medium*) (Hot drink: *Medium*)>	0.15
<(Fat dairy: *Medium*)>	0.32	<(Fat dairy: *High*) (Hot drink: *Low*)>	0.13
<(Fat dairy: *High*)>	1.46	<(Fat dairy: *High*) (Hot drink: *Medium*)>	0.05
<(Hot drink: *Low* Fat dairy: *Low*)>	0.13	<(Hot drink: *Low* Fat dairy: *Low*) (Hot drink: *Medium*)>	0.15
<(Hot drink: *Low* Fat dairy: *Medium*)>	0.32	<(Hot drink: *Low* Fat dairy: *Medium*) (Hot drink: *Medium*)>	0.15
<(Hot drink: *Medium* Fat dairy: *Low*)>	1.07	<(Hot drink: *Medium* Fat dairy: *Low*) (Hot drink: *Medium*)>	0.2
<(Hot drink: *Medium* Fat dairy: *Medium*)>	0.25	<(Hot drink: *Medium* Fat dairy: *Medium*) (Hot drink: *Medium*)>	0.2
<(Hot drink: *Low*) (Hot drink: *Low*)>	0.5	<(Hot drink: *Low* Fat dairy: *Low*) (Hot drink: *Low*)>	0.32
<(Hot drink: *Low*) (Hot drink: *Medium*)>	1.5	<(Hot drink: *Low* Fat dairy: *Medium*) (Hot drink: *Low*)>	1.48
<(Hot drink: *Medium*) (Hot drink: *Low*)>	0.2	<(Hot drink: *Medium* Fat dairy: *Low*) (Hot drink: *Low*)>	0.25
<(Hot drink: *Medium*) (Hot drink: *Medium*)>	0.4	<(Hot drink: *Medium* Fat dairy: *Medium*) (Hot drink: Low)>	0.5
<(Fat dairy: *Low*) (Hot drink: *Low*)>	0.32		

on the dataset shown by Table 20.3. To show the differences and changes in results, Table 20.5 presents the results of applying fuzzy prefixspan algorithm on the dataset shown by Table 20.5.

Tables 20.4 and 20.5 have been mined from the same basic transactions. It is clear that sequences of Table 20.4 are more general with higher support values. In this example if minimum support is equal to 1, then in the first case, the results will be

<(Hot drink: *Low*)>,
<(Fat dairy: *Low*)>,
<(Fat dairy: *High*)>,
<(Hot drink: *Medium* Fat dairy: *Low*)>,
<(Hot drink: *Low*)(Hot drink: *Medium*)>,
<(Hot drink: *Low* Fat dairy: *Medium*)(Hot drink: *Low*)>

Table 20.5 Output sequences found by applying the fuzzy prefixspan algorithm on *Table* 1 [12]

Sequences	Support	Sequences	Support
<(Cream: *Low*)>	0.25	<(Cream: *Low*) (Fruit juice: *High*)>	0.25
<(Cream: *Medium*)>	0.75	<(Cream: *Medium*) (Fruit juice: *Medium*)>	0.5
<(Tea: *Low*)>	0.75	<(Cream: *Medium*) (Fruit juice: *High*)>	0.5
<(Tea: *Medium*)>	0.25	<(Coffee: *Low*) (Fruit juice: *Low*)>	0.5
<(Coffee: *low*)>	0.75	<(Coffee: *Low*) (Fruit juice: *Medium*)>	0.5
<(Coffee: *Medium*)>	0.25	<(Tea: *Low* Cream: *Low*) (Fruit juice: *Medium*)>	0.25
<(Butter: *Medium*)>	0.75	<(Tea: *Low* Cream: *Low*) (Fruit juice: *High*)>	0.25
<(Butter: *High*)>	0.25	<(Tea: *Low* Cream: *Medium*) (Fruit juice: *Medium*)>	0.5
<(Fruit juice: *Low*)>	0.5	<(Tea: *Low* Cream: *Medium*) (Fruit juice: *High*)>	0.5
<(Fruit juice: *Medium*)>	1	<(Tea: *Medium* Cream: *Low*) (Fruit juice: *Medium*)>	0.25
<(Fruit juice: *High*)>	0.5	<(Tea: *Medium* Cream: *Low*) (Fruit juice: *High*)>	0.25
<(Tea: *Low* Cream: *Low*)>	0.25	<(Tea: *Medium* Cream: *Medium*) (Fruit juice: *Medium*)>	0.25
<(Tea: *Low* Cream: *Medium*)>	0.75	<(Tea: *Medium* Cream: *Medium*) (Fruit juice: *High*)>	0.25
<(Tea: *Medium* Cream: *Low*)>	0.25	<(Coffee: *Low* Butter: *Medium*) (Fruit juice: *Low*)>	0.5
<(Tea: *Medium* Cream: *Medium*)>	0.25	<(Coffee: *Low* Butter: *Medium*) (Fruit juice: *Medium*)>	0.5
<(Coffee: *Low* Butter: *Medium*)>	0.75	<(Coffee: *Low* Butter: *High*) (Fruit juice: *Low*)>	0.25
<(Coffee: *Low* Butter: *High*)>	0.25	<(Coffee: *Low* Butter: *High*) (Fruit juice: *Medium*)>	0.25
<(Tea: *Low*) (Fruit juice: *Medium*)>	0.5	<(Coffee: *Medium*) (Fruit juice: *Low*)>	0.5
<(Tea: *Low*) (Fruit juice: *High*)>	0.5	<(Coffee: *Medium*) (Fruit juice: *Medium*)>	0.5
<(Tea: *Medium*) (Fruit juice: *Medium*)>	0.25	<(Coffee: *Medium* Butter: *Medium*) (Fruit juice: *Low*)>	0.25
<(Coffee: *Medium*, Butter: *Medium*)>	0.25	<(Coffee: *Medium* Butter: *Medium*) (Fruit juice: *Medium*)>	0.25
<(Coffee: *Medium* Butter: *High*)>	0.25	<(Coffee: *Medium* Butter: *High*) (Fruit juice: *Low*)>	0.25
<(Tea: *Medium*) (Fruit juice: *High*)>	0.25	<(Coffee: *Medium* Butter: *High*) (Fruit juice: *Medium*)>	0.25
<(Cream: *Low*) (Fruit juice: *Medium*)>	0.25		

but in the second case the result will be

<(Fruit juice: Medium)>

lonely.

6 Conclusion and Future Work

This paper introduced a novel algorithm to mine sequences of more general items and concepts. This algorithm works based on the similar mental concepts and uses the fuzzy prefixspan algorithm and gives more general results as output sequences. Moreover, the proposed method was able to find the sequences which might be hidden when no mental similarity was considered. There are many future works in this case such as: for mining more effective sequences, time restriction can be considered in algorithm. Only events happen in a special range of time can be considered as a pattern. Moreover, change detection will be performed to discover more general changes.

References

1. Faloutsos C, Ranganathan M, Manolopoulos Y (1994) Fast subsequence matching in time-series database. In: Proceedings of the ACM SIGMOD international conference on management of data, Minneapolis, pp 419–429
2. LeBaron B, Weigend AS (1998) A bootstrap evaluation of the effect of data splitting on financial time series. IEEE Trans Neural Networks 9(1): 213–220
3. Rodr'ıguez J Alonso C, B¨ostrom H (2000) Learning first order logic time series classifiers. In: Cussens J, Frisch A (eds) Proceedings of the 10th international workshop on inductive logic programming. Springer, pp 260–275
4. Agrawal R, Srikant R (1994) Fast algorithms for mining association rules in large databases. In: 20th international conference on very large data bases, Santiago de Chile, pp 487–499
5. Chang CY, Chen MS, Lee CH (2002) Mining general temporal association rules for items with different exhibition periods. IEEE international conference on data mining, Maebashi City, pp 272–279
6. Lee CH, Chen MS, Lin CR (2003) Progressive partition miner: an efficient algorithm for mining general temporal association rules. IEEE Trans Knowledge Data Eng 15:1004–1017
7. Chen G, Wei Q (2002) Fuzzy association rules and the extended mining algorithms. Fuzzy Sets Syst 147:201–228
8. Agrawal R, Srikant R (1995) Mining sequential patterns. In: Yu PS, Chen ASP (eds) 11th International Conference on Data Engineering (ICDE'95). IEEE Computer Society Press, Taipei, pp 3–14
9. Agrawal R, Srikant R (1995) Mining sequential patterns: generalizations and performance improvements, Research Report RJ 9994, IBM Almaden Research Center, San Jose, pp 81–93
10. Srikant R, Agrawal R (1996) Mining sequential patterns: generalizations and performance improvements. In: Proceedings of the 5th international conference on extending database technology, Avignon, pp 327–332

11. Kuok CM, Fu A, Wong MH (1998) Mining fuzzy association rules in databases, SIGMOD Record 27:41–46
12. Gholizadeh M, Pedram MM, Shanbehzadeh J Fuzzy sequence mining for similar mental concepts, lecture notes in engineering and computer science. In: Proceedings of the international multiconference of engineers and computer scientists 2011, IMECS 2011, Hong Kong, 16–18 March 2011, pp 362–367
13. Zaki M (2001) SPADE: an efficient algorithm for mining frequent sequences. Mach Learn J 42(1/2):31–60
14. Yu C-C, Chen Y-L (2005) Mining sequential patterns from multidimensional sequence data. IEEE Trans Knowledge Data Eng 17(1): 136–140, ISSN 1041–4347
15. Zhao Q, Bhowmick S (2003) Sequential pattern mining: a survey, Technical report, center for advanced information systems, School of Computer Engineering, Nanyang Technological University, Singapore
16. Nancy PL, H-J Chen, W-H Hao, H-E Chueh, C-I Chang (2007) Mining negative fuzzy sequential patterns. In: Proceedings of the 7th WSEAS international conference on simulation, modelling and optimization, pp. 111–117
17. Dong G, Pei j (2007) Sequence data mining. Springer, New York
18. Jian P (2004) Mining sequential patterns by pattern-growth: the prefixspan approach, IEEE Trans Knowledge Data Eng 16(10):12–22
19. Hou X, Gu J, Shen X, Yan W (2005) Application of data mining in fault diagnosis based on ontology. In: Third international conference on information technology and applications (ICITA'05), Sydney

Chapter 21
A Comparative Study of Density-based Clustering Algorithms on Data Streams: Micro-clustering Approaches

Amineh Amini and Teh Ying Wah

Abstract Clustering data streams is a challenging problem in mining data streams. Data streams need to be read by a clustering algorithm in a single pass with limited time, and memory whereas they may change over time. Different clustering algorithms have been developed for data streams. Density-based algorithms are a remarkable group in clustering data that can find arbitrary shape clusters, and handle the outliers as well. In recent years, density-based clustering algorithms are adopted for data streams. However, in clustering data streams, it is impossible to record all data streams. Micro-clustering is a summarization method used to record synopsis information about data streams. Various algorithms apply micro-clustering methods for clustering data streams. In this paper, we will concentrate on the density-based clustering algorithms that use micro-clustering methods for clustering and we refer them as density-micro clustering algorithms. We review the algorithms in details and compare them based on different characteristics.

Keywords Data streams • Density-based clustering • Micro-cluster

1 Introduction

In recent years, mining data streams became more prominent due to a large number of applications that generate data streams such as network intrusion detection, weather monitoring, emergency response systems, stock trading, electronic business, telecommunication, planetary remote sensing, and Web site analysis.

With the applicability of data streams, clustering data streams have received more attention in data mining research. In clustering data streams, the aim is to cluster the

A. Amini (✉) • T.Y. Wah
Department of Information System, Faculty of Computer Science and Information Technology, University of Malaya (UM), 50603 Kuala Lumpur, Malaysia
e-mail: amini@siswa.um.edu.my; tehyw@um.edu.my

S.I. Ao et al. (eds.), *Intelligent Control and Innovative Computing*, Lecture Notes in Electrical Engineering 110, DOI 10.1007/978-1-4614-1695-1_21,

data streams continuously hence there is an up-to-date clustering of all objects seen so far.

Data streams posed additional challenges to clustering [18] such as:

- Single pass clustering: data arrive continuously so the clustering has to be done in a single pass over the data.
- Limited time: for real time application, the clustering algorithm has to be done in the limited time.
- Limited memory: data streams are infinite but the clustering algorithm needs to perform with the limited memory.
- Number and size of clusters: according to data streams' characteristics, the number, and the shape of clusters are unknown in advance.
- Evolving data: the algorithm has to consider that the data streams may change over the time.
- Noisy data: the clustering algorithm has to handle the noise in data stream since it affects the clustering.

Some of the clustering algorithms which are developed for data streams are distance-based. As such, they are suitable to find ellipsoid-shaped clusters, or at best convex clusters. However, for non-convex clusters, these methods have trouble finding the true clusters, since two points from different clusters may be closer than two points in the same cluster. Nevertheless, within one cluster, there must be enough intermediate points such that we can reach from one end of the cluster to other end [13].

Another group of developed clustering algorithms for data streams are density-based. Density-based clustering algorithms are developed based on density notion of clusters. They are designed to discover clusters of arbitrary shape and to handle outliers. Indeed, in these clustering algorithms the high density area is separated from the low one. Density is defined as the number of points within a specified radius [13]. A density-based cluster is a set of density-connected objects that is maximal with respect to density-reachability. Every object not contained in any cluster is considered to be noise [16].

In clustering data streams, we cannot save all the incoming data objects due to the limited memory. Micro-clustering is a method in stream clustering, which is used to record summary information about the data objects in the streams. Several algorithms are developed [2, 10, 18, 21] that use micro-clusters for their clustering.

The main objective of this paper is to review and compare the density-based clustering algorithms specifically developed for data streams, as well as using micro-clusters for saving synopsis information about the clusters; we call them as density micro-clustering algorithms.

The rest of the paper is organized as follows. Section 2 surveys related work. Section 3 provides an overview of the density micro-clustering based algorithms on data streams. Section 4 presents discussion, and Section 5 concludes our study.

2 Related Work

Recently, data stream clustering has attracted many researchers and they have published several articles which surveyed various algorithms in data stream clustering [1, 9, 14, 15].

In [11], the authors reviewed theoretical foundations of data streams and discussed about the mining techniques for them. The paper is a general review on mining data streams; however, they only discussed about some distance-based algorithms in the clustering part.

Aggarwal et al. in [1] surveyed the different methods in data streams. This book has a special chapter for clustering data stream with micro-cluster method. Nevertheless, they did not discuss about the density-based algorithms developed for data streams based on the micro-cluster approach.

A complete review on clustering data is proposed in [4]. This paper discussed about the application of clustering algorithms as well. However, it does not discuss about density-based clustering.

Anil in [5] surveys the challenges in clustering data and gives some idea about the semi-supervised clustering, ensemble clustering, and large scale data clustering. However, it only discusses about the density-based clustering and neither on data streams nor micro-cluster methods.

In [8] the authors discuss about density-based clustering algorithms on data streams however it only reviews the grid-based clustering.

Recently, researchers in [7] overview several density-based clustering algorithms using micro-clustering methods for data streams. This paper is an extension of the aforementioned paper.

2.1 Density-based Clustering Algorithms

Density-based clustering algorithms are developed for finding arbitrary shape clusters in large and noisy data sets. They consider the clusters as dense areas, which are separated by low-density areas. Outstanding density-based clustering algorithms include:

- DBSCAN (Density-Based Spatial Clustering of Applications with Noise) [12], grows clusters according to a density-based connectivity analysis.
- OPTICS (Ordering Points To Identify the Clustering Structure) [6], extends DBSCAN to produce a cluster ordering by setting different parameters.
- DENCLUE (DENsity-based CLUstEring) [17], clusters objects based on a set of density distribution functions.

DBSCAN is the most applicable density clustering that has been adopted for data stream which clusters the data as follows:

DBSCAN algorithm has two important parameters: ε-neighbourhood and MinPts. ε-Neighbourhood is a neighbourhood within a radius ε of a given object and the MinPts is the minimum number of points in the ε-neighbourhood. For clustering data, at first DBSCAN checks all the ε-neighbourhood of each point. If the ε-neighbourhood of a point has more than MinPts, the data point is considered as a core object and this core object forms a new cluster. After that, the algorithm finds all the data points which are density reachable objects and it forms this core object and form the cluster. The algorithm performs repeatedly until no new point is added to any cluster [16].

However, these algorithms are not applicable for data streams. They are developed for large or spatial databases. We will discuss later in this paper how these algorithms are adopted for clustering data streams.

2.2 *Micro-cluster*

Micro-clustering is a technique in clustering which saves synopsis information about the data items in data streams. Micro-cluster, which was first introduced in [2], is a temporal extension of cluster feature vector [25]. According to [25] and [2], the micro-cluster is defined as follows:

Definition 1 (Micro-Cluster). A micro-cluster for a set of d-dimensional points is defined as $(\overleftarrow{CF2^x}, \overleftarrow{CF1^x}, CF2^t, CF1^t, n)$, the entries are defined as follows:

$\overleftarrow{CF1^x}$: The linear sum of data points
$\overleftarrow{CF2^x}$: The squared sum of data points
$CF2^t$: The sum of squares of timestamps $T_{i1} \ldots T_{in}$
$CF1^t$: The sum of timestamps $T_{i1} \ldots T_{in}$
n: Number of data points

The micro-cluster for a set of points C is denoted by $\overleftarrow{CFT(C)}$.

3 Denisty Micro-clustering Algorithms on Data Streams

Clustering algorithms, which are developed based on micro-clustering methods, are categorized as distance-based [2, 18] and density-based. In this paper, we focus on the density-based clustering algorithms which apply the micro-clustering techniques. We call them as density micro-clustering algorithms.

One of the well-known designs for clustering data stream is two-phase clustering, which is introduced by Aggarwal et al. [2]. The two-phase clustering separates the clustering process into online and offline components. In this online-offline way, the online phase captures synopsis information from the data stream, and the offline phase generates clusters on the stored synopsis information. In the following sections, four remarkable algorithms, i.e. *DenStream, C-DenStream, rDenStream,* and *SDStream,* are explored and compared and their pros and cons are discussed in a separate section. Most of these algorithms are based on the two-phase clustering.

In the aforementioned algorithms the micro-cluster concept is extended as potential micro-cluster and outlier micro-clusters in order to keep difference between clusters and outliers.

Potential micro-cluster and outlier micro-cluster are defined as follows. Before defining these new concepts, we need to describe the core-micro-cluster.

Definition 2 (Core-micro-cluster (c-micro-cluster)). A c-micro-cluster is defined as $CMC(w, c, r)$ for a group of close points $p_{i1} \ldots p_{in}$ with timestamps $T_{i1} \ldots T_{in}$.

- $w = \sum_{j=1}^{n} = f(t - T_{ij})$, is the weight and $w \geq \mu$.
- $c = \frac{\sum_{j=1}^{n} f(t-T_{ij}) p_{ij}}{w}$, is the center.
- $r = \frac{\sum_{j=1}^{n} f(t-T_{ij}) dist(p_{ij},c)}{w}$, $r \leq \varepsilon$ is the radius.

$dist(p_{ij}, c)$ is *Euclidean distance* between point p_{ij} and the center c.

Definition 3 (Potential c-micro-cluster (p-micro-cluster)). p-micro-cluster at the time t for a group of close points $p_{i1} \ldots p_{in}$ with timestamps $T_{i1} \ldots T_{in}$ is defined as $\left(\overrightarrow{CF^1}, \overrightarrow{CF^2}, w\right)$.

- $w = \sum_{j=1}^{n} f(t - T_{ij})$, is the weight and $w \geq \beta\mu$. β is a parameter to determine the threshold of the outlier related to *c-micro-clusters* $(0 < \beta < 1)$.
- $\overleftarrow{CF^1} = \sum_{j=1}^{n} f(t - T_{ij}) p_{ij}$, is the weighted linear sum of the points.
- $\overleftarrow{CF^2} = \sum_{j=1}^{n} f(t - T_{ij}) p_{ij}^2$, is the weighted squared sum of the points.

Definition 4 (Outlier Micro-cluster (o-micro-cluster)). The definition of o-micro-cluster is similar to p-micro-cluster. It is defined as $\left(\overrightarrow{CF^1}, \overrightarrow{CF^2, w, t_0}\right)$.

However, it considers $t_0 = T_{i1}$ related to the creation time of *o-micro-cluster*.

The general framework for density micro-clustering is shown in Fig. 21.1 (adopted from [23]).

3.1 Denstream

DenStream [10] is a density-based clustering algorithm for evolving data streams which has two-phase clustering framework. In the online phase, it keeps the micro-clusters and in the offline it performs DBSCAN clustering on the micro-clusters. DenStream introduces two new concepts for the first time. Besides, it considers a buffer for the outliers that are likely to be changed to the real clusters. The algorithm

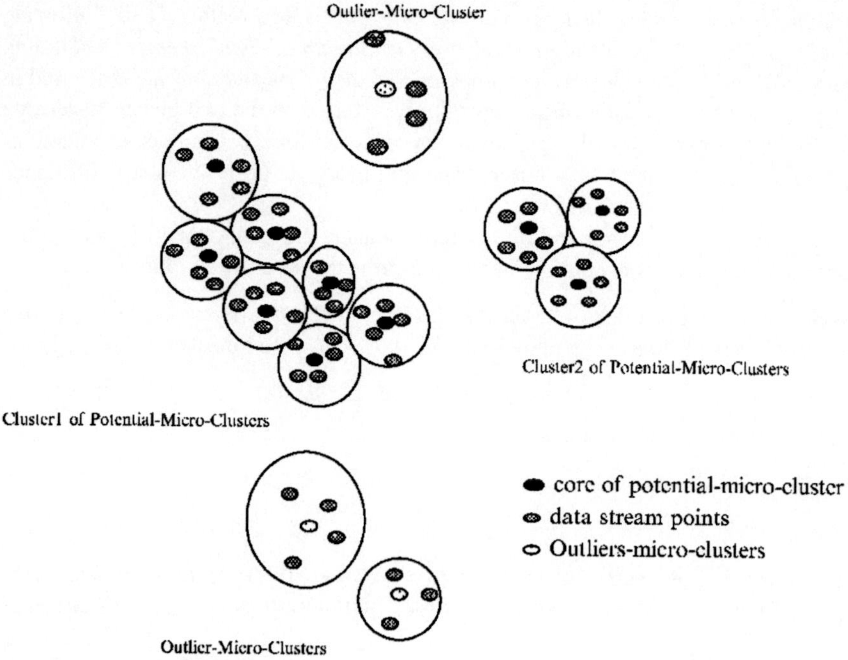

Fig. 21.1 Micro-clusters in DenStream

considers weights for each data point in the micro-cluster and uses fading function [3] to define the weight of the micro-cluster.

DenStream performs the clustering on the data streams as follows:

- **Online phase (Retaining Micro-Clusters)**: this phase captures the potential and outlier-micro-clusters. The online phase performs the DBSCAN clustering on initial points and then defines the potential and outlier micro-clusters. It puts all the outlier-micro-clusters in a buffer, which is called outlier buffer. When a new point arrives, the data point is merged with the existing potential or outlier micro-clusters as follows:
  - If the radius of the new data point to the existing potential-micro-cluster is below or equal to the c-micro-cluster threshold radius then the algorithm merges the new point to the nearest potential-micro-cluster.
  - Otherwise, the algorithm merges it with the nearest outlier-micro-cluster
  - If the radius of the new point to the existing outlier is less than the threshold radius. The weight of outlier-micro-cluster is checked if it is higher than the threshold weight; it probably changes into the potential micro-cluster, therefore, the outlier micro-cluster is removed from the outlier buffer and a new potential-micro-cluster is created.

  – Otherwise, the new outlier-micro-cluster is created with the new point and placed in the outlier buffer. It may be an outlier or the seed of a new cluster.

- **Offline phase (Perform Clustering)**: In this phase each potential micro-cluster is considered as a virtual point and the DBSCAN clustering is used for generating the clustering results.

DenStream has a technique for recognizing the real clusters from outliers. The algorithm monitors the potential and outlier micro-cluster weight in a specified time intervals. If the potential micro-cluster weight is less than the threshold weight, it means that it has not received any data for a long time and the potential micro-cluster should be altered to an outlier micro-cluster. Therefore, the potential-micro-cluster is discarded. The algorithm checks the weight of outlier micro-cluster as well. If the outlier-micro-cluster weight is below the lower limit of threshold, it is a real outlier and it will be omitted. Otherwise, the outlier-micro-cluster is considered as a potential one, which will be changed to a potential-micro-cluster.

The execution time of *DenStream* is evaluated on data stream with various dimensionalities and different number of natural clusters. The authors show that the quality of clustering is higher than CluStream.

3.2 rDenStream

In [19], researchers develop a three step clustering algorithm for data streams that is called rDenStream. The algorithm learns from the discarded outliers to improve the accuracy of the clustering. Since the two-phase algorithms process the data stream in the time window model, the knowledge points of one pattern is probable to be separated into different subsets, as a result some knowledge points are lost. Therefore, rDenStream adds one more step to DenStream which is called the retrospect. In this step, a classifier is made from discarded outliers to be used for the new data points to improve the accuracy of the clustering.

In *rDenStream* the discarded micro-clusters are kept in an outside temporary memory, giving them a new chance to attend the clustering and improve the accuracy of clustering. The three steps of *rDenStream* are described as follows:

- Micro-clustering phase: categorizes the data in the same time window into the potential micro-clusters and outliers micro-clusters. Recognizing the outlier and potential micro-cluster and merging the new points is the same as the DenStream. The outlier-micro-clusters are kept in a buffer that is called historical outlier buffer. Only the potential-micro-clusters are used for the next phase.
- Macro-clustering phase: in this phase, all the potential-micro-clusters from different subsets are placed in one set and uses the DBSCAN algorithm to get the final result.

- Retrospect phase: this phase uses the result of the macro clustering and builds classifier. After that, it uses the classifier to learn the outlier micro-clusters in the historical outlier buffer.

rDenStream is applicable for applications with large amounts of outliers. The experimental result of rDenStream shows that it has a better performance rather than *DenStream* in the initial phase. The time complexity and memory usage are more than *DenStream* for processing and recording the historical outlier buffer respectively.

3.3 C-DenStream

Authors in [21], developed a density-based clustering algorithm with constraints for data streams. The algorithm is called C-DenStream, which is an extension of static semi-supervised clustering for data streams. In semi-supervised clustering methods, the background knowledge guides the clustering process. Instance-level-constraints are one form of background knowledge. They refer to instances that must belong to the same cluster (Must-Link constraints) and those that must be assigned to different clusters (Cannot-link constraints) [24].

C-DenStream adopts the DenStream algorithm and changes it in two ways:

- Instead of using DBSCAN in the offline phase, the algorithm uses C-DBSCAN [22] to include constraints.
- It has micro-cluster transformer level in which the instance level constraints are converted to micro-cluster level constraint.

The algorithm has three steps, which are defined as follows:

Step 1: Initial clustering is generated with C-DBSCAN using the instance-level constraints. These clusters are then transformed into potential-micro-cluster, and initial instance-level constraints are converted to micro-cluster-constraints.

Step 2: The micro-clusters and micro-constraints are retained. Micro constraints are kept in a constraint matrix which stores constraints between micro-clusters, as well as their weighting. When a new data point and constraints arrive, the weight of micro-clusters and micro-cluster constraints are updated. It keeps the potential micro-cluster and outlier micro-cluster similar to DenStream and micro constraints of potential micro-clusters as well.

Step 3: At the request of the user, it generates final clustering and micro-cluster-level constraints using C-DBSCAN algorithm.

Ruiz et al. compare their algorithm performance with DenStream and show that it leads to a better performance result because it converts the domain information in the form of constraint to micro-cluster level constraint.

3.4 SDStream

In [20], the authors propose an algorithm titled SDStream for density clustering of data streams over sliding window. The algorithm considers the distribution of the most recent data stream in which the data points are not accommodated in sliding window length will be discarded. The main idea in the sliding window is to perform detailed analysis over the most recent data items and over summarized versions of the old ones [11]. SDStream is a two-phase clustering in which the online phase maintains potential micro-cluster the same way as DenStream and the offline phase performs the density clustering on the data streams. In this algorithm, the micro-clusters are kept in the form of *Exponential Histogram of Cluster Feature* (*EHCF*) [26]. Exponential histogram is an approach for generating synopsis data structure based on the sliding window. According to [26], *EHCF* concept is defined as follows:

Definition 5 (Exponential Histogram of Cluster Feature (EHCF)). EHCF is a cluster feature based on sliding window model. In EHCF, only the most recent N records are considered at any time. Every bucket in an EHCF is a temporal cluster feature (TCF) for a set of records. TCF in the EHCF is a temporal extension of cluster feature in the sliding window model, which is defined as: $\left(\overleftarrow{CF2^x}, \overleftarrow{CF1^x}, t, n\right)$. t is the time stamp of the most recent record.

The clustering procedure in SDStream is described as follows:

Online Phase

This phase records the potential and outlier micro-clusters in the main memory. As data points arrive, they are assigned to the nearest potential and outlier micro-cluster like DenStream's online procedure. The algorithm considers a threshold for the number of micro-clusters in the main memory. If the number of micro-clusters in the main memory exceeds the threshold limit then either two neighbored micro-clusters will be merged or an outdated outlier micro-cluster will be deleted. If the attribute t of the micro-cluster TCF does not belong to the length of the sliding window, it is considered as an outdated micro-cluster. The merging of two micro-clusters is completed by joining the two EHCF of them [26].

Offline Phase

When the clustering request arrives, the potential micro- clusters are considered as virtual points and DBSCAN algorithm is applied to get the final clustering result on them.

The authors show that *SDStream* has a better cluster quality compared to *CluStream* [2].

4 Discussion

In this section, we will discuss the advantages and disadvantages of density-micro clustering algorithms that have been introduced in the previous sections.

The *DenStream* algorithm considers weights for data points, so the outdated data is eliminated, which is very useful in data streams. It also saves time since it does not merge data into a micro-cluster and then defines the data as outlier. In fact, it determines the real outliers before merging.

rDenStream is built on *DenStream*; however, it can handle the outliers very well with high accuracy. On the other hand, it has higher time complexity compared to *DenStream*, since it processes the historical outlier buffer. In addition, it needs more memory space for saving the historical buffer.

SDStream uses the EHCF synopsis data structure that can track the cluster evolution better while consuming much less memory. EHCF provides a flexible framework for analyzing the cluster evolution. This algorithm is applicable for the applications in which the distributions of the most recent data streams are important.

C-DenStream uses background knowledge for guiding clustering and putting constraint on micro-cluster. It prevents the formation of the clusters that do not follow the applications' semantics. For instance, geographical objects such as houses separated by a river may not be assigned to the same cluster. Therefore, it is applicable in real applications.

All of these algorithms have their pros and cons. They cluster data streams from different perspectives. While some of them emphasize the handling of outliers, others ignore it. Some are more accurate, though they have a high complexity. Choosing the time window for processing data stream is also different in these algorithms; while some of them consider the whole data stream, the others only use the most recent data streams. Consequently, choosing these algorithms depends on what we want from the algorithms, such as low time complexity, high accuracy, the distribution of our data, and others. Table 21.1 summarizes the density micro-clustering algorithms with the important characteristic.

5 Conclusion

Clustering data streams places additional constraints on clustering algorithms. Data streams require algorithms to make a single pass over the data with bounded memory, limited processing time, whereas the stream may be highly dynamic, and evolve over time. Several clustering algorithms are introduced for data streams that are distance-based and cannot handle the interwoven clusters. Besides, saving the whole data streams is impossible, due to their infinite characteristics. Consequently, micro-clustering technique is introduced to record a summary of data.

Table 21.1 Summarization of Density Micro-clustering Algorithms

Algorithm name	Year	Objective	Phases	Density-based algorithm	Data structure	Time complexity
DenStream	2006	Clustering evolving data stream	Two-phases 1-Micro clustering 2-Macro clustering	DBSCAN	Potential micro-cluster, Outlier micro-cluster	$O(n)$
C-DenStream	2009	Clustering data streams with constraints	Three-phases 1-Convert the constraint to micro-cluster constraints 2-Micro clustering 3-Macro clustering	C-DBSCAN	Potential micro-cluster, Outlier micro-cluster, micro-cluster constraint	$O(n)$
rDenStream	2009	Improving the accuracy of clustering algorithm	Three-phases 1-Micro clustering 2-Macro clustering 3-Retrospect phase	DBSCAN	Potential micro-cluster, Outlier micro-cluster	$O(n)+$ time for processing the historical buffer
SDStream	2009	Density-based clustering over sliding window	Two-phases 1-Micro clustering 2-Macro clustering	DBSCAN	EHCF, TCF Potential micro-cluster, Outlier micro-cluster	Unknown

We explore four density-based clustering algorithms which use micro-clustering method. These algorithms utilize the density-based clustering because of their ability to find any shape clusters and micro-clusters as a general summarization of incoming data streams for solving data mining problems on streams.

The algorithms are two-phase, online and offline, in which the online phase maintains the micro-clusters and the offline phase generates the final clusters based on various kinds of density-based clustering algorithms.

As a future work, we will implement all these algorithms and compare them based on the cluster quality on a single dataset.

References

1. Aggarwal CC (ed) (2007) Data streams—models and algorithms. Springer, New york, USA
2. Aggarwal CC, Han J, Wang J, Yu PS (2003) A framework for clustering evolving data streams. In: Proceedings of the 29th international conference on very large data bases, VLDB Endowment, Berlin, Germany, pp 81–92
3. Aggarwal CC, Han J, Wang J, Yu PS (2004) A framework for projected clustering of high dimensional data streams. In: Proceedings of the thirtieth international conference on very large data bases VLDB Endowment, Toronto, Canada, pp 852–863
4. Anil KJ, Murty MN, Flynn PJ (1999) Data clustering: a review, ACM Comput Surveys 31: 264–323
5. Anil KJ (2008) Data clustering: 50 years beyond K-means, Pattern Recogn Lett 31(8):651–666
6. Ankerst M, Breunig MM, Kriegel H-P, Sander J (1999) OPTICS: ordering points to identify the clustering structure, SIGMOD Records 28:49–60
7. Amini A, Teh YW (2011) Density micro-clustering algorithms on data streams: a review, lecture notes in engineering and computer science: proceedings of the international multi-conference of engineers and computer scientists 2011, IMECS 2011, Hong Kong, 16–18 March 2011
8. Amini A, Teh YW, Saybani MR, Aghabozorgi SR (2011) A study of density-grid based clustering algorithms on data streams. In: Proceedings of the 8th international conference on fuzzy systems and knowledge discovery, Shanghai, pp 410–414
9. Babcock B, Babu S, Datar M, Motwani R, Widom J (2002) Models and issues in data stream systems. In: Proceedings of the twenty-first ACM SIGMOD-SIGACT-SIGART symposium on principles of database systems, PODS 2002, New York, pp 1–16
10. Cao F, Ester M, Weining Q, Aoying Z (2006) Density-based clustering over an evolving data stream with noise. In: SIAM conference on data mining, SIAM, Bethesda, Maryland, USA, pp 328–339
11. Elena I, Suzana L, Dejan G (2007) A survey of stream data mining. In: Proceedings of 8th national conference with international participation, ETAI, Ohrid, Republic of MACEDONIA, pp 19–21
12. Ester M, Kriegel H-P, Sander J, Xu X (1996) A density-based algorithm for discovering clusters in large spatial databases with noise. In: Proceedings of 2nd international conference on knowledge discovery and data mining (KDD), AAAI Press, Portland, Oregon, pp 226–231
13. Gan G, Ma C, Wu J (2007) Data clustering: theory, algorithms, and applications (ASA-SIAM series on statistics and applied probability). Society for Industrial and Applied Mathematics (SIAM), Philadelphia, Pennsylvania
14. Gaber MM, Zaslavsky A, Krishnaswamy S (2010) Data stream mining, data mining and knowledge discovery handbook, pp 759–787

15. Gaber MM, Zaslavsky A, Krishnaswamy S (2005) Mining data streams: a review. SIGMOD Record 34:18–26
16. Han J (2005) Data mining: concepts and techniques. Morgan Kaufmann, San Francisco
17. Hinneburg A, Keim DA (1998) An efficient approach to clustering in large multimedia databases with noise. In: Proceeding of 4th international conference on knowledge discovery & data mining, New York City, NY, pp 58–65
18. Kranen P, Assent I, Baldauf C, Seidl T (2011) The ClusTree: indexing micro-clusters for anytime stream mining. Knowl Inf Syst **29**(2): 249–272
19. Li-xiong L, Jing K, Yun-fei G, Hai H (2009) A three-step clustering algorithm over an evolving data stream. In: Proceedings of IEEE international conference on intelligent computing and intelligent systems (ICIS), Shanghai, China, pp 160–164
20. Ren J, Ma R, Ren J (2009) Density-based data streams clustering over sliding windows. In: Proceedings of the 6th international conference on fuzzy systems and knowledge discovery (FSKD), IEEE, Tianjin, China
21. Ruiz C, Menasalvas E, Spiliopoulou M (2009) C-DenStream: using domain knowledge on a data stream. In: Proceedings of the 12th international conference on discovery science, Springer, Berlin, pp 287–301
22. Ruiz C, Spiliopoulou M, Menasalvas E (2007) C-DBSCAN: density-based clustering with constraints. In: Proceedings of the international conference on rough sets fuzzy sets data mining and granular computing, Springer, Berlin, Heidelberg, pp 216–223
23. Tasoulis DK, Ross G, Adams NM (2007) Visualizing the cluster structure of data streams. In: Proceedings of the 7th international conference on intelligent data analysis, IDA, Springer, Berlin, pp 81–92
24. Wagstaff K, Cardie C, Rogers S, Schrodl S (2001) Constrained k-means clustering with background knowledge. In: Proceedings of the eighteenth international conference on machine learning, ICML, San Francisco, pp 577–584
25. Zhang T, Ramakrishnan R, Livny M (1996) BIRCH: an efficient data clustering method for very large databases. In: Proceedings of the 1996 ACM SIGMOD international conference on management of data, Montreal, Quebec, Canada, pp 103–114
26. Zhou A, Cao F, Qian W, Jin C (2008) Tracking clusters in evolving data streams over sliding windows. Knowledge Inform Syst 15:181–214

Chapter 22
A New Clustering Algorithm for Noisy Image Retrieval

Morteza Analoui and Maedeh Beheshti

Abstract The paper concerns an open problem in the area of content based image retrieval (CBIR) and presents an original method for noisy image data sets by applying an artificial immune system model. In this regard, appropriate feature extraction methods in addition to a beneficial similarity criterion contribute to retrieving images from a noisy data set precisely. The results show some improvement and resistance in the noise tolerance of content based image retrieval in a database of various images.

Keywords Artificial immune system • Content based noisy image retrieval, Fuzzy linking histogram • Similarity criterion

1 Introduction

The significant growth of the Internet and high availability of large number of images in various white-black or color types encourage us to finding a solution for image retrieval in order to tackle the difficulties of considerable increasing image data sets. In this regard each innovated technique of CBIR plays crucial role for retrieving images precisely. Although currently implemented CBIR systems

---

M. Analoui
School of Computer Engineering, Iran University of Science & Technology, Tehran, Iran

University of Science and Technology (IUST), Narmak, Tehran, Iran
e-mail: analoui@iust.ac.ir

M. Beheshti (✉)
School of Computer Engineering, Iran University of Science & Technology, Tehran, Iran

University of Science and Technology (IUST), Tehran, Iran
e-mail: maedehbeheshti@comp.iust.ac.ir; maedehbeheshti79@gmail.com

S.I. Ao et al. (eds.), *Intelligent Control and Innovative Computing*, Lecture Notes in Electrical Engineering 110, DOI 10.1007/978-1-4614-1695-1_22,

achieved a lot of improvements to correctly retrieve proper images, encountering noisy images in image data sets is another difficulty in content based image retrieval requiring reach a compromise.

Noise filtering (reduction) in images is a classical and prevailing task in the subject of Image processing and recognition but it is not a certain solution [1]. In addition to performance, precision and speed the CBIR methods have to provide a certain level of noise resistance, at least as far as the standard noise which is typical for the process of image retrieval. EFIRS (effective and fast image retrieval system) is an example of that systems which has been developed at the Bulgarian Academy of Sciences (BAS) for the needs of the Patent Office of Republic of Bulgaria (PORB), and specifically for their vast IDBs of trademark images [1–3].

Impulse noise is caused by malfunctioning pixels in camera sensors, faulty memory locations in hardware, or transmission in a noisy channel. Two common types of impulse noise are the salt-and-pepper noise and the random valued noise. For images corrupted by salt-and-pepper noise (respectively random-valued noise), the noisy pixels can take only the maximum and the minimum values (respectively any random value) in the dynamic range. There are many works on the restoration of images corrupted by impulse noise [4]. In this paper we try to apply artificial immune clustering algorithm to show its effectiveness in noisy image retrieval. In other words, in addition to appropriate feature extraction our concentration for similarity criterion in noisy image data sets is on a robust clustering. Natural immune system is a sustainable and powerful defense system that exhibits many signs of intelligence and cognitive learning [5, 6]. Through CBIR systems like classic histogram, developed artificial immune systems based on immune mechanism provides evolutionary learning mechanisms of unsupervised learning, self-organizing, etc, and combines with some merits of learning systems such as classifier, neural network and so on. It has strong capacity of processing robustness information, and provides a new capacity of solving complex problem. Artificial immune system has successfully been used in some fields such as image recognition, etc. [7].

2 Feature Extraction and Spatial Fuzzy Linking Histogram

The proposed method of fuzzy linking histogram [8,9] uses a small number of bins produced by linking the triplet from the L*a*b* color space into a single histogram by means of a fuzzy expert system. The L*a*b* color space was selected because it is a perceptually uniform color space which approximates the way that humans perceive color and perform better than other color spaces in various retrieval tests performed in the laboratory for this exact purpose [8–10].

In this paper, we have enhanced this method by adding spatial feature to fuzzy linking histogram method. We call the enhanced method "Spatial Fuzzy linking histogram". In L*a*b*, L* stands for luminance, a* represents relative greenness-redness and b* represents relative blueness-yellowness. All colors and gray levels can be expressed using a combination of three L*, a* and b* components. In fuzzy linking histogram a* component is subdivided into five regions representing green, greenish, the middle of the component, reddish and red [8, 9]. The b* component is subdivided into five regions representing blue, bluish, the middle of the component, yellowish and yellow [8, 9]. The L* component is subdivided into only three regions: dark, dim, bright areas [8, 9]. We have tried to enhance this method by inserting spatial information in horizontal and vertical position. In this way X and Y position axes are defined as two more inputs to the fuzzy linking system. X and Y give us more information about the spatial position of a pixel in color image. The X component is subdivided into three regions representing left, middle and right in horizontal axis. The Y component is subdivided into three regions representing up, middle and down in vertical axis. The spatial fuzzy linking of five components (L*, a*, b*, X, Y) is made according to at least 40 fuzzy rules which leads to output of the system. Figure 22.1a–e shows the fuzzification of the five mentioned inputs through triangular membership function and Fig. 22.1f shows one output through trapezoidal membership function. The output spatial fuzzy linking histogram is divided to 10 bars which approximately indicate black, dark gray, red, brown, yellow, green, blue, cyan, magenta and white colors. The Mamdani type of fuzzy inference is used in which the fuzzy sets from the output MFs of each rule are combined through the aggregation operator which is set to max and the resulting fuzzy set is defuzzified to produce the output of the system [8, 9, 11]. The implication factor which determines the process of shaping the fuzzy set in the output MFs based on the results of the input MFs is set to min and the OR and AND operators are set to max and min, respectively.

3 Image Noise and Artificial Immune Systems

The random variation of brightness or color information in images causes image noise produced by the sensor and circuitry of a scanner or digital camera [12]. Image noise can also originate in film grain and in the unavoidable shot noise of an ideal photon detector. Image noise is generally attended as an unexpected by-product of image capture. There are different types of noises, amplifier noise (Gaussian noise) that the standard model of it is additive, independent at each pixel and independent of the signal intensity. Salt and pepper noise or "impulsive" noise sometimes called spike noise. An image containing salt-and-pepper noise will have dark pixels in bright regions and bright pixels in dark regions. Dealing with large image data sets which are sometimes containing noisy images make us provide some scalable clustering techniques. In this regard some clustering techniques assume that clusters are hyper-spherical, clean of noise, similar in size and span the whole data space [6, 13]. Robust clustering algorithms have recently been proposed to handle noisy data [13, 14]. In [15], we proposed a new similarity criterion for CBIR based on a new unsupervised AIS learning approach, called TECNO-CBIR. The proposed method is based on another clustering algorithm which is called tracking evolving

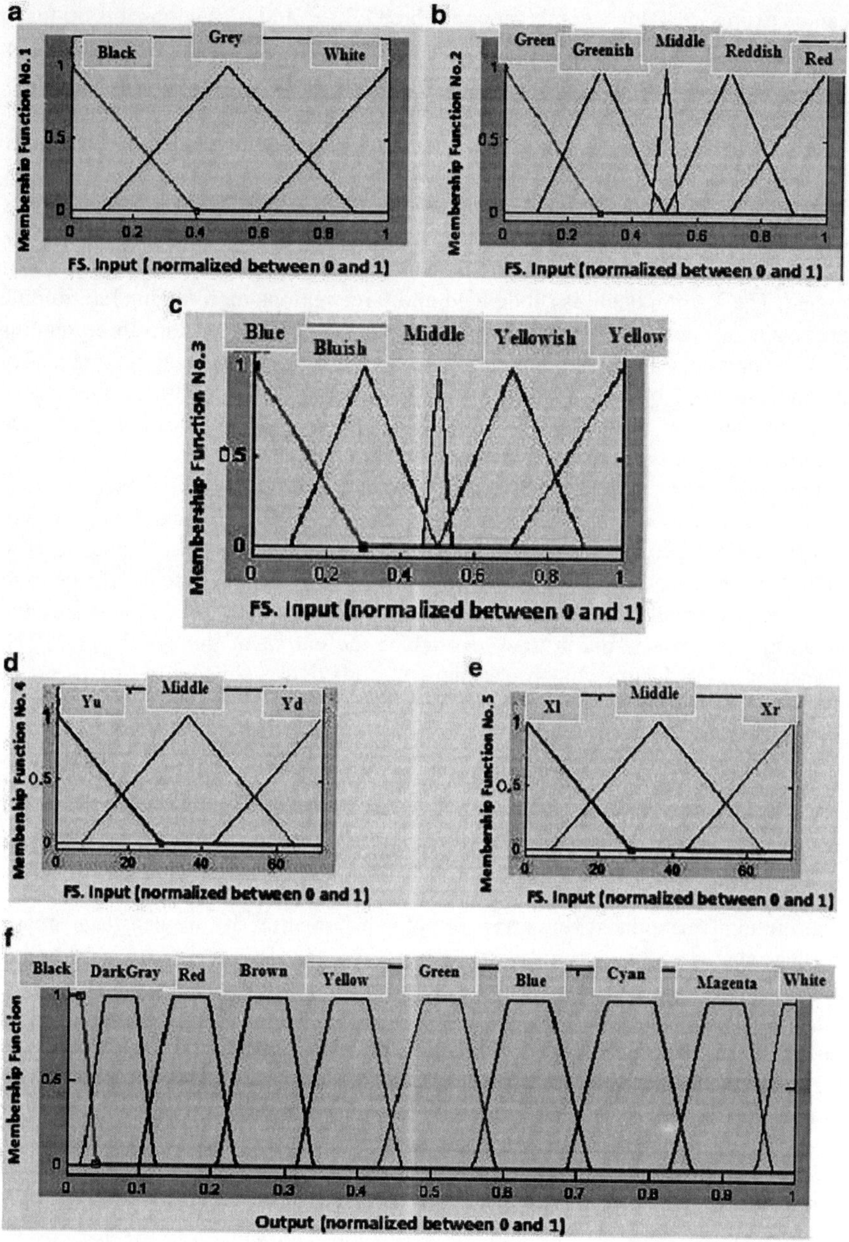

Fig. 22.1 **(a–e)** Fuzzy system input membership function of L*, a*, b*, Y and X. **(f)** Output of the fuzzy system

clusters in noisy streams. This approach improves learning abilities, scalability, robustness and automatic scale estimation [6, 13, 14, 16]. In comparison with some existing scalable clustering algorithms in areas like, insensitivity to initialization, robustness to noise, time complexity, required number of clusters, handling evolving clusters and robust automatic scale estimation, they proved that TECNO-Streams is more powerful than the other models. In Tecno-CBIR the input image feature vector is considered as an antigen. According to (22.1) the ith antibody represents a soft influence zone with the size proportional to $\sigma^2_{Ab_i,Ag_j}$, that can be interpreted as a robust zone of influence [6, 13, 14, 16].

$$IZ_i = \{Ag_j \in AG | w_{Ab_i Ag_j} \geq w_{\min}\} \tag{22.1}$$

When Ag_j has been presented to Ab_i pure stimulation and optimal scale can be updated incrementally using the (22.2) and (22.3) respectively:

$$P_{Ab_i,Ag_j} = \frac{\exp^{-\frac{1}{\tau}} W_{Ab_i,Ag_{j-1}} + w_{Ab_i Ag_j}}{\sigma^2_{Ab_i,Ag_j}} \tag{22.2}$$

$$\sigma^2_{Ab_i,Ag_j} = \frac{\exp^{-\frac{1}{\tau}} \sigma^2_{Ab_i,Ag_{j-1}} W_{Ab_i,Ag_{j-1}} + w_{Ab_i Ag_j} d^2_{Ab_i Ag_j}}{25\left(\exp^{-\frac{1}{\tau}} W_{Ab_i,Ag_{j-1}} + w_{Ab_i Ag_j}\right)} \tag{22.3}$$

According to the noisy image data set some crucial parameters such as optimal scale or $w_{\min}$ have been changed precisely. In order to create antigen and antibody vectors Ag_j, $j = 1, \ldots, P_{\mathrm{A}}$ (P_{A} = maximum population of antigens) and $AB = [Ab_1, Ab_2, \ldots, Ab_N]^T$ have been defined respectively. Because of time consuming and computational resources only 50 × 50 size images have been used in this implementation. Inevitably, according to the spatial fuzzy linking histogram, the input feature vector contains 2,500 elements. Each time just one antigen from an image data set (Ag_j) is presented to the immune network, each of which is selected by random. With each presentation the stimulation and the scale measures will be updated and the antigen index, j, will increase with the time. It indicates time differentiation for entering antigen into the immune network. In Tecno-CBIR, antibodies are the dynamic weighted B-cells (D-W-B-Cell) and represent an influence zone over the domain of discourse consisting of the training image data set [16]. In this experiment the images have been selected from Internet sites such as http://utopia.duth.gr/~konkonst and http://www.cs.washington.edu/research/imagedatabase. All the images in these sites are free and their subjects are related to the area of arts, sports, animals, culture and so on. In order to make an appropriate data set containing a variety of clear images and noisy ones, we tried to create some extra noisy images by inserting salt and pepper noise of density 0.03, 0.05, 0.15, brightened and blurred images through linear filtering [9]. Obviously, Image deblurring (or restoration) is an old problem in image processing, and it continues to attract the attention of researchers and practitioners alike. A number of real-world problems from astronomy to consumer imaging find applications for image restoration algorithms. Plus, image restoration is an easily visualized example of a larger class of inverse problems that arise in all kinds of scientific, medical, industrial and theoretical problems. Besides that, it's just fun to apply an algorithm to a blurry image and then see immediately how well you did [12]. Using mathematical descriptions like deconvlucy function is a solution to deblur an image but time consuming and high memory allocation in order to run these mathematical models every time in each image data set is one of its drawbacks. In this regard due to declining the mentioned model's repercussions we try to get two beneficial methods of feature extraction and similarity criterion together and show the evaluation results as well.

3.1 Cloning and Memory Construction

One important immune mechanism of defense is to reproduce those cells capable of recognizing and binding with antigens. The cellular reproduction in the immune system is based on cloning (mitosis), i.e. the creation of offspring cells that are copies of their parent cells subject to mutations. This proliferation will result in the production of a Clone of cells of the same type. To avoid preliminary proliferation of D-W-B-Cells, and to encourage a diverse repertoire, new D-W-B-Cells do not clone before they are mature (their age, Age_i, exceeds a lower limit $\mathrm{Age}_{\min}$). Due to the mutations, the cells within a clone are all similar but present slight differences that are capable of recognizing the antigen that triggered the immune response. A selective mechanism guarantees that those offspring cells (in the clone) that better recognize the antigen, which elicited the response, are selected to have long life spans; these cells are named memory cells. Figure 22.2 shows that the cloning and memory creation operations have been made using at least one of the existent antibodies stimulation for an input antigen. Antibody cloning and memory construction are made with respect to their stimulation levels and their age condition. In this regard, the $\mathrm{Ag}_{\mathrm{clone}i}$ is calculated based on (22.4) and the condition of $Age_{\min} \leq Age_i \leq Age_{\max}$ should be realized. In this experiment we have considered $k_{\mathrm{clone}} = 0.97$, $\mathrm{Age}_{\min} = 2$ and $\mathrm{Age}_{\max} = 10$. As it is illustrated in Fig. 22.2, when an antigen enters into the immune network for the first time, it creates a memory of external agent based on cloning the cells with the most stimulated ones and creating a memory of external agent that facilitates faster recognition of the same cells when they re-enter to the network. Even after disappearance of the external antigen, b-cells co-stimulate each other and in this way sustaining each other. Thus a memory network of images will be constructed. With memory construction, image retrieval speed will increase. It is obvious that the speed factor has an important role in image retrieval.

When the B-cell population size (N_{B}) exceeds a prespecified maximum ($N_{\mathrm{B\,max}}$), the B-cells are sorted in ascending order of their stimulation levels, and the top ($N_{\mathrm{B}} - N_{\mathrm{B\,max}}$) B-cells (with lowest stimulation) are killed. New B-cells

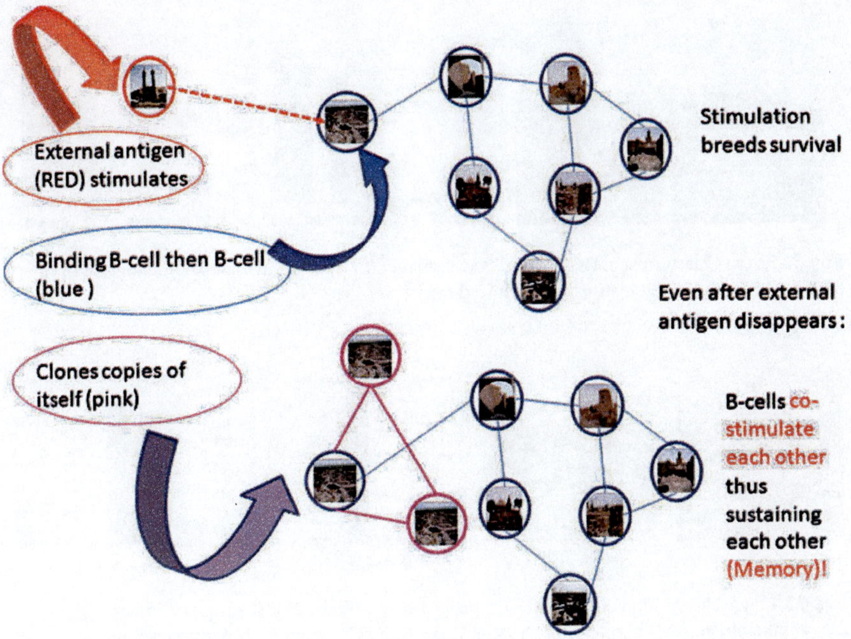

Fig. 22.2 Cloning procedure

($Age_i < Age_{min}$) are compensated to be able to compete with more mature cells in the immune network by temporarily (for the purpose of sorting) scaling their stimulation level to the network's average stimulation level.

$$Ag_{clonei} = K_{clone} \frac{stimulation_i}{\sum_{r=1}^{P_B} stimulation_r} \tag{22.4}$$

4 Results and Evaluation

In our proposed Tecno-CBIR approach the retrieval begins when a new image enters into the system. In this case, the system tries to find the closest images from the clusters to the new one [15]. This method is actually true for noisy image data sets [9]. In this paper, in order to make a comparison between different noisy images, two image prototypes with two related noisy ones (blurred and salt and pepper) have been presented. In Fig. 22.3a–d, two query images and the respective resulting spatial fuzzy histograms for original and noisy images (blurred through a filter which approximates the linear motion of a camera) have been presented. Because of making a comparison between Tecno-CBIR results for noisy images and histogram-based retrieval system results, we only choose three image datasets.

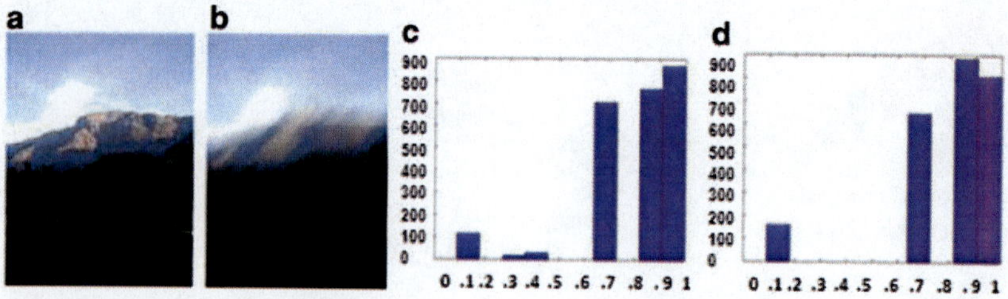

Fig. 22.3 (**a**) Query image. (**b**) Blurred query image. (**c**) Spatial fuzzy linked histogram of query. (**d**) Spatial fuzzy linked histogram of blurred query

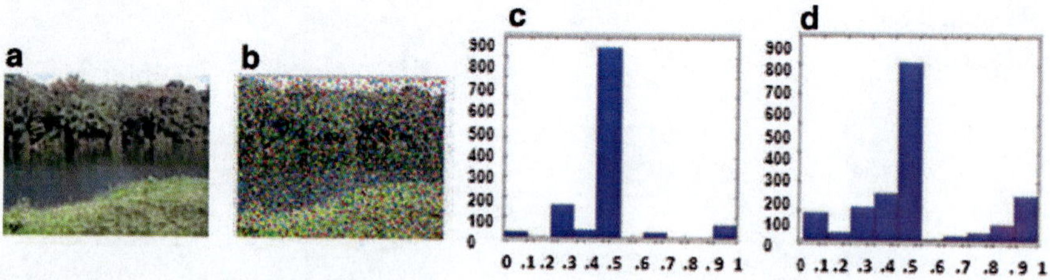

Fig. 22.4 (**a**) Query image. (**b**) Salt and pepper noise of density 0.15 query image. (**c**) Spatial fuzzy linked histogram of query. (**d**) Spatial fuzzy linked histogram of blurred query

The presented images are shown as the representative prototypes of similar semantic content images in database (green, cat, traditional building, etc.). According to Fig. 22.3c and d the spatial fuzzy linking histogram results for original image and noisy one are almost similar. One can easily notice the dominant colors in each of the images. In the first image Fig. 22.3a, bin 7, 9 and 10 (Fig. 22.3c) are activated because of the green trees, blue and white sky, magenta mountains. After applying Spatial fuzzy linking histogram on the noisy image, the output histogram shows the same activated histogram in Fig. 22.3d.

In Fig. 22.4a–d, two other query images and the respective resulting spatial fuzzy histograms for original and noisy images (salt & pepper noise) have been presented. According to Fig. 22.4c and d the spatial fuzzy linking histogram results for original image and noisy one are a bit similar. One can easily notice the dominant colors in each of the images. In the first image Fig. 22.4a, bin 3 and 5 (Fig. 22.4c) are activated because of the yellow grass and green trees. After applying spatial fuzzy linking histogram on the noisy image, the output histogram shows the same activated histogram in Fig. 22.4d. Salt and pepper noise is one of the Image independent noise which can often be described as an additive noise model, where the recorded image $\mathrm{f}(i, j)$ is the sum of the true image $\mathrm{s}(i, j)$ and the noise $\mathrm{n}(i, j)$. In this regard because of some colored data addition to Fig. 22.4 or because of the insertion of the noise $\mathrm{n}(i, j)$, second, fourth, ninth and tenth columns in Fig. 22.4d increased. They are symbols of red, dark gray, brown and magenta which have been inserted to the original pic.

Fig. 22.5 The 21 retrieved images from data set 1. Overall, two images wrongfully retrieved

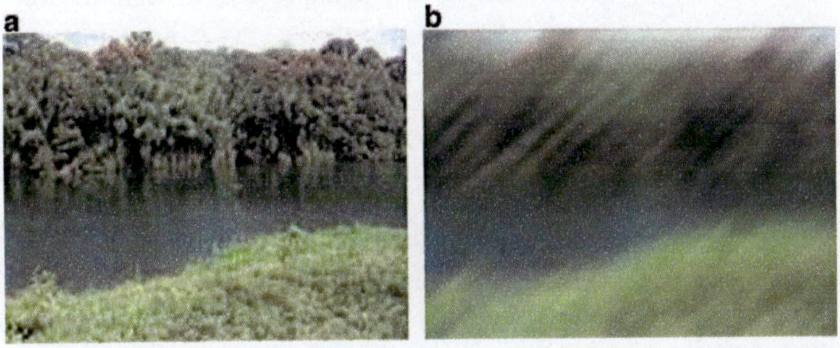

Fig. 22.6 (**a**) Query image. (**b**) Blurred query image

The Tecno-CBIR algorithm starts with k-means technique so early clustering will be done over B-cells population. In order to show the effectiveness of an appropriate similarity criterion, three more tasks were executed to test the robustness of the system on various noisy images. First of all, salt and pepper of density 0.15 was produced and random noise was inserted to the query image, then the brightness of the images was increased and decreased, and finally the images were blurred (Fig. 22.3b) by a filter. The filter becomes a vector for horizontal and vertical motions. K. Konstantinidis et al. [8] claimed the accuracy percentages for fuzzy linking histogram were decreased 5–10% when noisy images have been added to the data sets, which has strongly direct effect on Tecno-CBIR tests because of using their feature extraction approach. But after evaluations we found that Tecno-CBIR similarity approach is more effective and sustainable than histogram-based

Fig. 22.7 The three retrieved images from data set 1. All of the images correctly retrieved

similarity solution. Figure 22.5 shows some correct retrieved images when one noisy image Fig. 22.6b acts as a query. Also it is true for retrieving images from a dataset which is a mixture of original and noisy ones. It is interesting that all the retrieved images for original query image Fig. 22.6a and noisy one Fig. 22.6b are the same. In other words, this kind of noisy or corrupted query image did not have any strong drawback on our CBIR system. In this regard, the most crucial problems with noisy image datasets happen with salt and pepper noise [9]. Such images in one case loose some appropriate features, so extracted spatial fuzzy histogram for them concentrates on one or two bar which makes a lot of difficulties for Tecno-CBIR systems. In other case there is a lot of unimportant extra information in salt and pepper noisy images which cause some disabilities in CBIR systems. Although Fig. 22.4b and its equivalent diagram Fig. 22.4d prove the above claim about salt and pepper noisy images, the acquired results in Fig. 22.7 is nearly satisfying. This figure shows three correct retrieved images for salt and pepper noisy query image Fig. 22.8b which all of them are correctly retrieved. Against the minority of the retrieved images the performance of the CBIR system in front of this kind of noise is good. In order to show Tecno-CBIR effectiveness for noisy images a diagram of precision/recall comparison (Fig. 22.9) has been constructed. According to the diagram, three data sets' output of Tecno-CBIR for noisy images shows appropriate function of the proposed system for noisy datasets in retrieving 20 images. It represents also a comparison between Tacno-CBIR action on dataset 1 and histogram-based function again on dataset 1. As the diagram shows the outputs of two systems are alike but by comparison of Dataset1 and histogram-based lines we found out that Tecno-CBIR is more sustainable than histogram-based for selecting as a similarity criterion in retrieving noisy image. At the end of two representative lines for dataset 1 (Fig. 22.9) by Tecno-CBIR and histogram-based, the first one shows a linear characteristic in retrieving noisy images which sometimes has some up and down in retrieval. But another line (histogram-based) has a decreasing characteristic specially after 16th retrieval. In other words, we could eliminate the histogram-based downward trends in retrieving content-based noisy images and change it to a stable state through proposed similarity criterion.

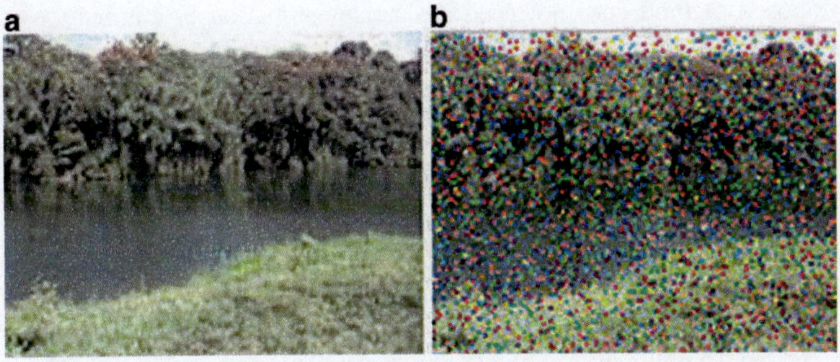

Fig. 22.8 (**a**) Query image. (**b**) Salt and pepper noise of density 0.15 query image

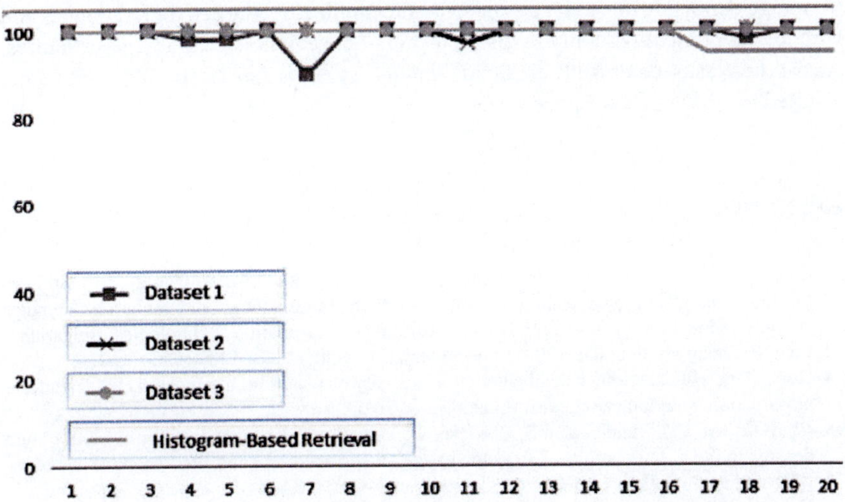

Fig. 22.9 Comparison of the two methods Tecno-CBIR (3 datasets) & histogram-based (1 dataset)

5 Conclusion

In this paper, a new spatial fuzzy linking approach along with an effective similarity criterion based on immune systems applied on noisy data sets. The proposed fuzzy linking histogram and Tecno-CBIR system, each of which had appropriate results in content-based image retrieval and previous results from last experiences proved it [15]. In this regard, in order to prove their robustness and consistency in more dynamic environments we tried to show their response on the noisy data sets. Fortunately the evaluation results on noisy data sets are satisfying. In comparison with usual methods tackling with noisy images and previously try to detect and eliminate image noise with image noise estimation, Laplacian convolution, edge detection, local variance computation and statistical averaging, our proposed method does not need any noise estimation or elimination so it is faster. In this way, all necessary mathematical equations for noise estimation have been eliminated. We only use a fuzzy linking histogram in order to extract feature vectors then the extracted features will be given to the Tecno-CBIR system to be classified. According to the results and evaluation outcomes, not only there is no drop in the acquired results but also there is an improvement from downward trends in histogram-based method to a steady one in proposed solution. The most difficulties happen in salt & pepper noisy environment but as a final conclusion, in comparison with histogram-based method, the proposed method even performs better for salt & pepper noisy images. In this paper, an amalgamation of two robust artificial intelligence methods like fuzzy algorithms and artificial immune approaches provide us with a high performance, reliable and fast CBIR system. That is because, many high efficiency works show that Fuzzy logic is appropriate for image denoising due to it can deal with the nonlinear nature of digital images and with the inherent uncertainty in distinguishing between noise and image structures. Also other tasks dealt with artificial immune systems (specially Tecno-Streams) resulted in a high performance system.

References

1. Dimov D, Marinov A (2006) Geometric-morphological method for artifact noise isolation in the image periphery by a contour evolution tree. In: International conference on computer systems and technologies—CompSysTech, Institute of Information Technologies, Bulgarian
2. Long F, Zhang Hj, Feng DD (2001) Fundamentals of content-based image retrieval.
3. Home Page of Fuzzy Image Segmentation, University of Waterloo, Content by: H.R. Tizhoosh, http://pami.uwaterloo.ca/tizhoosh/fip.htm
4. Vijay Kumar VR, Manikandan S, Ebenezer D, Vanathi PT, Kanagasabapathy P (2007) High density impulse noise removal in color images using median controlled adaptive recursive weighted median filter. IAENG Int J Comput Sci 14(1):31–34
5. Cohen I (2000) Tending Adam's garden. London NW1 7DX,Uk
6. Nasraoui O, Cardona C, Rojas C, Gonzalez F (2003) Mining evolving user profiles in noisy web clickstream data with a scalable immune system clustering algorithm. In: Proc. of WebKDD 2003–KDD Workshop on Web mining as a Premise to Effective and Intelligent Web Applications, Washington DC, p 71
7. Wei D, Zhan-sheng L, Xiaowei W (2007) Application of image recognition based on artificial immune in rotating machinery fault diagnosis. IEEE, Wuhan
8. Konstantinidis K, Gasteratos A, Andreadis I (2005) Image retrieval based on fuzzy color histogram processing Opt Commun 248:375–386
9. Analoui M, Beheshti M (2011) Content-based image retrieval using artificial immune system (AIS) clustering algorithms, lecture notes in engineering and computer science: proceedings of the international multiconference of engineers and computer scientists 2011, IMECS 2011, Hong Kong, 16–18 March 2011
10. Zimmerman HJ (1987) Fuzzy sets, decision making and expert systems. Kluwer Academic, Boston
11. www.mathworks.com, MATLAB and Simulink for Technical Computing (2008)
12. Wikipedia-Image noise, the free encyclopedia, http://en.wikipedia.org/wiki/Image_noise

13. Nasraoui O, Uribe CC, Coronel CR (2003) Tecno-streams: tracking evolving clusters in noisy data streams with a scalable immune system learning model. IEEE computer society, Melbourne
14. Nasraoui O, Rojas C (2003) From static to dynamic web usage mining: towards scalable profiling and personalization with evolutionary computation. Invited Paper in Workshop on Information Technology, Rabat, Morocco, March 2003
15. Analoui M, Beheshti M, Mahmoudi MT, Jadidi Z (2010) Tecno-streams approach for content-based image retrieval. In: Proceedings of the world congress on nature and biologically inspired computing, IEEE, Fukuoka
16. Nasaroui O, Gonzalez F, Dasgupta D (2002) The fuzzy artificial immune system: motivations, basic concepts, and application to clustering and web profiling. IEEE, Honolulu

Chapter 23
Page as a Meeting Place: Web Search Augmented with Social Communication

Yuya Matsui, Yukiko Kawai, and Jianwei Zhang

Abstract When searching for information, a user generally uses search engines on the web. Often search results do not satisfy all users because they may have different needs and varying levels of knowledge. We develop a search system that combines the merits of searching and social communication for improving user experience on the web. This system includes a page ranking algorithm based on the analysis of a hyperlink structure and a social link structure, and a communication interface attached to a page that allows real-time users to communicate with each other. By our system, users can quickly search not only for popular web pages but also for other users currently accessing them.

Keywords Web search • Social communication • Real-time processing • Link analysis

1 Introduction

Search engines, although commonly used in daily life, have some disadvantages. Long complex queries can result in poor search results, or the results may not correspond to the user's expectations or knowledge level. If a user comes across unknown or confusing content on a page and he or she is not an expert on the topic, more searches must be conducted and more, possibly irrelevant, pages must be read in order to understand the meanings of the content.

On the other hand, communication of human knowledge through SNS services can overcome the problems described above as users can freely ask questions and receive help. Other knowledgeable or experienced users can guide users, provide

Y. Matsui (✉) • Y. Kawai • J. Zhang
Kyoto Sangyo University, Motoyama, Kamigamo, Kita-Ku, Kyoto-City 603-8555, Japan
e-mail: i1054056@cc.kyoto-su.ac.jp; kawai@cc.kyoto-su.ac.jp; zjw@cc.kyoto-su.ac.jp

S.I. Ao et al. (eds.), *Intelligent Control and Innovative Computing*, Lecture Notes in Electrical Engineering 110, DOI 10.1007/978-1-4614-1695-1_23,

missing information, or explain the content of web pages. However, the response time in traditional SNS-based search is much longer than the one of web search engines as the replies depend on other users' behavior. Furthermore, the data coverage is much lower.

We have developed a system [7] that combines the merits of both the web search and social communication. It can be used to quickly search not only for web pages but also for other users currently accessing those pages ("real-time users"). Our augmented web search with real-time user communication gathers both web pages and user information, and smoothly connects the users through the accessed pages.

Most pages do not provide any means for contact with their authors or communication with other visitors. While some pages invite user feedback or contain contact information, this kind of communication is inefficient and often takes considerable time. Our system solves this problem by enabling the user to freely communicate with other users currently accessing an arbitrary page. In short, our system has the following characteristics:

1. Our ranking method is based on both hyperlinks and social links to find currently popular pages.
2. Each hyperlink on a page is annotated by the number of users currently accessing its corresponding linked page for offering popularity-based navigation.
3. The communication interface attached to a page enables users to talk with other users in real time, or browse the past communication logs asynchronously.
4. Users are allowed to highlight and share important information of a page, so that other users can immediately find the key portion of the page.

In this paper, we describe our augmented web search system and its implementation. Section 2 presents related work and Sect. 3 gives an overview of the system. Section 4 explains the page ranking algorithm and Sect. 5 presents the real-time communication function. Section 6 describes the implementation of our system. Section 7 presents the experimental evaluation. We conclude with a summary of the key points and a mention of future work in Sect. 8.

2 Related Work

There have been several studies on search and social communication [1, 3, 14]. For example, a recently proposed approach for combining search and social network [3] searches not only for web pages but also for experts. The user can send queries using a communication tool such as email. Another approach [14] identifies popular pages on the basis of users' browsing histories. The result is a SERP with links to websites frequently visited by other users with similar queries. Although these approaches may be able to locate expert users for each type of content, they do not solve the fundamental problems described in Sect. 1.

Other approaches are to use community-based recommendations [8, 9, 12]. For example, Pickens et al. [9] and Morris and Horvitz [8] propose a small community of searchers to do the collaborative filtering. These searchers search for target information or web pages by collaboratively improving the queries. However, a purpose-built interface is needed for real-time processing, and the searchers cannot use people outside their community in the collaboration process.

The detection of user characteristics and their application to news and auction sites have also been studied [4, 6, 11]. The detection analysis uses the history of a user's behavior such as the click history. However, this analysis cannot detect an appropriate expert for a particular page or query.

The use of chat for user communication is another kind of approaches [5, 10, 13]. These services support connections between users, and users accessing the same page can chat. However, because only communication service is provided, the users have to find interesting and popular pages for connecting with other users.

Our augmented web search system not only identifies other users accessing the same page but also provides high-quality information by analyzing the structure of hyperlinks and real-time user links. Furthermore, it shows the number of users currently accessing each page, and the users can communicate with each other.

3 System Overview

The flow of augmented web search with social communication is as follows:

1. After a user submits a query, he or she receives a SERP ranked by the values of the hyperlinks and the social links (Fig. 23.1).

   The pages are first ranked based on the analysis of the hyperlink structure, and then re-ranked based on the analysis of the social links. The hyperlinks on the SERP are also annotated by the number of users currently accessing the corresponding page.
2. When the user clicks a hyperlink on the SERP, the corresponding page is shown with a communication window on the right for chat (Fig. 23.2).

   The users currently accessing this page are represented by avatars. The user can directly ask another real-time user a question by using the communication window. The communication window also has logs of previous communication, i.e., the history of communication between users, which can provide the answers to previous, possibly similar, questions. The important information can also be highlighted by a user and shared to other users, which can help them efficiently detect the key portion of the page.
3. Our system also provides each hyperlink on the accessed page with the number of users currently accessing its corresponding page, so that the user can recursively find and communicate with more users through the links (Fig. 23.3).

   Using our system, a user may reach a communication with a user with an appropriate level of knowledge by following the link structure (Fig. 23.4).

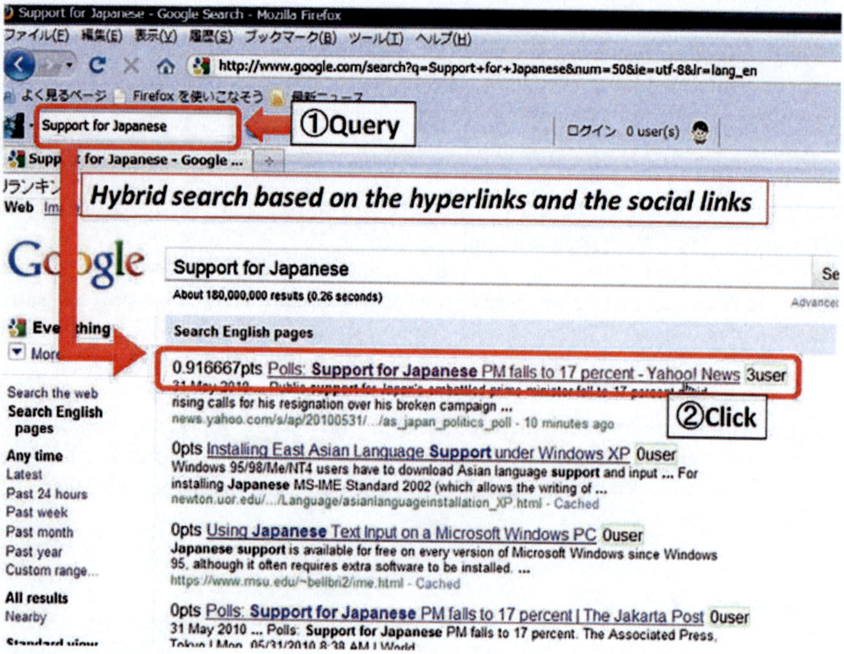

Fig. 23.1 Search results with the numbers of access users

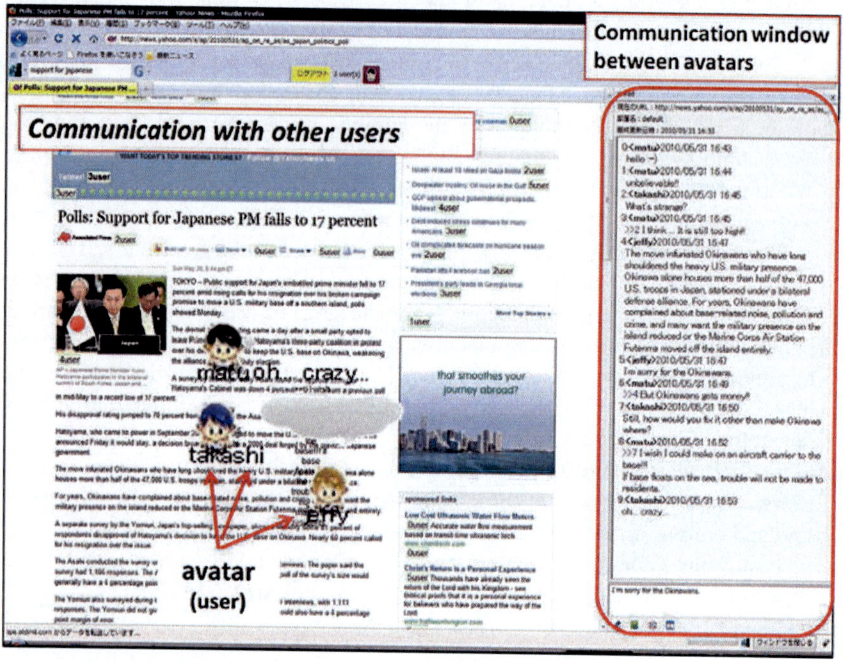

Fig. 23.2 Web page with a communication window

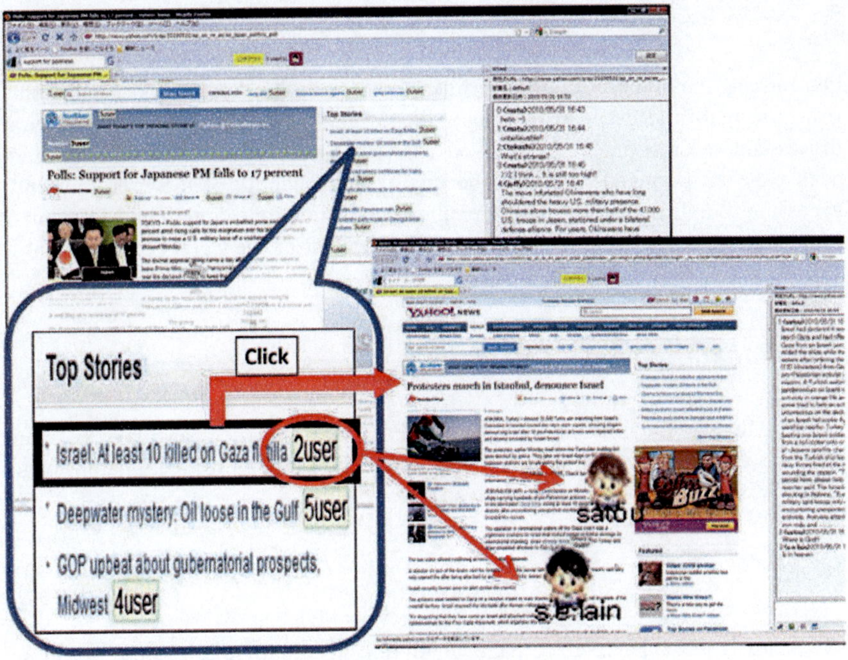

Fig. 23.3 Recursive user communication through deeper links

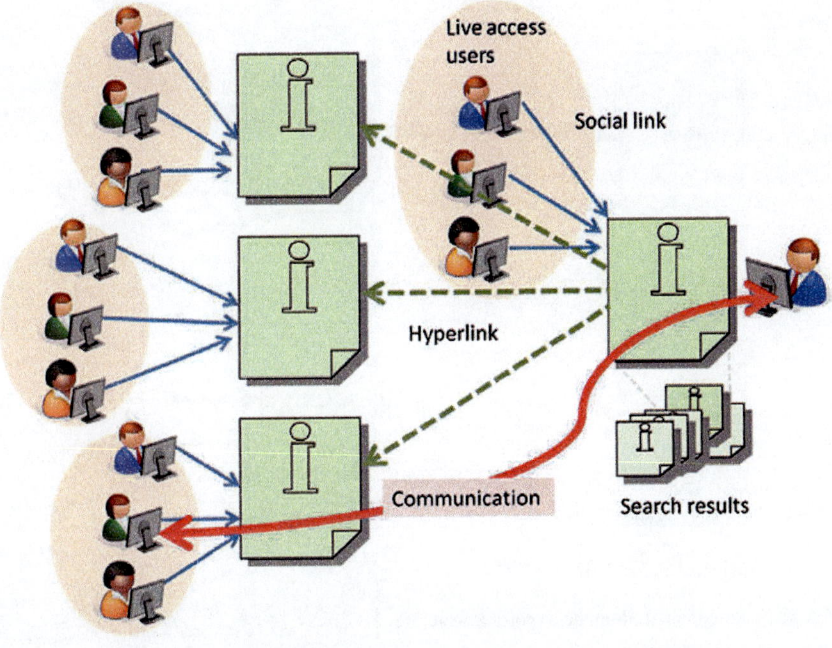

Fig. 23.4 Flow of finding a suitable expert

4 Ranking Based on Hyperlinks and Social Links

Our ranking method uses both hyperlinks and social links. Social links are the links from the real-time users to the pages. Page ranking has two major steps. First, the weight of each page is calculated considering the hyperlink structure. Then, each page is re-ranked based on the quality and quantity of the users currently accessing it. In this section, first we explain the method for calculating the weight of hyperlinks, and then describe the method for calculating the weight of social links.

4.1 Ranking Based on Hyperlinks

Our ranking method is based on the idea of PageRank [2]. Each page has a pagerank deriving values equally divided by the number of outlinks from its parent pages (Fig. 23.5). The pagerank value is given by

$$PR_p = \sum_{q \in parent(p)} PR_q / N_q \tag{23.1}$$

where p is a target page, q is a page which links to p, $parent(p)$ represents the set of q, PR_p and PR_q represent the pagerank values of p and q respectively, and N_q is the number of outlinks from q.

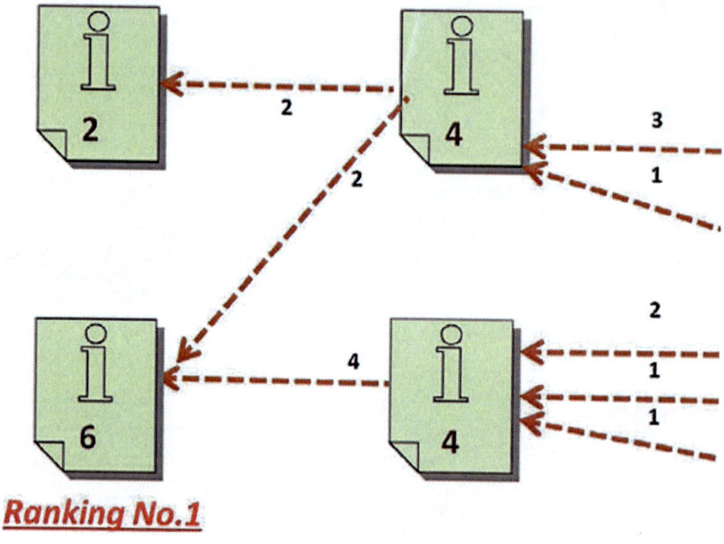

Fig. 23.5 Ranking based on the hyperlink structure

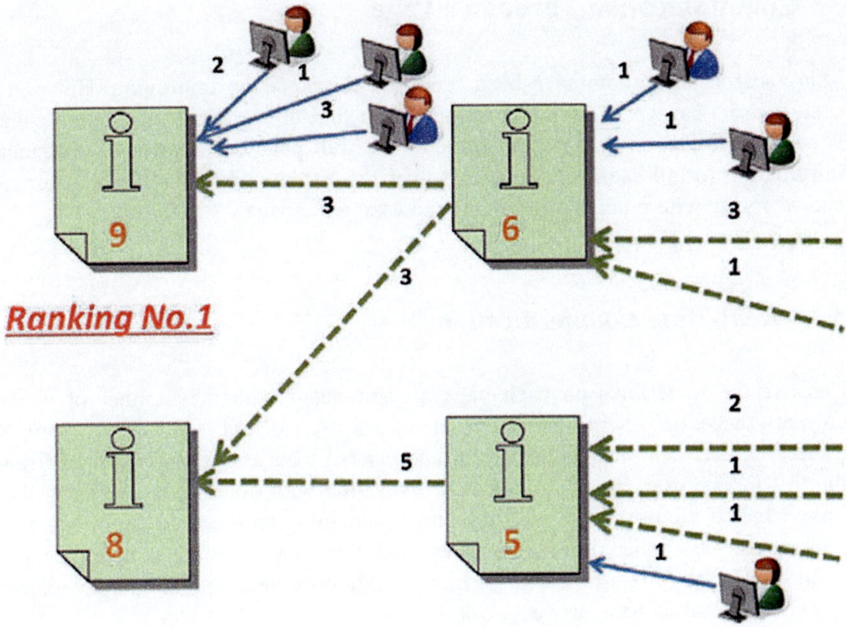

Fig. 23.6 Ranking based on the social link structure

4.2 Ranking Based on Social Links

As described above, our ranking method uses the hyperlink and social link structure, and is based on the PageRank algorithm. Each page has two values: one is the weight of the hyperlinks calculated using Equation 23.1, and the other is the weight of the social links (Fig. 23.6). The weight of the social links includes the weights based on the quality and quantity of users currently accessing a page, and the weights derived from the page's parents. Each page has a UR (UserRank) weight value given by

$$UR_p = \sum_{u \in user(p)} W_u + \sum_{q \in parent(p)} UR_q / N_q \tag{23.2}$$

where u is a user currently accessing page p, $user(p)$ represents the set of those users accessing p, W_u is the quality of user u, q is a parent page of p, UR_p and UR_q are the UR weight values of p and q respectively, and N_q is the number of outlinks from q. W_u is initiated as 1. Considering that a user's effect decreases as the access time increases, we calculate W_u by the following equation:

$$W_u = \begin{cases} 1 - (T_{now} - T_{access})/T & \text{if } (T_{now} - T_{access}) < T \\ 0 & \text{if } (T_{now} - T_{access}) \geq T \end{cases} \tag{23.3}$$

where T is the useful-life of users, T_{now} is now, and T_{access} is the time when the page was accessed.

5 Communication Through a Page

Many search algorithms have been proposed for precision improving. However, even if users can get a good SERP, they often find information on a linked page is confusing. This is because the content of a web page does not have sufficient information for all users. Our system solves this problem by enabling the user to communicate with other users currently accessing the page.

5.1 *Real-Time Communication*

Because the hyperlinks on each page are annotated with the number of users currently accessing them, the system enables a user to find even more real-time users. The user can thus connect with other users by accessing pages deeper from the SERP. The user can communicate in real time with other users accessing the same page. If the user has a question about something on the page, he or she can immediately ask other users about it by typing a query into the communication window. The other real-time users see the question in the communication window and can respond with an answer.

If a page is being accessed by a large number of users and its communication window has a large number of dialogues, it is difficult for users to find an expert with an appropriate level of knowledge. To solve this problem, we design the system by allowing users to freely create a "room" in which real-time users can talk with respect to a specific topic.

5.2 *Asynchronous Communication*

For some unpopular pages, the number of users accessing them may be small, even zero. In this case, few or no other users can be communicated with in real time. To solve this problem, our system is also designed to support asynchronous communication. The server maintains a communication log for each page, and the log is made available to the current users of that page. The log is presented on the basis of content. A user searching this log may find the answer to a previous query similar to his or her current question.

5.3 *Text Highlighting*

A user can highlight text that he or she considers important by selecting it and clicking a popup command, and other users can efficiently detect the important parts of the page. The highlighting remains on the page, and the system can automatically scroll down and present the highlighted text after the page is accessed. We also plan to make use of the highlighting for improving the ranking of pages. If there are many highlighted text in a page, the page is ranked higher.

6 Implementation

We implemented our augmented web search system. It consists of the server side and the client side.

6.1 *Server Side*

The server side receives and processes the client requests. It was built using Apache Tomcat 6.0.18 and Java Servlet on JDK 1.6. The servlet can perform parallel processing for multiple requests because it makes a "user thread" for each request. The servlet consists of two parts: a search servlet and a communication servlet.

6.1.1 Search Servlet

The search servlet extracts the query from line of request. It then retrieves web pages by using the query. Several search engines (Google, Yahoo!, Bing, etc.) are supported in this implementation. After the servlet gets the search results' URLs, it retrieves information about the URLs: the quality and quantity of current users from the database. Finally, it sorts the search results by PR values, extracts the top ones, and re-ranks them based on UP values.

6.1.2 Communication Servlet

The user thread on the communication servlet receives the requests from the clients and parses them. Table 23.1 shows the requests supported by the user thread, the corresponding action, and the information sent to the other clients and to the requesting client. The user thread sends commands to the database objects and room operation objects in accordance with Table 23.1. Most user requests are stored as chat room objects. The chat room data includes the URL, the room name, the communication log, the user information (avatar name, etc.), and the server response. The user thread operates the chat rooms by using the room operation class. Asynchronous communication was implemented by unblocking the I/O using Commet technology. With this setup, the server can send the server status to the clients in real time. The communication logs are stored in the server database.

Table 23.1 Requests that the server can process

Request	Action	Info sent to other users	Info sent to requesting users
Log in	Add user data to room and DB	Notification	Other users' data and chat log
Log out	Delete user data from room and DB	Notification	–
Chat write	Store query in room log	Message	–
Chat read	Store reply in room log	–	–
Get room info	Provide room data	–	Room data
Change room	Delete user data from room and DB	Notification	New room's user data and chat log
Change user name	Change user name	Notification	–
Share text	–	Highlighted text	–

Table 23.2 Requests that the clients can submit

Request (command format)	Info sent
RoomsInfo	Room name and number of users accessing the same URL
UsersInfo	Data on users in the same room
Logs	Room chat log
Enter	Data on newly entered users
Exit	Data on users recently logged out
Message	Message in new comment, data on user who wrote it
NewRoom	Data on newly created room
ChangeName	Modified user data
ShareText	Locations of highlighted text

6.2 Client Side

To use this search and communication system, a user needs to simply install a plug-in. The plug-in was developed using the expanded functions of Firefox, a cross-platform browser. The browser interface was programmed using XUL (extensible user-interface language), which is an expanded version of XML (extensible markup language), and the development was programmed using JavaScript.

6.2.1 Basic Operation

A user connects to the server using an asynchronous communication program running in an Ajax script. After the user logs in to the server, the client program sends the user's requests to the server. Table 23.2 shows the supported client requests. The client program continuously polls the read buffer. If the server is to reply immediately after receiving a request, request with polling is used. This polling operation reduces network traffic because the client need not periodically send requests to confirm the server's up-to-date. We implemented a toolbar supporting five basic operations. The user can (1) perform an augmented web search, (2) login and logout for communicating with other users, (3) get the information of the number of users accessing the current page, (4) set the avatar display on or off, and (5) configure his or her properties such as the name and the color of input text, etc.

6.2.2 Highlighting of Important Information

As described above, the highlight function can be used to both indicate important information and draw the attention of other users. The user can highlight the target text by, for example, left-clicking, dragging and right-clicking to select the interested parts of the page. This highlighting helps subsequent users to identify the important information because it remains on the page, and the augmented web search system can automatically scroll down and present the highlighted text after the page is accessed.

6.2.3 Ranking by the Number of Current Accesses

When a user gives a query, our system ranks the search results related to the query. When no query is given, our system also provides a function for ranking all the pages in the access log. An example of this ranking is shown in Fig. 23.7. The URL at the top has the most users currently accessing it. The pages with higher ranks are considered popular pages because they are more frequently accessed by users. By accessing such ranking page, a user can find the popular topics and connect to more people.

7 Evaluation

7.1 Ranking Effect

We compared the ranking given by a conventional search engine and the one given by our system. During the period from April 1 2010 to April 30 2010, there were three million access users and sixteen million queries.

We selected eight popular queries from the stored data and calculated spearman correlation of the top 20 ranking between the conventional search engine and our system. The result is shown in Table 23.3. A value close to 1 means a high correlation, and 0 means a low correlation. As we can, the relatively long term topics have high correlation and the temporarily hot topics have low correlation.

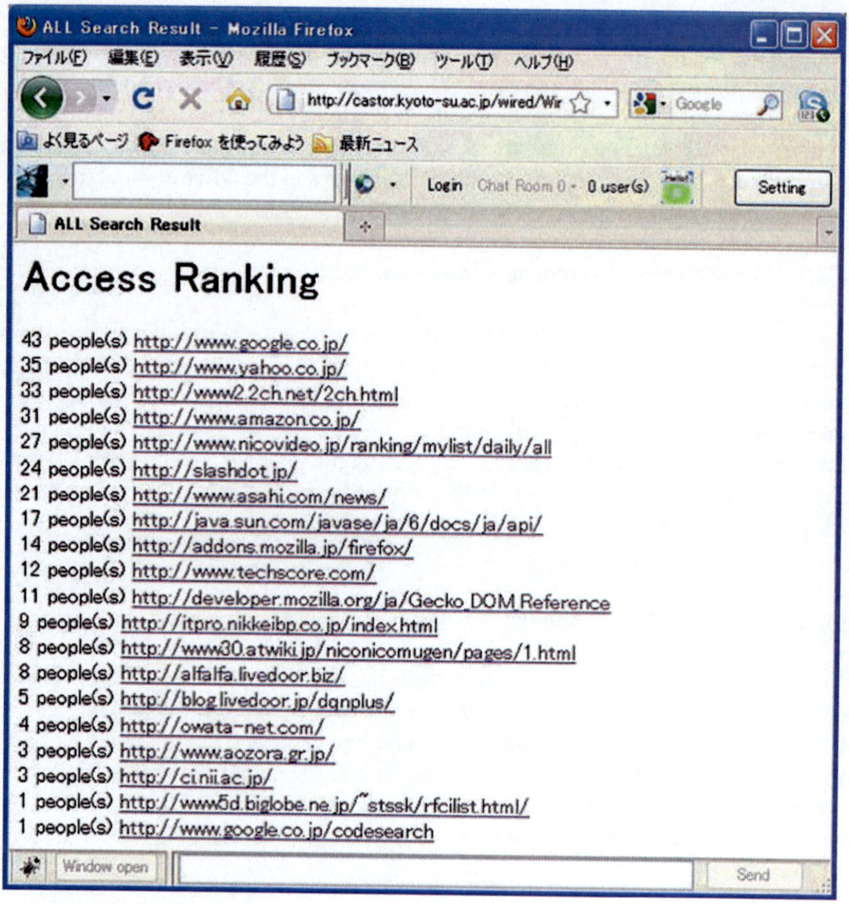

Fig. 23.7 Example of ranking based on current accesses

Table 23.3 Spearman correlation

Query	ρ	query	ρ
Sportsman A	0.04	Idol D	0.55
Idol B	0.2	Song E	0.62
Ipad	0.28	Idol F	0.77
Singer C	0.54	Bean diet	0.93

Figure 23.8 shows the ranking given by the conventional search engine and the ranking by our system are different. For viewability, we only plot the top 10 ranking for three queries. For example, for the query "Sportsman A", the page with rank 7 given by the conventional search engine is ranked as rank 2 by our system. This indicates that ranking pages based on the information of access users is necessary and useful.

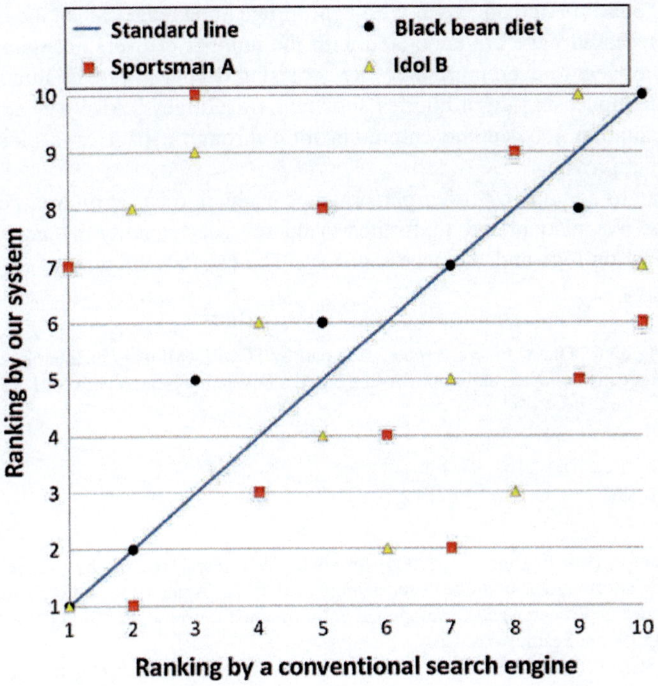

Fig. 23.8 Ranking by a conventional search engine vs. ranking by our system

7.2 *Communication Effect*

We published the client plug-in, and designed a web questionnaire to investigate how helpful the communication through a page was. Testees were asked to install the plug-in, search for web pages, browse the ranked web pages, chat with others if they like, and finally answer the questionnaire. We received the answers from 28 testees. The numbers of the testees, who thought the communication was "helpful", "somewhat helpful", "neither of them", "somewhat helpless", and "helpless", were 2, 9, 10, 4, and 3, respectively. The percentage of the testees giving the evaluations of "helpful" and "somewhat helpful" was about 40%, the percentage of "neither of them" was 35%, and the percentage of "somewhat helpless" and "helpless" was 25%. This indicates the communication through a page is helpful but the system should be further improved.

8 Conclusion

We developed an augmented web search system with social communication. The hyperlinks are used to obtain search results, and the social links are used to re-rank the results. A user can communicate with other users currently accessing the same page, and possibly find an expert with appropriate knowledge. Since the hyperlinks on each accessed page are annotated with the number of users accessing them, a user can connect to even more users by accessing deeper pages. Evaluation of our ranking method using eight queries showed its usefulness. Also a questionnaire-based evaluation showed the communication through a page was effective and should be improved.

We plan to conduct a further performance evaluation based on a larger number of testees. We also intend to further evaluate users' quality by analyzing the communication logs and the access history. The capacity of the system will also be expanded.

Acknowledgment This work was supported in part by SCOPE (Ministry of International Affairs and Communications, Japan) and implemented based on the joint research with NTT Resonant Inc.

References

1. Agichtein E, Brill E, Dumais S (2006) Improving web search ranking by incorporating user behavior information. In: ACM: Proceedings of the 29th Annual International ACM SIGIR Conference on Research and Development in Information Retrieval, SIGIR 2006, 6–11 August, 2006, pp 19–26, Seattle, WA, USA
2. Brin S, Page L (1998) The anatomy of a large-scale hypertextual web search engine. Comput Network 30(1–7):107–117
3. Chang EY (2009) Confucius and gitsh intelligent disciples. In: ACM: Proceedings of the 18th ACM Conference on Information and Knowledge Management, CIKM 2009, 2–6 November, 2009, pp 3–3, Hong Kong
4. Goel A, Munagala K (2009) Hybrid keyword search auctions. In: ACM: Proceedings of the 18th International Conference on World Wide Web, WWW 2009, 20–24 April, 2009, pp 221–230, Madrid, Spain
5. Greenberg S, Rounding M (2001) The notification collage: Posting information to public and personal displays. In: ACM: Proceedings of the SIG-CHI on Human Factors in Computing Systems, CHI 2001, 31 March - 5 April, 2001, pp 515–521, Seattle, WA, USA
6. Jatowt A, Kawai Y, Tanaka K (2008) What can history tell us? towards different models of interaction with document histories. In: ACM: Proceedings of the 19th ACM Conference on Hypertext and Hypermedia, Hypertext 2008, 19–21 June, 2008, pp 5–14, Pittsburgh, PA, USA
7. Matsui Y, Kawai Y, Zhang J (2011) Hybrid web search with social communication service. In: Lecture Notes in Engineering and Computer Science: Proceedings of The International MultiConference of Engineers and Computer Scientists 2011, IMECS 2011, 16-18 March, 2011, pp 687–692, Hong Kong
8. Morris MR, Horvitz E (2007) Searchtogether: An interface for collaborative web search. In: ACM: Proceedings of the 20th Annual ACM Symposium on User Interface Software and Technology, UIST 2007, 7–10 October, 2007, pp 3–12, Newport, Rhode Island, USA
9. Pickens J, Golovchinsky G, Shah C, Qvarfordt P, Back M (2008) Algorithmic mediation for collaborative exploratory search. In: ACM: Proceedings of the 31st Annual International ACM SIGIR Conference on Research and Development in Information Retrieval, SIGIR 2008, 20–24 July, 2008, pp 315–322, Singapore
10. Ranganathan A, Campbell RH, Ravi A, Mahajan A (2002) Conchat: A context-aware chat program. IEEE Pervasive Comput 1(3):51–57

11. Richardson M, Dominowska E, Ragno R (2007) Predicting clicks: Estimating the click-through rate for new ads. In: ACM: Proceedings of the 16th International Conference on World Wide Web, WWW 2007, 8–12 May, 2007, pp 521–530, Banff, Alberta, Canada
12. Twidale MB, Nichols DM, Paice CD (1997) Browsing is a collaborative process. Inform Process Manag 33(6):761–783
13. Viegas FB, Donath JS (1999) Chat circles. In: ACM: Proceedings of the CHI99 Conference on Human Factors in Computing Systems, CHI 1999, 15–20 May, 1999, pp 9–16, Pittsburgh, PA, USA
14. White RW, Bilenko M, Cucerzan S (2007) Studying the use of popular destinations to enhance web search interaction. In: ACM: Proceedings of the 30th Annual International ACM SIGIR Conference on Research and Development in Information Retrieval, SIGIR 2007, 23–27 July, 2007, pp 159–166, Amsterdam, The Netherlands

Chapter 24
The Game of n-Player White-Black Cutthroat and Its Complexity

Alessandro Cincotti

Abstract The game of N-player White-Black Cutthroat is an n-player version of White-Black Cutthroat, a two-player combinatorial game played on graphs. Because of queer games, i.e., games where no player has a winning strategy, cooperation is a key-factor in n-player games and, as a consequence, n-player White-Black Cutthroat played on stars is PSPACE-complete.

Keywords Combinatorial game • Complexity • Cutthroat • n-player game

1 Introduction

The game of Cutthroat [14] is a combinatorial game played on a graph where each vertex has been colored black, white or green. Two players, Left and Right, move alternately. Left removes a black or a green vertex and Right removes a white or a green vertex. After a move, or at the beginning of play, any monochromatic, connected component is also removed from the graph.

For example, if the graph were a single edge with one white and one black vertex then Left moving first would win since he would leave just the white vertex, a monochromatic component which would then be removed. Similarly, Right would win moving first. A winning strategy has been found for White-Black Cutthroat, i.e., Cutthroat where the vertices will only be colored white or black, played on stars [15].

N-player White-Black Cutthroat is the n-player version of White-Black Cutthroat played on graph where each vertex has been labeled by an integer $j \in \{1, 2, \ldots, n\}$. The first player removes a vertex labeled 1. The second player removes a vertex

A. Cincotti (✉)
School of Information Science, Japan Advanced Institute of Science and Technology,
1-1 Asahidai, Nomi, Ishikawa 923-1292, Japan
e-mail: cincotti@jaist.ac.jp

S.I. Ao et al. (eds.), *Intelligent Control and Innovative Computing*, Lecture Notes in Electrical Engineering 110, DOI 10.1007/978-1-4614-1695-1_24,

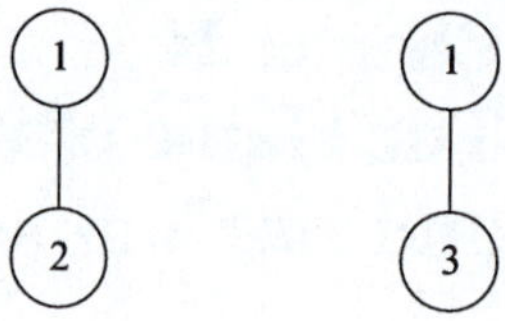

Fig. 24.1 A simple example of queer game in three-player White-Black Cutthroat

labeled 2. The other players move in similar way. After a move, or at the beginning of play, any connected component, where all the vertices have the same label, is also removed from the graph. Players take turns making legal moves in cyclic fashion (1-st, 2-nd, ..., n-th, 1-st, 2-nd, ...). When one of the players is unable to move, that player leaves the game and the remaining n-1 players continue playing in the same mutual order as before. The remaining player is the winner.

We briefly recall the definition of *queer* game introduced by Propp [17]:

Definition 1. A position in a three-player combinatorial game is called queer if no player can force a win.

Such a definition is easily generalizable to n players:

Definition 2. A position in an n-player combinatorial game is called queer if no player can force a win.

In the game of n-player Cutthroat, it is not always possible to determine the winner because of queer games, as shown in Fig. 24.1. In this case, no player has a winning strategy because if the first player removes the vertex on the left, then the third player wins but if the first player removes the vertex on the right, then the second player wins.

In two player games [1,2,11] players are in conflict to each other and coalitions are not allowed but in n-player games [3,5,12,13,18], when the game is queer, only cooperation between players can guarantee a winning strategy, i.e., one player of the coalition is always able to make the last move. As a consequence, to establish whether or not a coalition has a winning strategy is a crucial point.

In previous works, the complexity of different multi-player games has been analyzed [4, 6–9], included a n-player version of Cutthroat where every edge is labeled by a set of integers [10]. In this paper we show that, in White-Black Cutthroat, cooperation between a group of players can be much more difficult than competition and, as a consequence, n-player White-Black Cutthroat played on stars is PSPACE-complete.

2 The Complexity of n-Player White-Black Cutthroat

In this section we show that the PSPACE-complete problem of *Quantified Boolean Formulas* [16], QBF for short, can be reduced by a polynomial time reduction to n-player White-Black Cutthroat.

Let $\varphi \equiv \exists x_1 \forall x_2 \exists x_3 \ldots Q x_n \psi$ be an instance of QBF, where Q is $\exists$ for n odd and $\forall$ otherwise, and ψ is a quantifier-free Boolean formula in conjunctive normal form. We recall that QBF asks if there exists an assignment to the variables x_1, x_3, ..., $x_{2\lceil n/2 \rceil - 1}$ such that the formula evaluates to true. Without loss of generality, let us assume that the number of clauses is odd and each literal is contained in at least one clause because we can always introduce a dummy clause $(x_1, \overline{x_1})$.

Definition 3. A star S_m is a tree with one internal node and m leaves.

If n is the number of variables and k is the number of clauses in ψ, then the instance of n-player White-Black Cutthroat will have $n + k + 2$ players and $2n + k$ stars, organized as follows:

- For each variable x_i, with $1 \leq i \leq n$, we add two new stars. In the first one, the internal node is labeled i and there is a leaf for each clause containing x_i; in the second one, the internal node is labeled i and there is a leaf for each clause containing $\overline{x_i}$. These leaves are labeled j, with $n + 1 \leq j \leq n + k$.
- The last k stars are all S_2 star where the internal node is labeled $n + k + 2$ and the two leaves are labeled $n + k + 1$ and $j \in \{n + 1, n + 2, \ldots, n + k\}$.

Let us suppose that the first coalition is formed by $\lfloor n/2 \rfloor + 1$ players corresponding to the dominoes labeled 2, 4, ..., $2\lfloor n/2 \rfloor$, and $n + k + 1$. The second coalition is formed by the remaining players.

An example is shown in Fig. 24.2 where

$$\varphi \equiv \exists x_1 \forall x_2 \exists x_3 \forall x_4 (C_5 \wedge C_6 \wedge C_7)$$

and

$$C_5 \equiv (\overline{x_1} \vee x_2 \vee \overline{x_3})$$

$$C_6 \equiv (x_1 \vee \overline{x_2} \vee \overline{x_4})$$

$$C_7 \equiv (x_1 \vee x_3 \vee x_4)$$

The problem to determine the winning coalition is strictly connected to the problem of QBF, as shown in the following theorem.

Theorem 1. *Let G be a general instance of n-player White-Black Cutthroat played on stars. Then, to establish whether or not a given coalition has a winning strategy is a PSPACE-complete problem.*

Proof. We show that it is possible to reduce every instance of QBF to a set of stars G representing an instance of n-player White-Black Cutthroat. Previously we have described how to construct the instance of n-player White-Black Cutthroat, therefore we just have to prove that QBF is satisfiable if and only if the second coalition has a winning strategy.

If QBF is satisfiable, then there exists an assignment of x_i such that ψ is true with $i \in \{1, 3, \ldots, 2\lceil n/2 \rceil - 1\}$. If x_i is true, then the i-th player removes the vertex

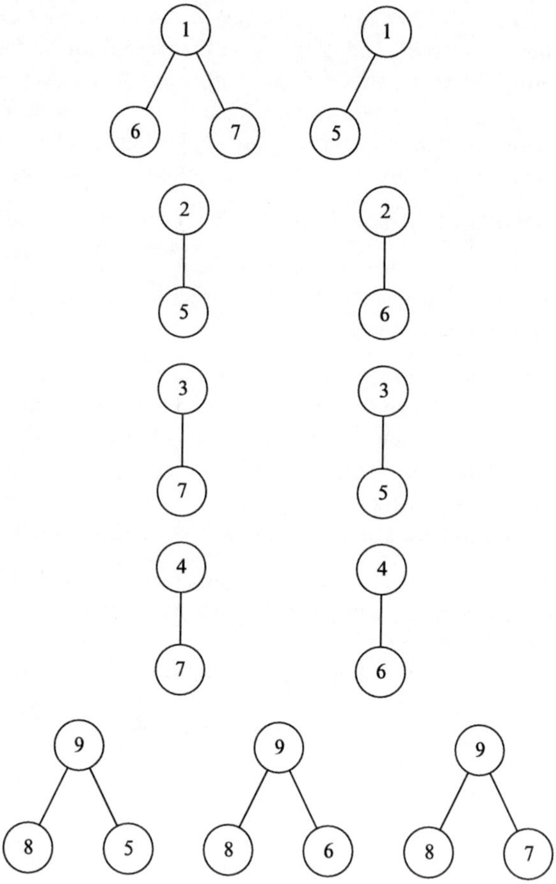

Fig. 24.2 The problem of QBF is reducible to the game of n-player White-Black Cutthroat played on stars

labeled i and, consequently, all the vertices corresponding to the clauses containing $\overline{x_i}$. If x_i is false, then the i-th player removes the vertex labeled i and, consequently, all the vertices corresponding to the clauses containing x_i.

Every clause contains at least a true literal, therefore the i-th player with $i \in \{n+1, n+2, \ldots, n+k\}$ can always remove one vertex from the star corresponding to that literal. The $n + k + 1$-th player removes a leaf from one of the k stars and the $n + k + 2$-th player can play in the same star. In the second turn, all the vertices labeled with $i \in \{1, \ldots, n + k\}$ will be removed and the $n + k + 2$-th player will be able to make the last move because now the number of stars is even. Therefore, the second coalition has a winning strategy.

Conversely, let us suppose that the second coalition has a winning strategy. We observe that the number of stars where the internal node is labeled $n+k+2$ must be odd before that the $n+k+2$-th player makes his/her first move, i.e., the $n+k+1$-th player cannot isolate any vertex labeled $n+k+2$. As a consequence, the i-th player

Fig. 24.3 A polynomial-space algorithm for n-player White-Black Cutthroat

```
Algorithm Check(G,C_0,C_1,p_i)
  j ← coalition (p_i);
  if ∄v ∈ V : label(v) = i then
      C_j ← C_j\{p_i};
      if C_j = ∅ then
          return 1 − j;
      else
          return Check(G,C_0,C_1, next(p_i));
      end
  else
      for all v ∈ V : label(v) = i do
          G' ← remove(G,v);
          if Check(G',C_0,C_1, next(p_i))= j then
              return j;
          end
      end
      return 1 − j;
  end
```

with $i \in \{n+1, n+2, \ldots, n+k\}$ must always be able to remove a vertex from the stars where the internal node is not labeled $n+k+2$, i.e., every clause has at least one true literal and QBF is satisfiable.

Therefore, to establish whether or not a coalition has a winning strategy in n-player White-Black Cutthroat played on stars is PSPACE-hard.

To show that the problem is in PSPACE we present a polynomial-space recursive algorithm to determine which coalition has a winning strategy as shown in Fig. 24.3.

Let us introduce some useful notations:

- $G = (V, E)$ is the graph representing an instance of n-player White-Black Cutthroat.
- p_i is the i-th player.
- C_0 is the set of current players belonging to the first coalition.
- C_1 is the set of current players belonging to the second coalition.
- `coalition`(p_i) returns 0 if $p_i \in C_0$ and 1 if $p_i \in C_1$.
- `label`(v) returns the label of the vertex v.
- `next`(p_i) returns the player which has to play after p_i.
- `remove`(G,v) returns the graph obtained after that the vertex v and any connected component, where all the vertices have the same label, have been removed from G.

Algorithm `Check` performs an exhaustive search until a winning strategy is found and its correctness can be easily proved by induction on the depth of the game tree.

Algorithm `Check` is clearly in PSPACE because the number of nested recursive calls is at most $|V|$ and therefore the total space complexity is $O(|V|^2)$. □

References

1. Albert MH, Nowakowski RJ, Wolfe D (2007) Lessons in play: an introduction to combinatorial game theory. AK Peters, USA
2. Berlekamp ER, Conway JH, Guy RK (2001) Winning way for your mathematical plays. AK Peters, USA
3. Cincotti A (2005) Three-player partizan games. Theor Comput Sci 332(1–3):367–389
4. Cincotti A (2009) On the complexity of n-player Hackenbush. Integers 9:621–627
5. Cincotti A (2010) N-player partizan games. Theor Comput Sci 411(34–36):3224–3234
6. Cincotti A (2010) On the complexity of n-player Toppling Dominoes. In: Ao SI, Castillo O, Douglas C, Feng DD, Lee J (eds) Lecture notes in engineering and computer science: Proceedings of the international MultiConference of engineers and computer scientists 2010, IMECS 2010, 17–19 March, 2010, Hong Kong. Newswood Limited, pp 461–464
7. Cincotti A (2010) On the complexity of n-player cherries. In: Hang Y, Desheng W, Sandhu PS (eds) Proceedings of the 2010 3rd IEEE international conference on computer science and information technology, 9-11 July, 2010, Chengdu. IEEE Press, pp 481–483
8. Cincotti A (2010) On the complexity of some map-coloring multi-player games. In: Ao SI, Castillo O, Huang H (eds) Intelligent automation and computer engineering, vol 52 of LNEE. Springer, pp 271–278
9. Cincotti A (2010) The game of n-player Shove and its complexity. In: Ao SI, Castillo O, Huang H (eds) Intelligent control and computer engineering, vol 70 of LNEE. Springer, pp 285–292
10. Cincotti A (2011) N-player Cutthroat played on stars is PSPACE-complete. In: Ao SI, Castillo O, Douglas C, Feng DD, Lee J (eds) Lecture notes in engineering and computer science: Proceedings of the international MultiConference of engineers and computer scientists 2011, IMECS 2011, 16–18 March, 2011, Hong Kong. Newswood Limited, pp 269–271
11. Conway JH (2001) On numbers and games. A K Peters, USA
12. Li SYR (1978) n-person nim and n-person moore's games. Intl J Game Theor 7(1):31–36
13. Loeb DE (1996) Stable winning coalitions. In: Nowakowski RJ (ed) Games of no chance. Cambridge University Press, Cambridge, pp 451–471
14. McCurdy SK (2003) Cherries and cutthroat: two games on graphs. Master's thesis, Dalhousie University
15. McCurdy SK (2005) Cutthroat, an all-small game on graphs. INTEGERS: Electron J Combin Number Theor 5(2):#A13
16. Papadimitriou CH (1994) Computational complexity. Addison-Wesley, MA
17. Propp JG (2000) Three-player impartial games. Theor Comput Sci 233(1–2):263–278
18. Straffin Jr PD (1985) Three-person winner-take-all games with mc-carthy's revenge rule. Coll Math J 16(5):386–394

Chapter 25
Combinatorial Problems With Closure Structures

Stefan Porschen

Abstract We consider a specific class of combinatorial search respectively optimization problems where the search space gives rise to a closure operator and essentially the hulls are the only relevant subsets that must be checked in a brute force approach. We suggest that such a closure structure can help to reduce time complexities. Moreover we propose two types of (structural) parameterizations of instance classes based on the closure property and outline how it could be used to achieve fixed-parameter tractability (FPT) characterizations. In this setting, three example problems are described: a covering problem from combinatorial geometry, a variant of the autarky problem in propositional logic, and a specific graph problem on finite forests.

Keywords Exact algorithmics • Closure operator • FPT • Combinatorial optimization • Computational complexity

1 Introduction

From the point of view of exact algorithmics it is desirable to get a structural insight into the search space of a combinatorial (optimization) problem in order to reduce the amount of search and therefore decrease time complexities. It turns out that there can occur situations in which a closure operator can be identified which may help to reduce the time complexity of an algorithm because the search region decomposes into equivalence classes corresponding to certain hulls.

In this paper we propose a structural perspective on a specific class of combinatorial problems over set systems. Namely those for which an appropriate closure

---

S. Porschen (✉)
Mathematics Group, Department 4, HTW Berlin, D-10313 Berlin, Germany
e-mail: porschen@htw-berlin.de

S.I. Ao et al. (eds.), *Intelligent Control and Innovative Computing*, Lecture Notes in Electrical Engineering 110, DOI 10.1007/978-1-4614-1695-1_25,

operator can be associated to the base set of each fixed instance of the problem. We develop the basic connections between the closures and the equivalence classes defined by the pre-images of a closure operator. Two classes of combinatorial problems based on such a closure structure are introduced. In the first one, we consider optimization problems such that the objective function is a class invariant with respect to the equivalence relation naturally defined by the closure operator; in this case the search space is the power set of the base set of every problem instance. On basis of such a closure structure it might be possible to reduce the computational time complexity of an algorithm, which is demonstrated for a rectangular covering problem in the Euclidean plane. In the second class of combinatorial problems proposed here, the subsets that are relevant in a brute force search approach, are the hulls of the closure operator which is defined in the search space, i.e., the base set of every problem instance (we call this a *closure property of the second kind*). An example of that type is provided in the context of the autarky problem in propositional logic.

Usually the parameterization of a combinatorial search problem is defined through the size of the solution [4]. Take for instance the vertex cover (VC) problem for a simple undirected graph. In its parameterized version, VC gets as input a graph $G = (V, E)$, and a positive integer k which serves as the parameter. Then one is asked for the existence of a VC of cardinality at most k. Clearly, VC is an NP-complete problem [6]. However, it is *fixed-parameter tractable* (FPT) in the sense that it can be solved by an exact algorithm of exponential worst-case time s.t. the exponential part depends only on the parameter k. The currently best bound is $O(1.2738^k + kn)$ [2] which in addition consumes polynomially bounded space only.

If a combinatorial search problem admits a closure property of the second kind, we propose a *structural parameterization* approach: the parameter of an instance class is defined as the maximal size of the corresponding hulls in the search space. We also propose another parameterization based on the number of the hulls. Such structural parameterizations might help to gain FPT instance classes of such a problem.

We discuss several examples of problems for which a closure property can be defined. The first one is an optimization problem that takes as input a finite set of points in the Euclidean plane and aims at finding a minimal rectangular covering subject to a certain objective function. We state the closure property and report how it can be used to reduce the time complexity of a dynamic programming optimization algorithm. Second, we address the autarky problem in propositional logic. The input instance here is a conjunctive normal form (CNF) formula and the question is, whether there is a subformula which can be satisfied independent of the remaining formula. A closure property closely related to the autarky problem is detected and a guide to a FPT characterization is proposed. Finally, we focus on a certain graph problem originally stemming from a falsifiability problem in propositional logic [5]. It takes as input a finite forest of rooted trees [7] and a mapping f assigning leaves to vertices which are considered as roots of subtrees. Then one is asked to find one leaf of every tree, s.t. none of these leaves is contained in the subtree rooted at the f-image of any other of these leaves. This problem is

NP-complete and belongs to the class FPT w.r.t. the number of trees in the forest as parameter [5]. Identifying a closure operator one can define a variant of this graph problem [12], for which an additional parameter is defined, namely the maximal hull size as stated above. A preliminary version of the material presented here has been published in Porschen [13].

2 Closures and Equivalences

A basic notion in this paper is the well-known closure operator defined on the power set 2^M of a given set M of finite size $n := |M|$. Recall that a closure operator $\sigma : 2^M \to 2^M$, by definition, has the following properties:

1. σ is *extensive*: $\forall S \subseteq M \Rightarrow S \subseteq \sigma(S)$
2. σ is *monotone:* $\forall S_1, S_2 \subseteq M : S_1 \subseteq S_2 \Rightarrow \sigma(S_1) \subseteq \sigma(S_2)$
3. σ is *idempotent:* $\forall S \subseteq M \Rightarrow \sigma(\sigma(S)) = \sigma(S)$

Easy examples are the identity mapping id on 2^M given by $\mathrm{id}(S) = S$, for every $S \in 2^M$, or the constant mapping $c_M(S) = M$, for every $S \in 2^M$. A more interesting example is given by the convex hull operator: Let $M \subseteq \mathbb{R}^2$ be a finite set of points in the euclidean plane. Then $\sigma(S) := \mathrm{conv}(S) \cap M$, for every $S \in 2^M$, defines a closure operator, as is easy to see, where $\mathrm{conv}(S)$ denotes the convex hull (cf. e.g. Boissonnat and Yvinec [1]) of all points in $S \subseteq \mathbb{R}^2$. Obviously, the convex hull operator is extensive and monotone. From $\mathrm{conv}(\mathrm{conv}(S) \cap M) = \mathrm{conv}(S)$ and $\mathrm{conv}(\mathrm{conv}(S)) = \mathrm{conv}(S)$ it directly follows that it is idempotent.

For convenience, we denote the image of a closure operator, i.e., the set of all closures, or *hulls* for short, by $\mathcal{H}_\sigma(M) := \sigma(2^M)$. If $|S| = k$ we call $\sigma(S) =: H(S)$ a k-hull, for $0 \leq k \leq n$, and we denote the collection of all k-hulls by $\mathcal{H}_\sigma^k(M)$. For short let $H(x) := \sigma(\{x\})$, for a single-element set $\{x\}$.

σ induces, w.r.t. $\subseteq$, a lattice on $\mathcal{H}_\sigma(M)$. For two elements $H(S_1), H(S_2) \in \mathcal{H}_\sigma(M)$ their supremum (resp. infimum) is given by

$$\sigma(S_1 \cup S_2) \quad (\text{resp.} \quad H(S_1) \cap H(S_2))$$

In the case $\sigma = \mathrm{id}$ this lattice coincides with the usual subset-lattice of M (cf. e.g. Birkhoff [14]).

Any closure operator on 2^M clearly defines an equivalence relation on 2^M by $S_1 \sim_\sigma S_2$ iff $\sigma(S_1) = \sigma(S_2)$. We refer $\sim_\sigma$ to as the *natural equivalence relation* of σ with classes $[S]_\sigma$. The next result tells us which conditions must be satisfied by an equivalence relation so that it gives rise to a closure operator as above.

Theorem 1. *Let M be a finite set, and let $\sim \subseteq 2^M \times 2^M$ be an arbitrary equivalence relation on 2^M. For each class $[S]$, $S \subseteq M$, define $W_{[S]} := \bigcup [S]$. Then $\sim$ satisfies*

(i) $\forall S \in 2^M : W_{[S]} \in [S]$, *and*
(ii) $\forall S_1, S_2 \in 2^M : S_1 \subseteq S_2 \Rightarrow W_{[S_1]} \subseteq W_{[S_2]}$,

if and only if

$$\sigma : 2^M \ni S \mapsto \sigma(S) := W_{[S]} \in 2^M$$

is a closure operator such that $(*)$: $\sigma(S_1) = \sigma(S_2) \Rightarrow S_1 \sim S_2$, *for all* S_1, $S_2 \in 2^M$.

Proof. First, let σ be a closure operator with property $(*)$. Then it is easy to see that even $\sigma(S_1) = \sigma(S_2) \Leftrightarrow S_1 \sim S_2$ holds for all $S_1, S_2 \in 2^M$. Since, by definition we have $\sigma(S) = W_{[S]}$ we obtain $\sigma(S) = \sigma(W_{[S]})$ because σ is extensive implying $W_{[S]} \in [S]$ which is (i). (ii) is a direct implication by the fact that σ is monotone.

Conversely, let $\sim$ be an equivalence relation satisfying both (i) and (ii). From (ii) it directly follows that σ as stated above is a monotone set mapping. For arbitrary $S \in 2^M$, we further have $S \subseteq W_{[S]} = \sigma(S)$ implying that σ is extensive. Property (i) means $S \sim W_{[S]}$. Therefore we have $S_1 \sim S_2 \Leftrightarrow W_{[S_1]} \in [S_2]$ and $W_{[S_2]} \in [S_1]$ $\Leftrightarrow W_{[S_1]} = W_{[S_2]} \Leftrightarrow \sigma(S_1) = \sigma(S_2)$, for arbitrary $S_1, S_2 \in 2^M$ which particularly implies $(*)$. Finally, we have that σ is idempotent, because (i) leads to $[W_{[S]}] = [S]$. Thus $\sigma(\sigma(S)) = \sigma(W_{[S]}) = \bigcup[W_{[S]}] = \bigcup[S] = \sigma(S)$, for every $S \in 2^M$. □

Closely related is the following assertion:

Theorem 2. *Let σ be an arbitrary closure operator on M. Then its natural equivalence relation $\sim_\sigma$, which is defined on 2^M by taking the preimages of σ as the equivalence classes, has the property*

$$W_{[S]_\sigma} := \bigcup[S]_\sigma = \sigma(S)$$

for every $S \subseteq M$.

Proof. For $S \in 2^M$ arbitrarily fixed, we have $\sigma(T) = \sigma(S)$ for every $T \in [S]_\sigma$. Hence, since $T \subseteq \sigma(T)$, we obtain

$$\bigcup_{T\in[S]_\sigma} T \subseteq \bigcup_{T\in[S]_\sigma} \sigma(T) = \sigma(S)$$

meaning $W_{[S]_\sigma} \subseteq \sigma(S)$. On the other hand, $\sigma(S) = \sigma(\sigma(S))$ implies $S \sim_\sigma \sigma(S)$. Hence, we obtain $\sigma(S) \in [S]_\sigma$ and therefore $\sigma(S) \subseteq \bigcup[S]_\sigma$ together yielding $W_{[S]_\sigma} = \sigma(S)$. □

In the next section we introduce two classes of combinatorial problems with a closure operator associated to every fixed problem instance.

3 Structural Parameters Based on Closure Properties

Consider a combinatorial problem Π such that each of its input instances I is associated to a base set $M(I)$. Let $F(I) := 2^{M(I)}$ denote the power set of the base set $M(I)$. Sometimes, but not in every case, $F(I)$ can be identified with the search space of the problem at hand. Examples for base sets are:

- Discrete point set in the plane in the input instance of a discrete geometry problem.
- Vertex set, or edge set in the input instance of a graph problem.
- Set of variables or expressions in the input instance of a logic problem.

Definition 1. Let Π be a combinatorial problem such that each instance $I \in \Pi$ admits a closure operator σ_I on $F(I)$.

(1) We say that a problem Π has a *closure property of the first kind*, if the search space is $2^{F(I)}$ and the objective is a class invariant w.r.t. the natural equivalence relation defined by σ_I.
(2) We say that Π has a *closure property of the second kind* if $F(I)$ is the search space and for deciding whether $I \in \Pi$, resp. for solving the search variant, it suffices to test all hulls given by σ_I.

Clearly, if a problem has a closure property of any kind, specifically we have an equivalence relation $\sim_{\sigma_I}$ on $F(I)$ defined by $\sigma(I)$. From Theorems 1 and 2 we know that the union of all elements in a given equivalence class belongs to the same class which thus is the supremum of all sets in that class w.r.t. the lattice defined by σ_I. Therefore these suprema (maximal hulls) give rise to a distinct family of class representatives.

The usual parameterization of a problem Π, is defined via the size of the solution [4]. As an example take the VC problem of a simple graph, for which a well-known FPT characterization exists in the following parameterized version:
Input: Graph $G = (V, E)$, $k \in \mathbb{N}$.
Problem: Find a VC of cardinality at most k in G or report that none exists.

In the following we shall define a structural parameterization by the cardinality of the image set of a closure operator, i.e., the number of all hulls. A useful observation is:

Lemma 1. *Let σ be a closure operator on M with at most k many 1-hulls, i.e., $|\mathcal{H}^1_\sigma(M)| = |\{H(x) := \sigma(\{x\}) : x \in M\}| \leq k$, for any fixed $k < n$, where n denotes the size of M. Then we have $|\mathcal{H}_\sigma(M)| \leq 2^k$.*

Proof. It suffices to show that for arbitrary $\emptyset \neq S \in 2^M$ it holds that $\sigma(S) = \sigma(\bigcup_{x \in S} H(x))$. Herewith the Lemma follows because $\bigcup_{x \in S} H(x)$ can be composed then out of at most k 1-hulls. Hence $\sigma(S)$ corresponds to a subset of $\{1, 2, \ldots, k\}$ implying the Lemma.

To prove the preceding assertion, let $\emptyset \neq S \in 2^M$ be arbitrary. Then there is $1 \leq p \leq n$ s.t. $|S| = p$; let $S = \{x_{i_1}, \ldots, x_{i_p}\}$. Then clearly $H(x_{i_q}) \subseteq \sigma(S)$ holds for $1 \leq q \leq p$. Therefore we obtain $\bigcup_{j=1}^p H(x_{i_j}) \subseteq \sigma(S)$ implying $\sigma(\bigcup_{q=1}^p H(x_{i_q})) \subseteq \sigma(S)$ because σ is monotone and idempotent. On the other hand, we clearly have $x_{i_q} \in H(x_{i_q})$, $1 \leq q \leq p$, thus $S \subseteq \bigcup_{q=1}^p H(x_{i_q})$; and since σ is monotone we obtain $\sigma(S) \subseteq \sigma(\bigcup_{q=1}^p H(x_{i_q}))$ yielding the assertion and the Lemma. □

Let us now propose a parameterization of instance classes of problems with closure structures based on the number of hulls.

Definition 2. Let Π be a problem with closure property of the second kind. In the *parameterization of* Π *by the number of hulls* we define the instance subclass according to fixed pre-defined parameter value k as

$$\Pi_k := \left\{ I \in \Pi : \exists j \leq |M(I)| : \sum_{i=1}^{j} |\mathcal{H}^i_{\sigma_I}(M(I))| \leq O(f(k)) \right\}$$

where $f : \mathbb{N} \to \mathbb{N}$ is an arbitrary fixed function, and $\mathcal{H}^j_{\sigma_I}(M(I)) := \{\sigma_I(S) : S \subseteq M(I), |S| = j\}$.

Observe that, according to Lemma 1, we specifically have that I is a member of Π_k if σ_I yields at most k 1-hulls.

Theorem 3. *Let* Π_k*,* $k \in \mathbb{N}$*, have a closure property of the second kind, and assume that for each* $I \in \Pi$ *we can check the membership in* Π_k *in polynomial time. Moreover assume that the subproblem corresponding to each hull is computable in polynomial time (or at least in FPT-time w.r.t.* k*). Then* Π_k *belongs to FPT w.r.t.* k*.*

Proof. The closure property of the second kind implies that we only have to check all hulls instead of all subsets of the search space $F(I)$. By assumption, the number of hulls in the instances of Π_k are bounded according to Definition 2. From the assumption $\sum_{i=1}^{j} |\mathcal{H}^i_{\sigma_I}(M(I))| \leq f(k)$ we particularly conclude that $|\mathcal{H}^1_{\sigma_I}(M(I))| \leq O(f(k))$. Hence, we have to check at most $2^{O(f(k))}$ many hulls of σ_I according to Lemma 1. Therefore, if the subproblem corresponding to each hull can be tested in polynomial time or in FPT-time w.r.t. k, we obtain an overall FPT-time w.r.t. k. □

Apart from the parameterization by the number of hulls, another one may be useful, namely that by the maximal size of all hulls: Let Π be a problem s.t. every instance admits a closure operator. Note that we do not necessarily require a closure property as defined above. For $k \in \mathbb{N}$ fixed, we define the instance class Π'_k by

$$\Pi'_k := \{I \in \Pi : \max\{|\sigma_I(S)| : S \in 2^{M(I)}\} \leq k\}$$

i.e., the collection of all Π-instances I such that the size of every hull is bounded by parameter k.

4 Several Example Problems

This section is devoted to illustrate that problems with closure properties occur in diverse areas. We start with a rectangular covering optimization problem which is identified to have a closure property of the first kind. Another problem lies in the

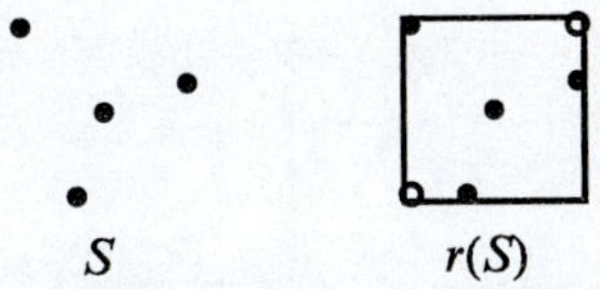

Fig. 25.1 *Black dots* represent points of S (*left*), *white dots* represent the diagonal points $z_u(S), z_d(S)$ of the rectangle $r(S)$ enclosing S (*right*)

area of propositional logic, and turns out to have a closure property of the second kind. Finally, a graph problem is discussed which is well-known to be in FPT [5]. Recently a variant of this problem has been studied [12], and an FPT classification together with a so-called *kernelization* [4] have been constructed. Here we discuss an underlying closure property of this problem variant on basis of the maximal hull size parameter.

4.1 Rectangular Coverings of Point Sets

We consider the following combinatorial optimization problem.
Given: Finite set M of points arbitrarily distributed in the euclidean plane. Objective function w on rectangular patches.
Problem: Find a covering R of M by rectangles such that $w(R) := \sum_{r \in R} w(r)$ is minimized.
Observe that a covering R is a selection of subsets $S \subseteq M$ each of which is covered optimally. So the search space, actually, is 2^{2^M} which by the closure structure is reduced to $2^{\mathcal{R}(M)}$ where $\mathcal{R}(M)$ is the set of all maximal hulls over all equivalence classes given by the corresponding closure operator as defined below.

The underlying closure structure of this problem relies on the concept of *base points* $b(S) = \{z_d(S), z_u(S)\}$ of every $S \in 2^M$ which are defined through: $x_d(S) := \min_{z \in S} x(z)$, $y_d(S) := \min_{z \in S} y(z)$ and $x_u(S) := \max_{z \in S} x(z)$, $y_u(S) := \max_{z \in S} y(z)$. The base points immediately yield the upper right and lower left diagonal points of the smallest rectangle $r(S)$ enclosing S as depicted in Fig. 25.1.

There is an equivalence relation on 2^M defined by $S_1 \sim S_2$ iff $b(S_1) = b(S_2)$, for $S_1, S_2 \in 2^M$ with classes $[S]$.
Defining

$$\sigma : 2^M \ni S \mapsto \sigma(S) := r(S) \cap M \in 2^M$$

and $\mathcal{R}(M) := \{S \subseteq M : \sigma(S) = S\}$, we have:

Theorem 4 ([10]). *$\sigma : 2^M \to 2^M$ is a closure operator; and there is a bijection between $\mathcal{R}(M)$ and $2^M / \sim$ defined by $S \mapsto [S], S \in \mathcal{R}(M)$. Moreover $\mathcal{R}(M)$ is of polynomial size in the variable $|M|$ and can be also be computed in polynomial time.*

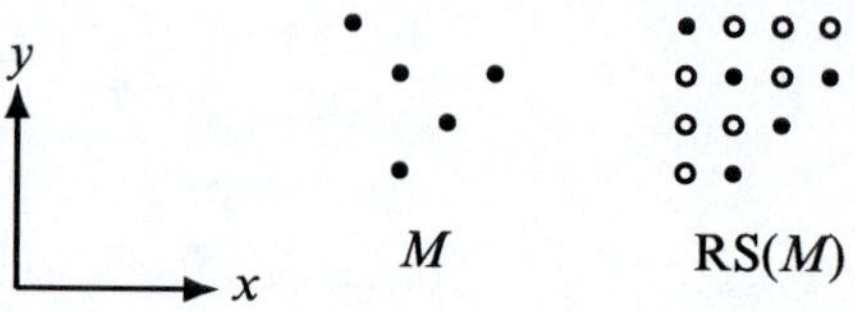

Fig. 25.2 A set M and its rectangular subset closure $\mathrm{RS}(M)$

Obviously the objective w_I is a class invariant because $w_I(r(T))$ has the same value on every $S \in [S]$. Hence we face a problem with closure property of the first kind. Using this closure structure, the time complexity of a dynamic programming approach for solving the covering problem can be decreased. Concretely, one can show the following results [10]: There exists a dynamic programming algorithm for solving this problem with time complexity $O(n^2 3^n)$, where n is the size of M. On behalf of the closure property in $F(I) = 2^{M(I)}$ this time complexity can be reduced to $O(n^6 2^n)$. Due to a more subtle closure property investigated in Porschen [9] one can establish in certain situations the slightly better bound of $O(n^4 2^n)$.

Roughly speaking this additional closure structure appears as follows using the concept of the rectangular subset closure $\mathrm{RS}(M)$ of a point set M introduced in Porschen [9]. By definition $\mathrm{RS}(M)$ is the smallest superset of M containing the base points of all subsets of M: $\mathrm{RS}(M) = M \cup \bigcup_{S \subseteq M} b(S)$.

As an example consider Fig. 25.2, where all *additional* base points, for a given set M, contained in the corresponding $\mathrm{RS}(M)$ are represented as white dots.

The rectangular subset closure gives rise to a closure operator defined for a fixed finite rectangular region $L \subset \mathbb{N} \times \mathbb{N}$ in the integer grid. Due to Porschen [9] one has that $\mathrm{RS} : 2^L \to 2^L$ in fact is a closure operator.

By the result above we can compute $\mathcal{R}(M)$ in polynomial time $O(p(|M|))$ where M is the set of input points, and p is a polynomial. Unfortunately, it is an open problem, of how to check the subproblems corresponding to each hull in polynomial time, or even in FPT-time. Therefore the question whether one can achieve a FPT-characterization of this problem w.r.t. the parameterization by the number of hulls is left for future work.

4.2 Autarkies in CNF Formulas

In this section we consider the propositional satisfiability problem (SAT) on CNF formulas [3]. It is convenient to regard a CNF formula C as a set of its clauses $C = \{c_1, \ldots, c_m\}$. By $V(C)$ we denote the set of all propositional variables occurring in C. The concept of autarky in the context of CNF-SAT was introduced in Monien and Speckenmeyer [8]. Roughly speaking, an *autark* set of variables can be removed from a CNF formula without affecting its satisfiability status. More precisely, given CNF formula C, we call a subset $U \subseteq V(C)$ an *autark set (of variables)*, iff there exists a (partial) truth assignment $\alpha : U \to \{0, 1\}$ satisfying the subformula $C(U)$ defined by $C(U) := \{c \in C : V(c) \cap U \neq \emptyset\}$. Removing $C(U)$ from C therefore

preserves the satisfiability status of the resulting formula compared to the original one. Consider the decision problem AUT:
Input: $C \in$CNF
Question: $\exists$ an autark set $U \subseteq V(C)$?
It is not hard to see that AUT is NP-complete. However, a basic open question is whether AUT is FPT regarding the traditional parameterization, namely w.r.t. the parameter k defining the maximum cardinality of an autark set in the input formula. In the following we propose an alternative approach based on a parameterization defined by a closure property: Given $C \in$ CNF, then for every $U \subseteq V(C)$ we define the set $\sigma_C(U) \subseteq V(C)$ as

$$\sigma_C(U) := V(C) - V(C - C(U))$$

We call $\sigma_C(U)$ the *autarky closure or autarky hull* of U (introduced as *variable hull* in Porschen [11]).

Lemma 2. *Given $C \in$ CNF, then $\sigma_C : 2^{V(C)} \to 2^{V(C)}$ as defined above is a (finite) closure operator.*

Proof. Obviously σ_C is extensive. Let $U_1, U_2 \subseteq V(C)$ with $U_1 \subseteq U_2$, then $C(U_1) \subseteq C(U_2)$, hence $V(C - C(U_2)) \subseteq V(C - C(U_1))$. Now suppose there is $x \in \sigma_C(U_1)$ and $x \notin \sigma_C(U_2)$, then by definition $x \in V(C - C(U_2))$ and therefore $x \in V(C - C(U_1))$ contradicting the assumption, thus (ii) holds. Finally, let $W := \sigma_C(U)$, for $U \in 2^{V(C)}$. We have $C(W) = C(U)$ since no variable of W occurs outside $C(U)$, yielding $\sigma_C(W) = W$ which means that σ is idempotent. □

Next, we have:

Lemma 3. *Given $C \in$ CNF.*

1). For $U_1, U_2 \subset V(C)$, $U_1 \sim U_2 : \Leftrightarrow C(U_1) = C(U_2)$ defines an equivalence relation on $2^{V(C)}$ with classes $[U]$.
2). The quotient space $2^{V(C)}/\sim$ is in 1:1-correspondence to $\{\sigma_C(U) : U \in 2^{V(C)}\}$.

Proof. The first part is obvious. For proving the second part we claim that for each $U_1, U_2 \in 2^{V(C)}$ holds

$$\sigma_C(U_1) = \sigma_C(U_2) \quad \Leftrightarrow \quad U_1 \sim U_2$$

from which 2.) obviously follows. Now $U_1 \sim U_2$ means $C(U_1) = C(U_2)$ implying $\sigma_C(U_1) = \sigma_C(U_2)$. For the reverse direction we observe that $C(\sigma_C(U)) = C(U)$, for each $U \subseteq V(C)$. Therefore, $\sigma_C(U_1) = \sigma_C(U_2)$ implies $C(U_1) = C(\sigma_C(U_1)) = C(\sigma_C(U_2)) = C(U_2)$ thus $U_1 \sim U_2$. □

The next result justifies the notion autarky closure:

Lemma 4. *For $C \in$ CNF and $U \subseteq V(C)$, we have that $\sigma_C(U)$ is autark if U is autark.*

Proof. Suppose U is autark, but $\sigma_C(U)$ is not autark. Because $U \sim \sigma_C(U)$, we have $C(U) = C(\sigma_C(U))$, hence any truth assignment $\alpha : U \rightarrow \{0, 1\}$ satisfying $C(U)$ also satisfies $C(\sigma_C(U))$, thus $\sigma_C(U)$ is an autark set. □

Observe that the last result tells us that instead of checking all subsets of $V(C)$ for autarky, it suffices to check the hulls only. Indeed, there can be left no autark set, because, if U is autark also $\sigma_C(U)$ is. And supposing no autarky hull W is autark, then there is no autark set at all, because otherwise its hull must have been checked positive for autarky. Therefore, we have a problem with a closure property of the second kind.

An autarky hull is called *free* (cf. Porschen [11]) if it does not contain any subhull.

Lemma 5 ([11]). *All free hulls of C are 1-hulls. There exists at most $|V(C)|$ free hulls of C.*

Free hulls have the computational property:

Lemma 6 ([11]). *A free hull $U \subset V(C)$ can be checked for autarky in linear time.*

Let $k \in \mathbb{N}$ be fixed, and suppose we were able to identify the instance class $\mathcal{C}(1,k) \subseteq$ CNF defined through the requirement that every $C \in \mathcal{C}(1,k)$ exhibits at most k 1-hulls. Then by Theorem 3 we would be able to achieve an FPT characterization for autarky testing in this class if one could check each 1-hull for autarky in FPT-time (not just the free 1-hulls). It is a future work task to investigate this question.

4.3 A Variant of the Shadow Problem

Next we consider a specific problem from algorithmic graph theory, for basic notions on that topic cf. e.g. Golumbic [7]. The problem is called the *shadow (independent set) problem* (SIS), and represents a falsifiability problem from propositional logic in terms of graph theory. SIS is given as follows:
Input: Finite forest F and a function, $f : L(F) \rightarrow V(F)$ from the set of all leaves into the set of all vertices of F, called the *shadow map*
Question: Does exist a set of $|F|$ many leaves, exactly one from each tree (called *transversal*) that are *mutually shadow independent*?
In this context, let the *shadow of a leaf* ℓ be the set of all leaves in the subtree rooted at the vertex $f(\ell)$. Then two leaves ℓ_1, ℓ_2 are called *shadow-independent* iff ℓ_1 is no element of the shadow of ℓ_2 and vice versa; this notion directly transfers to every set of leaves. SIS stems from the falsifiability problem of pure implicational formulas in propositional logic and is NP-complete.

A parameterized version of SIS was defined in Franco et al. [5] where the parameter is the number of trees in the forest F: For fixed $k \in \mathbb{N}$ let $\mathrm{SIS}_k := \{(F, f) \in \mathrm{SIS}: |F| \leq k\}$. Also a first FPT characterization of the time complexity of SIS_k was achieved in [5]. However, for SIS_k so far no kernelization has been constructed explicitly.

Now, in Porschen [12] a problem variant is proposed for which an FPT-bound with kernelization was provided. In fact this bound is based on the maximum closure parameter for a specific closure property which can be identified as follows: Consider the set $\hat{L}$ of all leaves in the forest which occur in at least one shadow. Then a closure operator σ on $2^{\hat{L}}$ is defined as follows: For $S \subseteq \hat{L}$ s.t. all leaves of S lie in the same tree, let $\sigma(S)$ be the set of leaves in the largest shadow containing the whole set S; in Porschen [12] this largest tree is called envelope. In the remaining case, S is distributed over more than one tree. Then $\sigma(S)$ is defined as the union over all these trees of the leaves in every envelope containing the corresponding fragment of S in the tree.

On basis of this closure property one can define a parameter resting on the maximum hull cardinality bound: $s_f := \max\{|\sigma(S)| : S \subseteq \hat{L}\}$. In fact this parameter corresponds to the largest shadow in F generated by f. The corresponding vector-parameter $\kappa := (k, s) \in \mathbb{N}^2$ is a pair, and one arrives at

$$\mathrm{SIS}_\kappa := \{(F, f, \kappa) : (F, f) \in \mathrm{SIS}, |F| \leq k, s_f \leq s\}$$

As in SIS_k the parameter component k controls the forest F, whereas the second parameter component s controls the shadow map f

The next result from Porschen [12] states that the problem variant SIS_κ is in the class FPT w.r.t. κ and moreover has a bound of the kernel form:

Theorem 5 ([12]). *Whether $(F, \sigma, \kappa = (k, s))$ belongs to SIS_κ can be decided in $O(k \cdot n^2 + [s \cdot \rho(k, s)]^3 3^k)$ time, where $n = |V(F)|$ and $\rho(x, y) := x(x-1)(y+1)$.*

Therefore the maximal hull parameterization indeed enables one to construct a kernelization for that variant of SIS.

5 Concluding Remarks and Open Problems

We introduced two classes of combinatorial problems based on closure structures associated to the corresponding search spaces. That perspective offers the proposal for a *structural* parameterization of such combinatorial problems which possibly could help to gain a FPT complexity characterization in case of NP-hardness. A guideline of how this could be achieved was outlined. Moreover we discussed three example problems and examined their closure structures as well as to what extent it could be used to decrease time bounds or to obtain FPT improvements, respectively.

There are left several open questions for future work: Does there exist a connection between the usual parameterization of problems according to the solution size and the structural one according to a closure structure? And/or, does it make sense to consider vector parameterizations: solution size and structural parameters simultaneously? Can one find other relevant examples for problems with closure properties. Finally, the question whether an FPT-characterization can be provided based on the number of hulls of the corresponding closure operator remained open for the first and second example problems.

References

1. Boissonnat JD, Yvinec M (1998) Algorithmic geometry, Cambridge University Press, Cambridge
2. Chen J, Kanj IA, Xia G (2010) Improved upper bounds for vertex cover. Theor Comp Sci 411:3736–3756
3. Cook SA (1971) The complexity of theorem proving procedures. In: Proceedings of the 3rd ACM Symposium on Theory of Computing, ACM, Ohio, USA, pp 151–158
4. Downey RG, Fellows MR (1999) Parameterized complexity. Springer, New York
5. Franco J, Goldsmith J, Schlipf J, Speckenmeyer E, Swaminathan RP (1999) An algorithm for the class of pure implicational formulas. Discrete Appl Math 96:89–106
6. Garey MR, Johnson DS (1979) Computers and intractability: a guide to the theory of NP-completeness. W. H. Freeman and Company, San Francisco
7. Golumbic MC (1980) Algorithmic graph theory and perfect graphs. Academic Press, New York
8. Monien B, Speckenmeyer E (1985) Solving satisfiability in less than 2^n steps. Discrete Appl Math 10:287–295
9. Porschen S (2005) On the rectangular subset closure of point sets. Proc ICCSA/CGA 2005, LNCS 3480:796–805
10. Porschen S (2006) Algorithms for rectangular covering problems. Proc ICCSA/CGA 2006, LNCS 3980:40–49
11. Porschen S (2007) A CNF formula hierarchy over the hypercube. Proc AI 2007, LNAI 4830:234–243
12. Porschen S (2009) An FPT-variant of the shadow problem with Kernelization. Proc ICCS 2009, 432–439, Hong Kong
13. Porschen S (2011) On problems with closure properties. Lecture notes in engineering and computer science: Proc IMECS 2011, pp. 258–262, 16–18 March, 2011, Hong Kong
14. Birkhoff G (1995) Lattice theory. American Mathematical Society, Providence, Rhode Island

Chapter 26
Decidable Containment Problems of Rational Word Relations

Wojciech Fraczak and Stéphane Hassen

Abstract We study a particular case of the inclusion problem for rational relations over words. The problem consists in checking whether a submonoid, M, is included in a rational relation, R. We show that if M is rational and commutative then the problem $M \subseteq R$ is decidable. In the second part of the paper we study the inclusion problem, $M \subseteq \downarrow R$, where M is a commutative submonoid and $\downarrow R$ is the prefix-closure of a rational word relation R. We describe an algorithm which solves the problem in a polynomial time, assuming that the number of tapes (arity of the word relation) is constant.

Keywords Formal language • Multi-tape automata • Rational relation • Inclusion

1 Introduction

Multi-tape finite automata play an important role in many areas of the theoretical computer science, computational linguistics, and even software engineering, [5, 10, 12]. However, unlike the usual (1-tape) automata which defines *rational languages*, the multi-tape automata defining *rational (word) relations* are "difficult". From the three major problems from language theory, namely *membership*, *equality*, and *inclusion*, for rational relations only the *membership* problem is decidable whereas *equality*, and thus *inclusion*, are not [13, 14].

In this paper (which revises and extends our previous work, [6]) we study the following inclusion problem for rational word relations. Let M be a rational submonoid of $(\Sigma^*)^d$ and $R \subseteq (\Sigma^*)^d$ a rational word relation, both given by finite

W. Fraczak (✉) • S. Hassen
Dépt d'informatique et d'ingénierie, Université du Québec en Outaouais, 101 rue Jean-Bosco, Gatineau, QC J8X 3X7, Canada
e-mail: fraczak@uqo.ca; shassen@univ-reunion.fr

S.I. Ao et al. (eds.), *Intelligent Control and Innovative Computing*, Lecture Notes in Electrical Engineering 110, DOI 10.1007/978-1-4614-1695-1_26,

multi-tape automata. We want to check if $M \subseteq R$. It is worth noting that already for $M = w^*$ (one generator) and $d = 1$ (one tape) the problem is co-NP-hard, [9, 15]. In general, the problem is undecidable, which follows directly from the undecidability of the *Post correspondence problem.* However, in the case when M is commutative, i.e., when $ab = ba$ for every a, b in M, the problem becomes solvable. Our proof of the decidability of the problem uses two known results. The first result, also proved in this paper in the case of the word relations, says that the intersection of a recognizable set with a rational set is rational [1, 7]. The second result we need is the well-known decidability result for Presburger arithmetic (see, e.g., Fischer and Rabin [4]). The combination of that two results yields an elegant (but high complexity) procedure for solving the problem.

In the second part of the paper we study a slightly different inclusion problem: $M \subseteq \downarrow R$, where $\downarrow R$ is the prefix-closure of a word relation R given by a finite multi-tape automaton A. We show that $\downarrow R$ is rational, and thus, by the result of the first part, the problem is decidable. However our construction of a finite automaton for $\downarrow R$ yields $|A|^{2^d}$ states, where $|A|$ is the number of states of A and d is the number of tapes (arity of the word relation). This amplifies the complexity of the already expensive procedure. Thus, we explore an alternative way to tackle the problem. Firstly, we observe that for any commutative submonoid M of $(\Sigma^*)^d$, there exists a multiword $w \in (\Sigma^*)^d$ such that $\downarrow w^* = \downarrow M$. Secondly, we propose a new algorithm which, assuming that d is a constant, solves $w^* \subseteq \downarrow R$ in a polynomial time with respect to the size of w and A. The motivation for considering the class of prefix-closed rational relations comes from our study of FIFO-channel systems [2, 11] as the semantics domain of programming languages.

2 Rational and Recognizable Word Relations

Let Σ be a finite set of *letters*, usually called *alphabet*. By Σ^* we denote the set of all *words* which are finite sequences over Σ, with the empty word denoted by ε.

A *multiword* of dimension $d \geq 1$, also called d-word, is a d-dimensional vector of words. The ith dimension of a multiword w will be referred to by $w[i]$. We introduce the notation (i, u), where $i \in \{1, \dots, d\}$ and $u \in \Sigma^*$, to describe the multiword having the empty word ε on each dimension but the i-th one which is u, i.e.:

$$(i, u)[k] \stackrel{\text{def}}{=} \begin{cases} u & \text{if } i = k \\ \varepsilon & \text{otherwise.} \end{cases}$$

For example, by $(2, aba)$ we describe multiword $(\varepsilon, aba, \varepsilon, \dots, \varepsilon)$.

The set $(\Sigma^*)^d$ of all d-words over an alphabet Σ constitute a monoid where the *concatenation* is defined component-wise

$$(u[1], \dots, u[d]) \cdot (v[1], \dots, v[d]) \stackrel{\text{def}}{=} (u[1]v[1], \dots, u[d]v[d])$$

and the neutral element, $\mathbb{1} = (\varepsilon, \ldots, \varepsilon)$, is the d-word with ε on each of its dimensions.

The *length* $|w|$ of a multiword w is its total number of letters. Also, if S is a set, by $|S|$ we denote its cardinality.

Subsets of $(\Sigma^*)^d$ are called *relations* on words. The rational operations (concatenation, union and Kleene-star) are defined as usual. Let $R, R' \subseteq (\Sigma^*)^d$:

$$RR' \stackrel{\text{def}}{=} \{uv \mid u \in R, v \in R'\}$$

$$R + R' \stackrel{\text{def}}{=} R \cup R'$$

$$R^* \stackrel{\text{def}}{=} \bigcup_{k \in \mathbb{N}} R^k, \text{ with } R^k \stackrel{\text{def}}{=} \begin{cases} \{\mathbb{1}\} & \text{if } k = 0 \\ RR^{k-1} & \text{if } k > 0 \end{cases}$$

A usual way to deal with words is by means of automata. To each language (1-dimensional relation) in the Chomsky hierarchy corresponds a class of finitely described devices called automata. Among these classes, there are *finite automata* accepting the regular languages. Based on a work of Rabin and Scott, Elgot and Mezei gave a straight generalization of the finite automata in the sense that the automata may still be described by finite graphs but work on several tapes.

A *multi-tape finite automaton* of dimension d, or d-tape automaton, is a tuple $A = (Q, \Sigma, d, \delta, I, F)$ where Q is a finite set of states, Σ an alphabet, $d > 0$ a number of tapes (dimensions), $\delta \subseteq Q \times (\Sigma^*)^d \times Q$ a finite transition relation, and $I, F \subseteq Q$ are sets of *initial* and *final* states, respectively.

An *execution* π *starting* in state q and *ending* in state p is a sequence of transitions $\pi = (q_0, u_0, p_0) \ldots (q_k, u_k, p_k)$ such that $q = q_0$, $p = p_k$, and for all $i \in \{1, \ldots, k\}$, $q_i = p_{i-1}$. We then say that the state p is *accessible* from q and that q is co-accessible from p. The *label* $\lambda(\pi)$ of execution π is the d-word $u_0 u_1 \cdots u_k$. If $p = q$ then π can be the empty sequence of transitions, describing the empty execution with label $\mathbb{1}$.

An execution starting in an initial state is called an *initial* execution. An initial execution ending in a final state is said to be *accepting*. The set of labels of all accepting executions defines a relation $R(A) \subseteq (\Sigma^*)^d$.

The family RAT of all relations which can be defined by a finite multitape automaton coincides with the closure of the finite relations by rational operations. The elements of family RAT are called *rational relations*.

Another important class of subsets of a monoid are *recognizable sets* which, in the case of word relations, is a strict subclass of rational sets. More precisely, a relation R is *recognizable* if there exists a finite monoid M and a monoid morphism $\phi : R \to M$ such that $R = \phi^{-1}(\phi(R))$. The family of the recognizable relations, denoted by REC, corresponds to all subsets of $(\Sigma^*)^d$ which can be represented as a finite union of direct products of the form $X_1 \times \cdots \times X_d$, where $X_1, \ldots, X_d$ are rational subsets of Σ^* (see, e.g., [3], where the result is attributed to Mezei).

3 Containment of Submonoids in Rational Relations

A word relation R is a *submonoid* if $R^* \subseteq R$.

Let M be a rational submonoid of $(\Sigma^*)^d$ and $R \subseteq (\Sigma^*)^d$ a rational word relation, both given by finite multi-tape automata. We want to check if $M \subseteq R$.

It is relatively easy to reduce an instance of the Post correspondence problem of size n (where n is the number of tiles) to an instance of the containment problem with $d = 2$ and where M has n generators. Therefore, in general, the containment problem is undecidable.

In what follows we will assume that M is commutative, i.e., $ab = ba$ for every a, b in M, and we will show that in this case the containment problem becomes decidable.

Given a relation $R \subseteq (\Sigma^*)^d$, we define $\mathbb{D}(R) \subseteq (\Sigma^*)^d$ as follows:

$$\mathbb{D}(R) \overset{\text{def}}{=} \{(u_1[1], u_2[2], \ldots, u_d[d]) \mid u_1, u_2, \ldots, u_d \in R\}.$$

Intuitively, a multiword w is in $\mathbb{D}(R)$ if for every dimension $i \in \{1, \ldots, d\}$ there exists a multiword $u_i \in R$ such that $u_i[i] = w[i]$. For example:

$$\mathbb{D}(\{(ab, b, \varepsilon), (b, b, a)\}) = \{(ab, b, \varepsilon), (ab, b, a), (b, b, \varepsilon), (b, b, a)\}.$$

Lemma 26.1. *If $M \subseteq (\Sigma^*)^d$ is a submonoid then so is $\mathbb{D}(M)$. Moreover, if M is commutative then $\mathbb{D}(M)$ is commutative, too.*

Proof. By definition $M \subseteq \mathbb{D}(M)$, thus $\mathbb{1} \in M$ implies $\mathbb{1} \in \mathbb{D}(M)$. If $w, w' \in \mathbb{D}(M)$ then for every dimension $i \in \{1, \ldots, d\}$ there exist $u_i, u_i' \in M$ such that $u_i[i] = w[i]$ and $u_i'[i] = w'[i]$. Clearly, $u_i u_i' \in M$ implies $ww' \in \mathbb{D}(M)$, and $u_i u_i' = u_i' u_i$ implies $ww' = w'w$. □

Let $\psi : (\Sigma^*)^d \to (\{a\}^*)^d$ be a monoid morphism defined by:

$$\psi(w) \overset{\text{def}}{=} \left(a^{|w[1]|}, \ldots, a^{|w[d]|}\right).$$

For example, ψ maps (abb, b, ε) to (aaa, a, ε). Intuitively, for a given multiword w, $\psi(w)$ may be seen as the unary representation of lengths of the dimensions of w. As usual, for a relation R we write $\psi(R)$ to denote $\{\psi(w) \mid w \in R\}$.

Lemma 26.2. *For a relation R and a commutative monoid M we have*

$$M \subseteq R \iff \psi(M) \subseteq \psi(R \cap \mathbb{D}(M)).$$

Proof. $M \subseteq R \implies \psi(M) \subseteq \psi(R \cap \mathbb{D}(M))$ follows immediately from $M \subseteq \mathbb{D}(M)$ and the definition of ψ.

In order to prove the implication from right to left we observe that for all $u, v \in \mathbb{D}(M)$, we have:

$$\forall u, v \in \mathbb{D}(M)\ \psi(u) = \psi(v) \implies u = v. \tag{26.1}$$

Indeed it is true, since M is commutative and by Lemma 26.1, $uv = vu$. By the usual argument of combinatorics of words, u and v are powers of a same multiword. Thus they are equal since their lengths on every dimension are equal.

Now, the implication $\psi(M) \subseteq \psi(R \cap \mathbb{D}(M)) \implies M \subseteq R$ can be proved by contradiction. Suppose that there exists $u \in M$ such that $\psi(u) \in \psi(R \cap \mathbb{D}(M)$ and $u \notin R$.

Since $\psi(u) \in \psi(R \cap \mathbb{D}(M))$, there exists $v \in R \cap \mathbb{D}(M)$ such that $\psi(u) = \psi(v)$. By Lemma 26.1, we have $u \in \mathbb{D}(M)$, and by (26.1) we have $u = v$. Thus, we conclude that $u \in R$. □

Since, as we are going to see, $R \cap \mathbb{D}(M)$ is rational whenever R and M are rational, the characterization of the inclusion from Lemma 26.2 allows us to reduce the problem to an instance of the inclusion problem for rational relations over one letter alphabet. That follows from the two properties:

1. If R is rational then $\mathbb{D}(R)$ is recognizable, and
2. The intersection of a recognizable relation and a rational relation yields a rational relation.

Lemma 26.3. *If $R \subseteq (\Sigma^*)^d$ is rational then $\mathbb{D}(R)$ is recognizable.*

Proof. $\mathbb{D}(R)$ can be written directly in the "recognizable" form:

$$\mathbb{D}(R) = R[1] \times R[2] \times \ldots \times R[d],$$

where $R[i] = \{w[i] \mid w \in R\}$, for $i \in \{1, \ldots, d\}$. Obviously, every $R[i]$ is rational (one-tape) language. □

Lemma 26.4. *Let $rec \in \mathrm{REC}$ be a recognizable relation and $rat \in \mathrm{RAT}$ be a rational relation, both over $(\Sigma^*)^d$. Intersection $rec \cap rat$ yields a rational relation over $(\Sigma^*)^d$.*

Proof. The general property that "the intersection of a recognizable part with a rational part of an arbitrary monoid M yields a rational part of M" was already proved, e.g., in Berstel [1] and Gilmer [7]. Here we sketch a constructive proof in the case of word relations by building a finite multi-tape automaton accepting the relation $rec \cap rat$.

To compute the intersection of these two, it is not necessary to consider the whole rec relation. We show that the problem may be simplified by considering only the intersection of rat with a rational language (1 dimension).

The recognizable relation rec may be given as a finite union of direct products of recognizable subsets over Σ^*: $rec = \bigcup_{1 \leq j \leq k} r_1^j \times \ldots \times r_d^j$ for $k \geq 0$. By using standard properties of sets we have:

$$\begin{aligned} rat \cap rec &= rat \cap \Big(\bigcup_{1 \leq j \leq k} r_1^j \times \ldots \times r_d^j \Big) \\ &= \bigcup_{1 \leq j \leq k} rat \cap (r_1^j \times \ldots \times r_d^j). \end{aligned}$$

Since RAT is closed by union, we consider here only the case when $k = 1$. Let $\overline{(i, L)} \stackrel{\text{def}}{=} \{w \in (\Sigma^*)^d \mid w[i] \in L\}$, for $i \in \{1, \ldots, d\}$ and $L \subseteq \Sigma^*$. We have

$$rat \cap r_1 \times \ldots \times r_d = rat \cap \bigcap_{1 \leq i \leq d} \overline{(i, r_i)}.$$

To show that $rat \cap rec$ is a rational relation, we will show that it is the case for the relation $rat \cap \overline{(i, r)}$ for every dimension i and every rational language $r \subseteq \Sigma^*$.

Now that the problem has been simplified, let us deal with the automata. Firstly we build a d-tape automaton A accepting the relation $\overline{(i, r)}$ and, without loss of generality, we assume that $i = 1$.

Let r be the language accepted by $A_1 = (Q_A, \Sigma, 1, \delta_1, I_A, F_A)$. We build a finite d-tape automaton $A = (Q_A, \Sigma, d, \delta, I_A, F_A)$ with

$$\delta \stackrel{\text{def}}{=} \left\{ p \xrightarrow{(1,a)} q \mid p \xrightarrow{a} q \in \delta_1 \right\} \cup \left\{ p \xrightarrow{(i,a)} p \mid i \in \{2, \ldots, d\} \wedge a \in \Sigma \right\}.$$

Hence $R(A) = \overline{(1, r)}$ which is obviously a recognizable relation on $(\Sigma^*)^d$.

Given a d-tape automaton $A_{RAT} = (Q', \Sigma, d, \delta', I', F')$ for rat, i.e., $rat = R(A_{RAT})$, we build automaton $B = (Q, \Sigma, d, \Delta, I, F)$ for $rat \cap \overline{(1, r)}$ by the standard product construction:

- $Q = Q' \times Q_A$, i.e., the states are pairs of states, one of A_{RAT}, one of A;
- $I = I' \times I_A$, i.e., a state is initial if all of its components are initial in their respective automaton;
- $F = F' \times F_A$, i.e., a state is final if all of its components are final in their respective automaton;
- The transition function $\Delta \subseteq Q \times (\Sigma^*)^d \times Q$ is defined as follows:

$$\{((q', q_A), (i, a), (p', p_A)) \mid (q', (i, a), p') \in \delta' \wedge (q_A, (i, a), p_A) \in \delta\}.$$

□

The following is the consequence to the decidability results for the Presburger arithmetic.

Lemma 26.5. *Let P, R be two rational relations over one letter alphabet $\{a\}$. The inclusion problem $P \subseteq R$ is decidable.*

Proof. Since rational relations over one letter alphabet are semi-linear, they form an effective Boolean algebra where, in particular, the inclusion is decidable, see, e.g., Ginsburg and Spanier [8]. The checking for $P \subseteq R$ can be done by using the decidability procedure for Presburger arithmetic in double exponential time, [4]. □

We are ready to state our main theorem.

Theorem 26.6. *Let $M \subseteq (\Sigma^*)^d$ be a commutative rational submonoid and $R \subseteq (\Sigma^*)^d$ a rational relation. The inclusion problem $M \subseteq R$ is decidable.*

Proof. Let A_R, A_M be two automata for R and M, respectively.

1. From A_M compute an automaton for $\mathbb{D}(M)$ (Lemma 26.3).
2. Construct an automaton B for $R \cap \mathbb{D}(M)$, e.g., by using the construction from the proof of Lemma 26.4.
3. Construct automaton B' for $\psi(R \cap \mathbb{D}(M))$ from B by relabeling its transitions by ψ, i.e., a transition $(p \xrightarrow{w} q)$ in B becomes $(p \xrightarrow{\psi(w)} q)$ in B'. Similarly, construct an automaton for $\psi(M)$ from A_M.
4. Check whether $\psi(M) \subseteq \psi(R \cap \mathbb{D}(M))$, e.g., using Lemma 26.5.

By Lemma 26.2, the last step of the above procedure gives the answer to the original question, i.e., whether $M \subseteq R$. □

A direct corollary of the above theorem is that the containment problem $w^* \subseteq R$, for any multiword $w \in (\Sigma^*)^d$ and any rational relation $R \subseteq (\Sigma^*)^d$, is decidable since the one-generator submonoid, w^*, is rational and commutative.

4 Prefix-Closed Rational Relations

We say that a multiword u is a *prefix* of a multiword w if there exists a multiword v such that $uv = w$. If such a v exists, it is unique and it is denoted by $u^{-1}w$. The set of all prefixes of a multiword w is denoted by $\downarrow w$. For any relation $R \subseteq (\Sigma^*)^d$, by $\downarrow R$ we denote the prefix-closure of R, i.e.:

$$\downarrow R \stackrel{\text{def}}{=} \bigcup_{w \in R} \downarrow w.$$

The set of *active dimensions* of a multiword w, denoted by $d(w)$, is the set of all "non-empty" dimensions of w, i.e., $i \in d(w)$ if and only if $w[i] \neq \varepsilon$, for $i \in \{1, \ldots, d\}$.

Given a set D of dimensions, $D \subseteq \{1, \ldots, d\}$, and a d-word w, the multiword $w \setminus D$ is defined as:

$$(w \setminus D)[i] \stackrel{\text{def}}{=} \begin{cases} \varepsilon & \text{if } i \in D \\ w[i] & \text{otherwise} \end{cases}$$

Intuitively, $w \setminus D$ represents the *deactivation* of dimensions D in w. For example, we have: $w \setminus \{1, \ldots, d\} = w \setminus d(w) = \mathbb{1}$.

Lemma 26.7. *If $R \subseteq (\Sigma^*)^d$ is a rational relation then so is $\downarrow R$.*

Proof. Given an automaton $A = (Q, \Sigma, d, \delta, I, F)$ defining a rational relation $R = R(A)$, we build an automaton $\downarrow A$ such that $\downarrow R = R(\downarrow A)$. Without loss of generality we assume that in A the transitions are labeled by words of length at most 1.

We set $\downarrow A = (Q \times 2^d, \Sigma, d, \delta', I', F')$, where:

- 2^d denotes the set of all subsets of $\{1, \ldots, d\}$;
- Transitions are defined by:

$$\delta' = \left\{ (p, D) \xrightarrow{w \setminus D} (q, D) \mid (p, w, q) \in \delta \right\}$$
$$\cup \left\{ (p, D) \xrightarrow{\mathbb{1}} (p, D \cup \{i\}) \mid i \in \{1, \ldots, d\} \right\}.$$

- Initial states are: $I' = I \times \{\emptyset\}$;
- Final states are: $F' = F \times \{\{1, \ldots, d\}\}$.

Intuitively, $\downarrow A$ mimics the behavior of A, however, in $\downarrow A$, at any moment we may decide to stop reading any dimension i by taking a transition of the form $(p, D) \xrightarrow{\mathbb{1}} (p, D \cup \{i\})$. □

By the result of the previous section, Theorem 26.6, together with Lemma 26.7 we obtain the following result.

Corollary 26.8. *Let $M \subseteq (\Sigma^*)^d$ be a rational commutative submonoid and $R \subseteq (\Sigma^*)^d$ a rational relation. The inclusion problem $M \subseteq \downarrow R$ is decidable.*

The above theoretical result is of little use in practice because of the considerable blowup of the size of the input automaton (in order to produce its prefix-closure) which in turn is the input to the high complexity decision procedure as described in the previous section. Thus, we propose an alternative approach for solving the problem.

Lemma 26.9. *Let $R_1, R_2 \subseteq (\Sigma^*)^d$ be two relations. We have:*

$$R_1 \subseteq \downarrow R_2 \iff \downarrow R_1 \subseteq \downarrow R_2.$$

Proof. The right-to-left implication follows from the fact that $R_1 \subseteq \downarrow R_1$. The left-to-right implication holds since, by definition, if $x \in \downarrow R_2$ then $\downarrow x \subseteq \downarrow R_2$. □

Lemma 26.10. *Let $M \subseteq (\Sigma^*)^d$ be a commutative submonoid. There exists a multiword $w \in (\Sigma^*)^d$ such that:*

$$\downarrow M = \downarrow w^*.$$

Proof. We define w as follows:

$$w[i] \stackrel{\text{def}}{=} \begin{cases} \varepsilon & \text{if } u[i] = \varepsilon \text{ for all } u \in M \\ \sqrt[*]{u[i]} & \text{otherwise, i.e., } \exists u \in M, u[i] \neq \varepsilon \end{cases}$$

By $\sqrt[*]{x}$, for a word $x \in \Sigma^*$ we denote the shortest word z such that x is a power of z, i.e., there exists $i \geq 1$ such that $x = z^i$. Since M is commutative, the definition of $w[i]$ doesn't depend on the choice of u from M if $u[i] \neq \varepsilon$. □

Theorem 26.11. *Let $M \subseteq (\Sigma^*)^d$ be a commutative submonoid and $R \subseteq (\Sigma^*)^d$ a relation. There exists a multiword $w \in (\Sigma^*)^d$ such that:*

$$M \subseteq \downarrow R \iff w^* \subseteq \downarrow R.$$

Proof. Directly from Lemmas 26.9 and 26.10. □

In other words, as long as we are able to calculate the word w from the monoid M such that $\downarrow M = \downarrow w^*$ (Lemma 26.10), the problem of deciding whether $M \subseteq \downarrow R$ holds, is equivalent to the problem of deciding whether $w^* \subseteq \downarrow R$ holds (Lemma 26.9). Notice that finding such a w is easy for rational M, however w always exists even if M is not rational.

For the description of our procedure we need an auxiliary operation on multiwords. Given two multiwords u, w we define $u@w$, the *rotation* of w by u, as the following multiword:

$$u@w \stackrel{\text{def}}{=} u^{-1}(wu).$$

For example, $(a, aba, a)@(aba, ab, \varepsilon) = (baa, ba, \varepsilon)$.

Notice that $u@w$ is defined if and only if u is a prefix of wu. The rotation $u@\mathbb{1}$ is defined for any u, $u@\mathbb{1} = \mathbb{1}$. By $@w$ we denote the set of all rotations of w, i.e. $u \in @w$ if and only if there is a v such that $u = v@w$.

Proposition 26.12. g *Let $w \in (\Sigma^*)^d$ be a multiword. We have:*

1. $|\downarrow w| = \prod_{i=1}^{d} (|w[i]| + 1) \leq \left(\left\lceil \frac{|w|}{d} \right\rceil + 1\right)^d$, *and*
2. $|@w| \leq |\downarrow w|$.

Proof. The first statement follows from a simple count of all prefixes.

For the second statement the key point is to notice that the rotation operation is length preserving.

When $d = 1$, consider w to be written on a ring torus. Then the number $|@w|$ of rotations of w is the number of distinct words of length $|w|$ that can be read on this torus. When $|w| \leq 1$, there is only one such word, w itself. Hence $|@w| = 1$. Otherwise, the word read depends on which letter is considered to be the first. Hence $|@w| \leq |w|$. When $d > 1$, consider d tori, one torus for each dimension. For each dimension the previous reasoning holds and the number of different words read on the d tori is the number of rotations of w:

$$|@w| = \prod_{i=1}^{d} |@w[i]| \leq \prod_{i=1}^{d} |w[i]| \leq |{\downarrow}w|.$$

□

4.1 Deciding Procedure for $w^* \subseteq {\downarrow}R(A)$

The inputs are:

- A multiword $w \in (\Sigma^*)^d$, and
- A multi-tape finite automaton $A = (Q, \Sigma, d, \delta, I, F)$.

Without loss of generality we assume that A is co-accessible, i.e., for every state q there is a computation in A starting in q and ending in a final state.

1. We start by constructing a digraph whose vertexes are elements of $Q \times @w$, where Q is the set of states of automaton A and $@w$ denotes the set of all rotations of w; there is an edge labeled by $v \in (\Sigma^*)^d$ from vertex (p, u) to vertex (p', u'), i.e., $(p, u) \stackrel{v}{\to} (p', u')$, if and only if there exists a transition $p \stackrel{v}{\to} p'$ in A and $u' = v@u$. Since d is a constant, the graph can be constructed in $O(|@w||A|)$, where $|@w|$ is the number of rotations of w and $|A|$ is the size of the automaton (by size of an automaton we mean the number of states plus the number of transition plus the total length of transition labels).
2. We find all strongly connected components of the constructed graph accessible from a vertex (p, w), where p is an initial state in A and w is the input word. With every such strongly connected component C we associate all *its dimensions*: dimension i is attached to C if and only if there is an edge with source and destination being in C and labeled by u such that $u[i] \neq \varepsilon$, i.e., u reads at least a letter on the dimension i. Let $d(C)$ denotes the set of those dimensions.

   If there is C such that $d(w) \subseteq d(C)$ then the algorithm stops reporting success, i.e., $w^* \subseteq {\downarrow}R$.

   Otherwise, let $\mathcal{C} = \{C \mid d(C) \cap d(w) \neq \emptyset\}$, i.e., $\mathcal{C}$ is the set of those strongly connected components which read active dimensions of w.

   If $\mathcal{C}$ is empty, then the algorithm stops reporting failure, i.e., $w^* \not\subseteq {\downarrow}R$.
3. For every C from $\mathcal{C}$ we generate a new instance of the problem:

   $$(v \setminus d(C))^* \subseteq {\downarrow}R_p$$

   by choosing any vertex (p, v) from C; R_p is the rational relation defined by A but with a single initial state set to p, i.e., $R_p = R(B)$, where $B = (Q, \Sigma, d, \delta, \{p\}, F)$.

   Finally, $w^* \subseteq {\downarrow}R$ if and only if there exists at least one of the sub-problems which reports success (in Step 2).

Since the depth of the sub-problem generation is bounded by d (d being constant), the number of all sub-problems to consider is polynomial in $|A|$ and $|w|$.

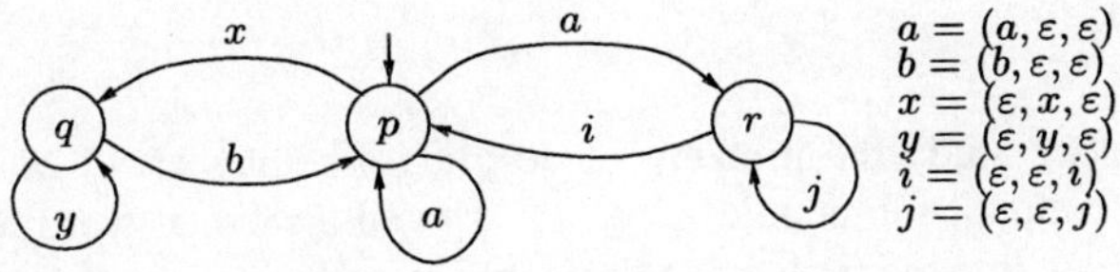

Fig. 26.1 A multi-tape relation R (all states are final)

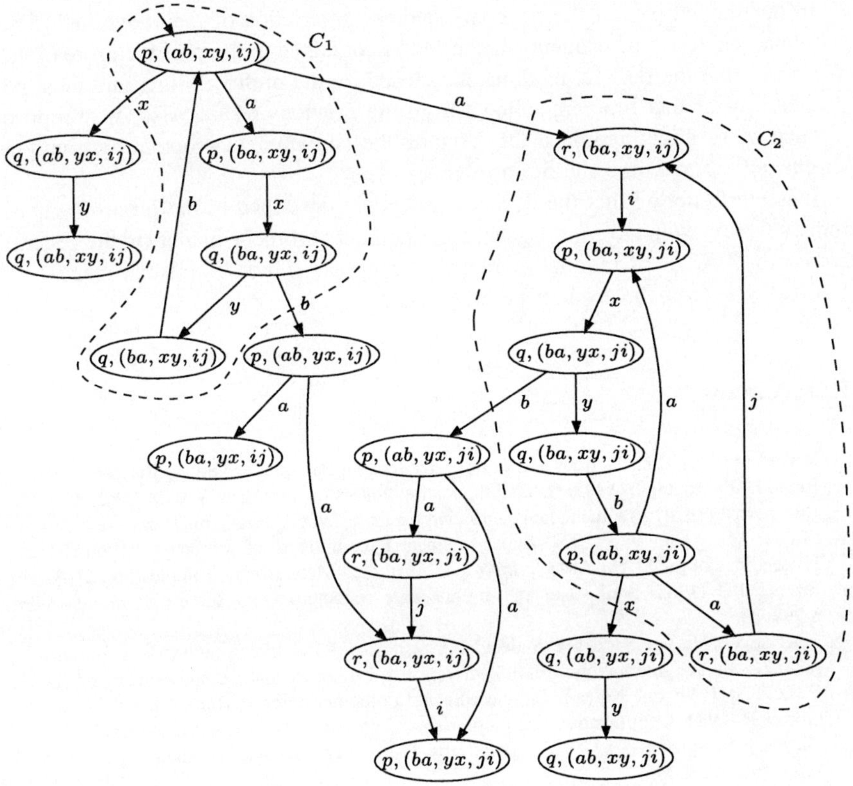

Fig. 26.2 Two strongly connected components, C_1 and C_2, reachable from the initial node "$p,(ab,xy,ij)$"

4.2 Example

Let us consider a ternary word relation R defined by the automaton of Fig. 26.1. In order to check whether $w^* \subseteq \downarrow R$, for $w = (ab, xy, ij)$, we construct the digraph as described in Step 1. The resulting graph, see Fig. 26.2, has two reachable strongly connected components, C_1 and C_2, with $d(C_1) = \{1,2\}$ and $d(C_2) = \{1,2,3\}$. Since $d(w) \subseteq d(C_2)$, see Step 2, the algorithm reports success, so we conclude that $w^* \subseteq \downarrow R$ effectively holds.

5 Conclusions

In this paper we study the problem of deciding whether $M \subseteq R$, for a given commutative rational submonoid $M \subseteq (\Sigma^*)^d$ and a rational word relation $R \subseteq (\Sigma^*)^d$. We show that the problem is decidable by reducing it to the same problem over unary alphabet, which is known to be decidable via the decidability of the Presburger arithmetic.

In the second part of the paper we address a variation of the above inclusion problem, where we investigate whether M is contained in the prefix-closure of R. We prove that the rational relations are closed by the prefix-closure, and thus the decidability of that problem follows from the previous result. Also, we propose a much more direct approach for deciding the containment problem by giving a simple polynomial time algorithm solving it.

It is worth noting that the decision procedure described in the proof of Theorem 26.6 may extend to a more general containment problem, namely to the general inclusion $P \subseteq R$ of rational relations, as long as relation P is such that $\mathbb{D}(P)$ fulfills (26.1) from the proof of Lemma 26.2.

References

1. Berstel J (1979) Transductions and context-free languages. Teubner Verlag, Stuttgart
2. Brand D, Zafiropulo P (1983) On communicating finite-state machines. J ACM, 30(2):323–342
3. Eilenberg S (1974) Automata, languages, and machines, vol A. Academic Press, New York
4. Fischer MJ, Rabin MO (1974) Super-exponential complexity of presburger arithmetic. In: Proceedings of the SIAM-AMS Symposium in Applied Mathematics, volume 7, pp 27–41
5. Fischer PC (1965) Multi-tape and infinite-state automata — a survey. Commun ACM 8:799–805
6. Fraczak W, Hassen S (2011) A decidable instance of the inclusion problem for rational relations. In The International MultiConference of Engineers and Computer Scientists 2011, IMECS 2011, Lecture Notes in Engineering and Computer Science, Hong Kong
7. Gilmer R (1984) Commutative semigroup rings. University of Chicago, Chicago, Illinois
8. Ginsburg S, Spanier EH (1966) Semigroups, Presburger formulas, and languages. Pacific J Math 16(2):285–296
9. Habermehl P, Mayr R (2000) A note on SLRE. Technical report, LIAFA - Universit Denis Diderot, 2, place Jussieu, Paris Cedex 05, France
10. Johnson JH (1986) Rational equivalence relations. Theor Comput Sci 47(3):39–60
11. Josephs MB (2004) Models for data-flow sequential processes. In: Abdallah AE, Jones CB, Sanders JW (eds) 25 Years communicating sequential processes. Lecture notes in computer science, vol 3525. Springer, Berlin, pp 85–97
12. Klarlund N (1998) Mona & fido: The logic-automaton connection in practice. In: Selected Papers from the11th International Workshop on Computer Science Logic, CSL '97, Springer, London, UK, pp 311–326
13. Rabin MO, Scott D (1959) Finite automata and their decision problems. IBM J Res Dev 3(2):114–125
14. Sakarovitch J (2003) Eléments de théorie des automates. Vuibert Informatique, Paris, France.
15. Stockmeyer LJ, Meyer AR (1973) Word problems requiring exponential time (Preliminary Report). In: Proceedings of the fifth annual ACM symposium on Theory of computing, STOC '73, ACM, New York, NY, USA, pages 1–9

Chapter 27
Pattern Matching Algorithm Using a Succinct Data Structure for Tree-Structured Patterns*

Yuko Itokawa, Masanobu Wada, Toshimitsu Ishii, and Tomoyuki Uchida

Abstract Two things are important in developing a fast, memory-efficient graph mining method that extracts characteristic graph structures from Web pages and other tree-structured data. One is tree patterns that can express the features of a graph structure and the other is data structures for tree patterns and for representing tree-structured data. In this paper, we first apply a depth-first unary degree sequence (DFUDS), which is one succinct data structure for an ordered tree, as a succinct data structure for tree patterns that express the features of a graph structure. We then propose a pattern matching algorithm that uses the DFUDS succinct data structure, to determine whether or not a given tree-structured data has features of tree pattern. We also implement the proposed algorithm on a computer and evaluate the algorithm by experiment. The results are reported and discussed.

Keywords Graph algorithm • Succinct data structure • Tree pattern matching • Tree structured data

---

*This is an extended and revised version of the paper presented at the International Multi-Conference of Engineers and Computer Scientists, Hong Kong, 2011.

Y. Itokawa (✉)
Department of Kansei Design, Hiroshima International University, 555-36, Kurose Gakuendai, Higashi Hiroshima, Hiroshima 739-2695, Japan
e-mail: y-itoka@he.hirokoku-u.ac.jp

M. Wada
Department of Information Sciences, Hiroshima City University 3-4-1, Ozuka-Higashi, Asa-Minami-Ku, Hiroshima 731-3194, Japan
e-mail: m-wada@ml.info.hiroshima-cu.ac.jp

T. Ishii • T. Uchida
Faculty of Information Sciences, Hiroshima City University 3-4-1, Ozuka-Higashi, Asa-Minami-Ku, Hiroshima 731-3194, Japan
e-mail: t-ishii@ml.info.hiroshima-cu.ac.jp; uchida@hiroshima-cu.ac.jp

S.I. Ao et al. (eds.), *Intelligent Control and Innovative Computing*, Lecture Notes in Electrical Engineering 110, DOI 10.1007/978-1-4614-1695-1_27,

1 Introduction

In recent years, with the rapid progress in networks and information technology, Web documents and other such material that does not have a clear structure and is referred to as semi-structured data have become innumerous. Semi-structured data that has a tree structure is called tree-structured data, and can be represented by an ordered tree. To extract useful information from tree-structured data, it is necessary to extract tree patterns that are common to tree-structured data. Moreover, to present efficient tree mining tools, pattern matching algorithms for determining whether or not given tree-structured data has features represented by given tree pattern are necessary to be efficient. Suzuki et al. [14] have proposed a matching algorithm for determining whether or not an edge-labeled ordered tree is generated by substituting all structural variables in given edge-labeled ordered tree pattern with arbitrary edge-labeled ordered trees. The aim of this paper is to present faster and more memory-efficient pattern matching algorithm for edge-labeled ordered tree patterns than Suzuki's.

To reduce the memory required to store an ordered tree, succinct data structures for ordered trees have been proposed [2–5,7,9,10,12]. As one succinct data structure for an edge-labeled ordered tree, Ferragina et al. [4] proposed the **xbw** transform and a path search algorithm for an xbw transformed edge-labeled tree.Using the xbw transform enables both compact storage of tree-structured data and fast path search. Itokawa et al. [6] proposed an efficient algorithm that involves successive compression and expansion of a huge set of tree-structured data.They also proposed an algorithm for more compact tree-structured data and a path-finding algorithm that is both efficient and fast. As a succinct data structure for an ordered tree, Benoit et al. [2] proposed a depth-first unary degree sequence (**DFUDS**) representation.The DFUDS representation uses a string of parentheses constructed by a depth-first traversal of all vertexes in which, if the index of a vertex is k, the k-th $\boxed{(}$ and its subsequent $\boxed{)}$ are output. By taking $\boxed{(}$ to be "0" and $\boxed{)}$ to be "1", the ordered tree representation can be handled as a bit string. Also proposed is supplementary data structure for an ordered tree represented by DFUDS that enables use of the *rank* and *select* operations to tour the tree in constant time. Almost of results about succinct data structures are from theoretical viewpoints. However, most recently, a few practical results are known [1].

In order to represent structural features of tree-structured data, Itokawa et al. [8] introduced a DFUDS representation for an edge-labeled ordered tree patterns based on a DFUDS representation of an ordered tree presented by Benoit et al. [2]. We also proposed an efficient matching algorithm for solving the membership problem for tree patterns having structural variables using the DFUDS representation of an edge-labeled tree as the data structure. The matching algorithm we described performs top down matching, but the method applied by Suzuki et al. [14] performs bottom up. In this paper, by proposing dynamic findopen operation used in the matching algorithm, we improve the matching algorithm presented by Itokawa et al. [8] to be memory-efficient. The results in this paper leads us to design fast and memory-efficient tree mining tools for tree-structured data. Adaptation of that algorithm to the TTSP graph forest representation proposed by Itokawa et al. [7] has application in the data mining of TTSP graphs.

This paper is organized as follows. In Sect. 2, we describe the tree structure that we deal with in this paper and the tree pattern which represents the structural features of edge-labeled ordered trees. In Sect. 2, we briefly describe the DFUDS representation for ordered tree patterns and operations, including dynamic findopen operations proposed in this paper, on succinct data structures. In Sect. 3, we formulate the membership problem for edge-labeled trees and give a matching algorithm for solving it. In Sect. 4, the algorithm for solving the membership problem is implemented on a computer. The test results for the implementation are reported and discussed. Section 5 concludes the paper. This paper is an extended and revised version of our paper [8].

2 Preliminaries

2.1 Tree Patterns

Let Σ and χ denote finite alphabets with $\Sigma \cap \chi = \emptyset$. Let V_t be the set of vertexes and $E_t \subseteq V_t \times (\Sigma \cup \chi) \times V_t$ the set of edges. For edge $e = (u, a, v) \in E_t$, let character $a \in (\Sigma \cup \chi)$ be the *edge label* of e. In particular, when $a \in \chi$, a is a *variable label* and e is a *variable*. For variable $e = (u, x, v)$, u is the *parent port* of e and, v is a *child port* of e. $t = (V_t, E_t)$ is called an *edge-labeled ordered term tree* or simply *term tree* if t has only one vertex u whose coming degree is 0, $(V_t, \{\{u, v\} \mid (u, a, v) \in E_t\})$ is a rooted tree having u as its root and all children of all internal vertexes are ordered. The term trees of concern in this paper are assumed to have all mutually different variables. Term trees having no variables are simply *trees*. In Fig. 27.1, term tree p and trees t, g_0, g_1, g_2 are shown.

For term tree t and its vertex u, the tree consisting of u and all of its descendants is called a *subtree* of t and is denoted as $t[u]$. In the same way, for edge $e = (u, a, v)$ of t, $t[e]$ denotes the tree consisting of e and $t[v]$. For two children u' and u'' of vertex u of term tree h, $u' <_u^h u''$ denotes that in the ordering of the children of u, u' is lower than u''. For term trees $t = (V_t, E_t)$ and $f = (V_f, E_f)$, if bijection $\pi : V_t \to V_f$ that satisfies the following conditions (1)-(3) exists, then t and f are *isomorphic*, which is denoted as $t \cong f$.

1. For any $a \in \Sigma$, if and only if $(u, a, v) \in E_t$, $(\pi(u), a, \pi(v)) \in E_f$.
2. If and only if there is $x \in \chi$ for which $(u, x, v) \in E_t$, there is $y \in \chi$ for which $(\pi(u), y, \pi(v)) \in E_f$.
3. If and only if for vertexes u of t and the two children u' and u'' of u, $u' <_u^t u''$, $\pi(u') <_{\pi(u)}^f \pi(u'')$.

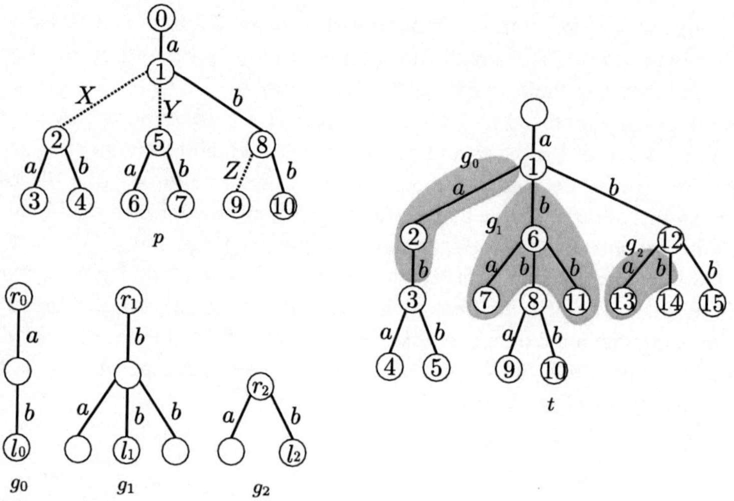

Fig. 27.1 Ordered term tree p and trees g_0, g_1, g_2, $t \cong p\theta$, where $\theta = \{X := [g_0, (r_0, l_0)], Y := [g_1, (r_1, l_1)], Z := [g_2, (r_2, l_2)]\}$

Let variable label $x \in \chi$ and let r be the root of a term tree g, and l the leaf of g. Then, $x := [g, (r, l)]$ is the *binding* of x and a finite set of variable label bindings is a *substitution*. A new term tree f can be obtained by applying substitution $\theta = \{x_0 := [g_0, (r_0, l_0)], \ldots, x_n := [g_n, (r_n, l_n)]\}$ to a term tree $g = (V_g, E_g)$ in the following way. A new term tree f can be obtained by regarding w_0 to be the same as r_i and w_1 to be the same as l_i in edge $e = (w_0, x_i, w_1)$ that has variable label x_i, and exchanging them for each $0 \le i \le n$. The resultant term tree f is denoted by $g\theta$. Tree t in Fig. 27.1 can be obtained by applying substitution $\theta = \{X := [g_0, (r_0, l_0)], Y := [g_1, (r_1, l_1)], Z := [g_2, (r_2, l_2)]\}$ to term tree p.

2.2 Succinct Data Structures for Tree Patterns

We explain the basic data structure for dealing with ordered term trees. In this paper, a word RAM with a word length of $\Theta(\log n)$ bits is used as the computation model. For sequence S of length n on alphabet $\mathcal{A}$, denote the i-th character $(0 \le i \le n-1)$ in S as $S[i]$. For each i, j $(0 \le i < j \le n-1)$, denote the sub-sequence from the i-th character to the j-th character of S as $S[i \ldots j]$. For sequence S of length n on alphabet $\mathcal{A}$, character c and natural number i $(0 \le i \le n-1)$, define a *rank* function and a *select* function as follows.

$rank_c(S, i)$ returns the number of occurrences of character c in sub-sequence $S[0 \ldots i]$.

$select_c(S, i)$ returns the position of the i-th character c from the beginning of S.

When the context makes it clear, S is omitted.

p's DFUDS representation: ((((a ((X a b ((Y a b ((b Z b
⓪ ① ② ③④ ⑤ ⑥⑦ ⑧ ⑨⑩

t's DFUDS representation : ((((a (a ((b a b (((b a ((b a b b (((b a b b
⓪ ① ② ③ ④⑤ ⑥ ⑦ ⑧ ⑨⑩⑪ ⑫ ⑬⑭⑮

Fig. 27.2 Edge-labeled DFUDS representations for term tree p and tree t

There are many succinct data structures for fast computation of the *rank* function and the *select* function [2,5,11,13]. One is the fully indexable dictionary *(FID)* [11]. The FID is an $n + o(n)$ bit data structure that allows the *rank* and *select* functions to be calculated in constant time for a bit sequence on $\{0, 1\}$ of length n with a word RAM model. The FID has a bit sequence of length n and a supplementary data structure. The supplementary data structure is a table that stores responses for all inputs in advance, considering the fine division of the character sequence as the problem of small size. The responses for any bit sequence of length $\frac{1}{2}\log n$ in a bit array of length n are at most $2^{\frac{1}{2}\log n} = \sqrt{n}$, so all responses can be stored in a $(\sqrt{n} \cdot \mathrm{polylog}(n))$bit table, which can be searched in constant time. That area can be reduced when there are few "1" in S. For a character sequence that contains m "1", there is a $\log \binom{n}{m} + O(n \log\log n/\log n) = m \log \frac{n}{m} + \Theta(m) + O(n \log\log n/\log n)$ bit data structure for which the *rank* and *select* functions can be calculated in constant time. If $m = O(n/\log n)$ bits, the size of that data structure becomes $O(n \log\log n/\log n)$. The *rank* and *select* functions can be expanded to a function for obtaining the number of occurrences and positions of a character sequence pattern. Consider the representation of an ordered tree of n vertexes that allows execution of the *rank* and *select* functions in constant time. There exist $\binom{2n+1}{n}/(2n+1)$ ordered trees of n vertexes, so the information theoretical lower bound of the data structure size is $2n - \Theta(\log n)$ bits.

Benoit et al. [2] proposed the *DFUDS* representation, which is a succinct data structure for ordered trees. The DFUDS representation for an ordered tree t of m edges is defined inductively as follows. The DFUDS representation of the tree consisting only one vertex is $\boxed{(}\boxed{)}$. The DFUDS representation of a t that has k subtrees $t_1, \ldots, t_k$ is a sequence of parentheses constructed by concatenating $k + 1$ $\boxed{(}$, one $\boxed{)}$, k DFUDS representations of $t_1, \ldots,$ and t_k in this order (here, the initial $\boxed{(}$ of the DFUDS representation of each subtree has been removed). The DFUDS representation is a sequence of balanced parentheses of length 2 m.

The DFUDS representation proposed by Benoit et al. [2] is a data structure for an ordered tree with no edge labels. In the DFUDS representation of Benoit et al. [2], $\boxed{)}$ must occupy the rightmost position for each vertex. Therefore, a hash function that returns the edge label that corresponds to that vertex for each $\boxed{)}$ makes a DFUDS representation of a tree possible. DFUDS representations for term tree p and tree t are shown in Fig. 27.2. For convenience in this example, a hash function that returns the edge labels that correspond to all of the $\boxed{)}$ has been executed.

The sequence of parentheses that is a DFUDS representation can be interpreted as the result of visiting all vertexes in pre-order and outputting k (for each vertex whose index is k the following one). Hence, the following theorem obviously holds.

Theorem 27.1 ([8]). *Given tree t of m edges, the edge-labeled DFUDS representation for t can be computed with $O(m)$.*

2.3 Operations on Succinct Data Structures

A DFUDS representation can be regarded as a bit sequence by replacing (with "0" and) with "1". Representing an ordered tree with a bit sequence DFUDS representation makes it possible to execute the *rank* and *select* functions in constant time. Furthermore, a supplementary data structure that allows execution of the following operations in constant time by using *rank* and *select* functions on DFUDS representation P has been proposed by Benoit et al. [2] and Jansson et al. [10].

findclose(x) return the position of the closing parentheses for an opening parenthesis in $P[x]$.

findopen(x) return the position of the opening parentheses for a closing parenthesis in $P[x]$.

enclose(x) return the position of the opening parentheses of the pair that most tightly encloses $P[x]$.

The three operations above can be used to express the following operations for touring the ordered tree t.

degree(x) return the number of children of vertex x.

child(x, i) return the position of the i-th child from the left of vertex x.

subtree(x) return the pair of the start and the end positions of the interval representing the subtree for which vertex x is the root.

id(x) return the visiting order of vertex x.

label(x) return the position of the closing parenthesis of the *id*(x)-th vertex.

child takes $O(i)$ time, but the other operations can be executed in constant time using an $o(n)$ bit supplementary data structure [2, 3, 12].

The *findopen* operation can be regarded as findclose operation for reverse sequence of an input DFUDS representation. The implementation of *findopen* operation based on such an idea is called a *naive findopen implementation*. The naive findopen implementation is simple, but supplementary data structure required to store a table for reverse sequence are needed. That is, supplementary data structures need twice memories. A sequence p of an input DFUDS representation is divided into equal-sized *blocks* (e.g., size $\lceil \log |p|/2 \rceil$). In this paper, we consider a *findopen* operation as follows. Operation *findopen* finds a block B including current position. If the block number n_B of B is marked as already lookup, then *findopen* lookups the table. Otherwise, *findopen* marks n_B as already lookup and adds the value of reversed sequence of B to the table. This implement of *findopen* operation based on the above new idea is called *dynamic findopen operation.*

3 Algorithm for Solving Membership Problem for Tree Patterns

$\mathcal{OT}_\Sigma$ denotes a set of edge-labeled ordered trees on Σ and $\mathcal{OTT}_{\Sigma\cup\chi}$ denotes a set of edge-labeled ordered term trees on $\Sigma \cup \chi$. In this section, we formulate the membership problem for $\mathcal{OTT}_{\Sigma\cup\chi}$ and propose a matching algorithm that uses a DFUDS representation as a data structure for solving that problem. Given edge-labeled ordered term tree $p \in \mathcal{OTT}_{\Sigma\cup\chi}$ and edge-labeled ordered tree $t \in \mathcal{OT}_\Sigma$, the problem of determining whether or not there exists a substitution θ for which $t \cong p\theta$ is referred to as the membership problem for $\mathcal{OTT}_{\Sigma\cup\chi}$.

Membership problem for $\mathcal{OTT}_{\Sigma\cup\chi}$

Instance: Edge-labeled ordered term tree $p \in \mathcal{OTT}_{\Sigma\cup\chi}$ and edge-labeled ordered tree $t \in \mathcal{OT}_\Sigma$
Problem: Determine if there exists a θ for which $t \cong p\theta$.

A matching algorithm Pattern_Matchingthat determines whether or not there exists a substitution θ for which $t \cong p\theta$ given edge-labeled ordered term tree $p \in \mathcal{OTT}_{\Sigma\cup\chi}$ and edge-labeled ordered tree $t \in \mathcal{OT}_\Sigma$, $t \cong p\theta$ is listed in Fig. 27.3.

By treating the DFUDS representation of an edge-labeled tree as a data structure, we can consider pattern matching for the character sequence that consists of $\boxed{(}$ and edge labels. We therefore execute the pattern matching from the beginning of the DFUDS representation in the touring order from the root of the tree. Given the DFUDS representation $P[0\ldots m-1]$ of edge-labeled ordered term tree p as the pattern and the DFUDS representation $T[0\ldots n-1]$ of edge-labeled ordered tree t as the text, matching begins with the first characters of p and t. If characters $P[i]$ and $T[j]$ match, matching is performed for the next characters. If characters $P[i]$ and $T[j]$ do not match, the following procedure is performed (line 4–line 14). The five pattern cases for which $P[i]$ and $T[j]$ do not match are shown in Table 27.1. Mismatch for which the condition of line 4 of the algorithm applies relates to Case 1 in Table 27.1, so edge-labeled ordered term tree p and edge-labeled ordered tree t are not the same. Accordingly, Pattern_Matching algorithm returns *false*. The first half of the condition of line 5 relates to Case 5 in Table 27.1; the second half of the condition is the case in which parent and child have sequential variables in edge-labeled ordered term tree p. In that case, it is determined whether on not subtrees $p[P.id(i)]$ and $t[T.id(j)]$ match. If they do not match, then Pattern_Matching algorithm returns *false*. Line 9 relates to Cases 2, 3 and 4 in Table 27.1, where, subtree $p[P.id(i)]$ is not found in subtree $t[T.id(j)]$ and Pattern_Matching algorithm returns *false*. If either character sequence p or t has

Algorithm Pattern_Matching

Input: DFUDS representations $P[0\ldots m-1]$ and $T[0\ldots n-1]$ of term tree p and tree t, respectively.

Output: true if there exists a substitution θ with $t \cong p\theta$, otherwise false.

```
1. i := 0, j := 0;
2. while i < m and j < n do {
3.   if P.label(i) ≠ T.label(j) then
     /* Case 1 */
4.     if P.label(i) ∈ Σ and T.label(j) ∈ Σ then return false;
     /* Case 5 */
5.     if P.label(i) ∈ Σ and T[j] is an open parenthesis then
6.       if Pattern_Matching(P[P.subtree(i)], T[T.subtree(j)]) is false
7.       then return false;
8.       else {i := P.subtree(i).second; j := T.subtree(j).second;}
     /* start of Cases 2,3,4 */
9.     else {
10.      if, for any k ∈ [0...T.degree(j)],
11.      Pattern_Matching(P[P.subtree(i)], T[T.subtree(T.child(j, k))])
12.      is false then return false;
13.      else {i := P.subtree(i).second; j := T.subtree(j).second;}
14.    }
     /* end of Cases 2,3,4 */
15.  i++, j++;
16. }
17. if m − i > 0 or n − j > 0 then return false;
18. return true;
```

Fig. 27.3 Algorithm Pattern_Matching

Table 27.1 Mismatch cases of $T[j]$ and $P[i]$

	*Case*1	*Case*2	*Case*3	*Case*4	*Case*5
$P[i]$	EL	VL	OP	VL	EL
$T[j]$	EL	EL	EL	OP	OP

EL, VL and OP denote edge label in Σ, variable label in χ and open parenthesis (, respectively

not been examined to the end (line 17) when the while statement ends, then there exists no substitution θ for which $t \cong p\theta$. If none of the above cases hold, then there exists a θ such that $t \cong p\theta$.

When DFUDS representations of edge-labeled ordered term tree p and edge-labeled ordered tree t in Fig. 27.2 are given, we illustrate the process of Pattern_Matching algorithm in Fig. 27.4. Since $P[0\ldots5]$ and $T[0\ldots5]$ are same, line 3 of Pattern_Matching algorithm is firstly executed at $i = 6$ and $j = 6$. Since $P[6] =$ (, $T[6] = a \in \Sigma$, lines 9 to 14 are executed. Then, since there exists the subtree $t[3]$ which is isomorphic to the subtree $p[2](= p[P.id(6)])$,

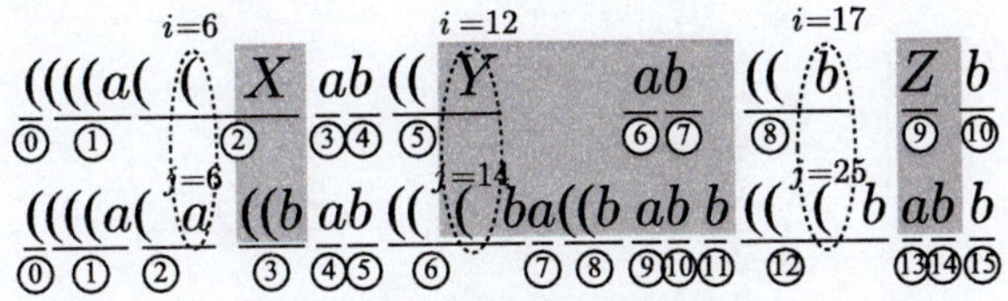

Fig. 27.4 Illustration of process of Pattern_Matching algorithm when DFUDS represents of p and t are given

after line 13 is executed, $i = 9$ and $j = 11$ are obtained. Next, $p[10\ldots 11]$ and $t[12\ldots 13]$ are same, and $P[12] = Y \in \chi$ and $T[6] = \boxed{(}$. Hence, lines 9 to 14 are executed again. Then, since there exists the subtree $t[8]$ which is isomorphic to the subtree $p[5](= p[P.id(12)])$, after line 13 is executed, $i = 14$ and $j = 22$ are obtained. Next, since $P[17] = b \in \Sigma$ and $T[25] = \boxed{(}$, line 5 is executed. when the subtree $p[(P.id(P.parent(17)), b, P.id(17))](= p[(1, b, 8)])$ of p and the subtree $t[(T.id(T.parent(25)), b, T.id(25))](= t[(1, b, 12)])$ of t are given, Pattern_Matching returns *true*. Hence, after line 8 is executed, $i = 19$ and $j = 29$ are obtained. Finally, Pattern_Matching algorithm exits while loop, returns *true*, and terminates.

The following theorem holds.

Theorem 27.2 ([8]). *The membership problem for $\mathcal{OTT}_{\Sigma\cup\chi}$ can be computed in $O(mn)$ time, where m and n are lengths of DFUDS representations of a given term tree and a given tree, respectively.*

4 Experiment and Discussion

We implemented Pattern_Matching algorithm, described in Sect. 3, in C++ on a computer equipped with a 3.16 GHz Intel XEON X5460 processor, main memory of 4.00 GB and running the Microsoft Windows Vista SP1 operating system. In this section, we describe the experimental setup, present the results, and discuss what was learned.

To avoid effect of the edge label count, we set the edge label count to 1. Let a be an edge label in Σ and r a real number from 0 up to 1. We artificially created a *pattern collection* $D^r_{\Sigma\cup\chi}(m)$ of one hundred each of term trees in $\mathcal{OTT}_{\Sigma\cup\chi}$ that has the edge label a, m edges including rm variables, and maximum degree 5, and a *data collection* $D_\Sigma(n)$ of one hundred each of trees in $\mathcal{OT}_\Sigma$ that has the edge label a, n edges and maximum degree 5. We remark that r denotes a *ratio* of the variable count for the edge count in a term tree in $D^r(m)$. From the design of Pattern_Matching algorithm, since Pattern_Matching algorithm may return *false* as soon as finds mismatch positions of given term tree p and tree t, we can see that, in general, the execution time of Pattern_Matching algorithm may not depend on the edge count of p and t even if quit huge data collections are used. Hence, we artificially created data collection $D_\Sigma(n)$ from $D^r_{\Sigma\cup\chi}(m)$ in the following way. For

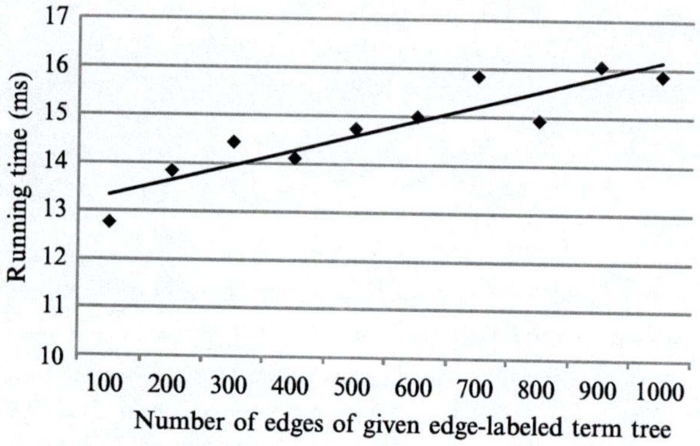

Fig. 27.5 Running times of Pattern_Matching algorithm for the edge count of given term tree

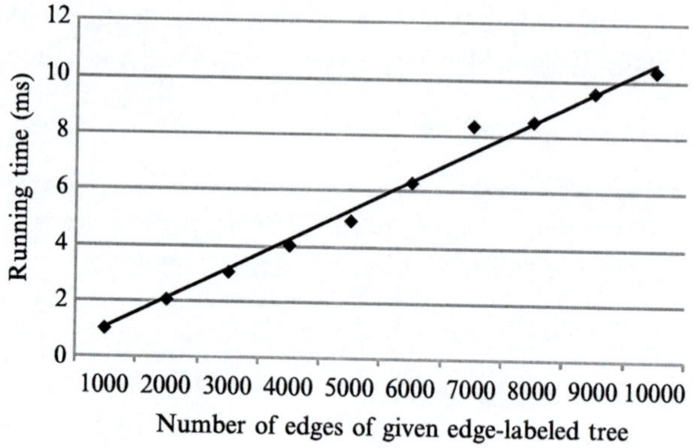

Fig. 27.6 Running times of Pattern_Matching algorithm for the edge count of given tree

each term tree p in $D^r_{\Sigma\cup\chi}(m)$, we add a tree t in $\mathcal{OT}_{\{a\}}$ so that t has edge counts of n and there exists a substitution θ with $t \cong p\theta$ to $D_\Sigma(n)$.

In order to show efficiency of Pattern_Matching algorithm, we evaluated Pattern_Matching algorithm by experiments using data collections and pattern collections. Let t be a tree selected in $D_\Sigma(10,000)$. Firstly, for each $m \in \{100, 200, 300, 400, 500, 600, 700, 800, 900, 1000\}$, when t and a term tree p in $D^{0.1}_{\Sigma\cup\chi}(m)$ are given, the average execution time of Pattern_Matching algorithm was shown in Fig. 27.5. When the edge count of given tree is fixed, we can demonstrate that Theorem 27.2 holds, because the matching time varies linearly with the edge count of given term tree. Conversely, let p be a term tree in $D^{0.1}_{\Sigma\cup\chi}(100)$. Secondly, by varying edge counts n from 1,000 to 10,000, when p and tree in $D_\Sigma(n)$ are given, the average execution time of Pattern_Matching algorithm was shown in Fig. 27.6. From Fig. 27.6, we can also demonstrate that Theorem 27.2 holds.

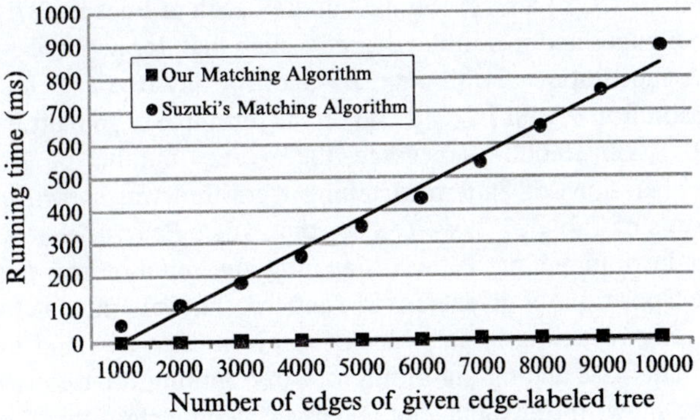

Fig. 27.7 Pattern_Matching algorithm vs. Suzuki's matching algorithm

Table 27.2 Reducing memories of Pattern_Matching algorithm

# edges		1,000	5,000	10,000
# blocks		37	169	333
# reversed blocks	Match trees	36.6	166.0	329.0
	Mismatch trees	9.2	75.2	181.7

edges, # blocks and # reversed blocks denote the numbers of edges, blocks and reversed blocks, respectively

Suzuki et al. [14] have proposed a matching algorithm for determining whether or not, given a term tree p and a tree t, there exists a substitution θ so that $t \cong p\theta$. The strategy of Suzuki's matching algorithm differs from ours. The matching algorithm we describe here performs top down matching, but the method applied by Suzuki et al. [14] performs bottom up. Let p be a term tree in $D^{0.1}_{\Sigma \cup \chi}(100)$ and t a tree in $D_{\Sigma}(n)$ obtained by varying edge counts n from 1,000 to 10,000. In order to show the advantage of Pattern_Matching algorithm for Suzuki's matching algorithm, we compared Pattern_Matching algorithm with Suzuki's matching algorithm for each data collection. The results are shown in Fig. 27.7. Figure 27.7 shows that Pattern_Matching algorithm is faster than Suzuki's matching algorithm. The reasons are follows. (1) If there exists no substitution θ with $t \cong p\theta$, Pattern_Matching algorithm returns *false* faster than Suzuki's matching algorithm. (2) Pattern_Matching algorithm are implemented using DFUDS as data structure. These experimental results leads us to give faster tree mining tools using Pattern_Matching algorithm proposed in this paper than using Suzuki's matching algorithm.

In order to show memory-efficiency of Pattern_Matching algorithm, we compared Pattern_Matching algorithm using naive findopen operation with Pattern_Matching algorithm using dynamic findopen operation, described in Sect. 2. Let p be a term tree in $D^{0.1}_{\Sigma \cup \chi}(100)$. For each $n \in \{1{,}000, 5{,}000, 10{,}000\}$, we artificially created new data collections $D^{+}_{\Sigma}(n, p)$ and $D^{-}_{\Sigma}(n, p)$ as follows.

$D_{\Sigma}^{+}(n, p)$ and $D_{\Sigma}^{-}(n, p)$ are sets of one hundred each of trees in $\mathcal{OT}_{\Sigma}$ that has the edge label a, n edges and maximum degree 5 such that, for any tree $s \in D_{\Sigma}^{+}(n, p)$, there exists a substitution τ with $s \cong p\tau$ and, for any tree $t \in D_{\Sigma}^{-}(n, p)$, there exists no substitution θ with $t \cong p\theta$. When the term tree p and a tree in $D_{\Sigma}^{+}(n, p)$ ($n \in \{1{,}000, 5{,}000, 10{,}000\}$) are given, the average number of reversed blocks for *findopen* operations of Pattern_Matching algorithm was shown in Table 27.2. From the results of Table 27.2, we can see that, since Pattern_Matching algorithm performs top down matching, Pattern_Matching algorithm returns false as soon as finds mismatch positions of given term tree and tree, but Pattern_Matching algorithm returns true after reversing almost blocks, if given term tree and tree match. In adaption of Pattern_Matching algorithm to a tree mining for tree structured data, this result also shows the advantage of Pattern_Matching algorithm.

5 Conclusion

We used the DFUDS representation as a data structure to formulate the membership problem for edge-labeled trees and introduced a matching algorithm that solves that problem in polynomial time. Moreover, we improved the above algorithm to memory-efficient algorithm by proposing a dynamic findopen operation. Evaluation experiments performed with computer implementation of the algorithm demonstrated its efficiency.

Itokawa et al. [7] analyzed a TTSP graph and proposed its representation as a series of edge-labeled ordered trees (*forest representation*). As applications of this research, we are considering adaptation of the DFUDS representation proposed here to succinct data structures of TTSP graphs based on forest representation proposed by Itokawa et al. [7] and the compressed tree proposed by Itokawa et al. [6]. Moreover, we are considering graph mining algorithms for semi-structured data using a succinct data structure.

References

1. Arroyuelo D, Cánovas B, Navarro G, Sadakane K (2010) Succinct trees in practice. In Proceedings of the Twelfth Workshop on Algorithm Engineering and Experiments (ALENEX), SIAM, 84–97
2. Benoit D, Demaine ED, Munro JI, Raman V (2005) Representing trees of higher degree. Algorithmica 43(4):275–292
3. Chiang Y-T, Lin C-C, Lu H-I (2005) Orderly spanning trees with applications. SIAM Journal on Computing, SIAM, 34(4):924–945
4. Ferragina P, Luccio F, Manzini G, Muthukrishnan S (2005) Structuring labeled trees for optimal succinctness, and beyond. In Proceedings of the 46th Annual IEEE Symposium on Foundations of Computer Science (FOCS'05), IEEE, 184–196
5. Geary RF, Rahman N, Raman R, Raman V (2006) A simple optimal representation for balanced parentheses. Theoretical Computer Science, ELSEVIER 368(3):231–246

6. Itokawa Y, Katoh K, Uchida T, Shoudai T (2010) Algorithm using expanded LZ compression scheme for compressing tree structured data. Lecture notes in electrical engineering. Springer, Berlin, pp 333–346
7. Itokawa Y, Miyoshi J, Wada M, Uchida T (2010) Succinct representation of ttsp graphs and its application to the path search problem. In Proceedings of the Sixth IASTED International Conference on Advances in Computer Science and Engineering IASTED, pp 33–40
8. Itokawa T, Wada M, Ishii T, Uchida T (2011) Tree pattern matching algorithm using a succinct data structure. In Lecture notes in engineering and computer science: Proceedings of the international multiConference of engineers and computer scientists 2011, IMECS 2011, 16–18 March, 2011, Hong Kong, pp 206–211
9. Jacobson G (1989) Space-efficient static trees and graphs. In Proceedings of the 30th Annual IEEE Symposium on Foundations of Computer Science (FOCS'89), IEEE, 549–554
10. Jansson J, Sadakane K, Sung W-K (2007) Ultra-succinct representation of ordered trees. In Proceedings of the eighteenth annual ACM-SIAM Symposium on Discrete Algorithms (SODA'07), Society for Industrial and Applied Mathematics, 575–584
11. Munro JI (1996) Table. In Proceedings of the 16th Conference on Foundations of Software Technology and Theoretical Computer Science (FSTTCS), 18–20 December, Hyderabad, India, Lecture Notes in Computer Science Vol. 1180, Springer, pp 37–42
12. Munro JI, Raman V (2001) Succinct representation of balanced parentheses and static trees. SIAM Journal on Computing, SIAM, 31(3):762–776
13. Pagh R (2001) Low redundancy in static dictionaries with constant query time. SIAM Journal on Computing, SIAM, 31(2):353–363
14. Suzuki Y, Inomae K, Shoudai T, Miyahara T, Uchida T (2003) A polynomial time matching algorithm of structured ordered tree patterns for data mining from semistructured data. In Proceedings the 12th international conference on inductive logic programming, ILP 2002, 9-11 July, Sydney, Australia, Lecture Notes in Computer Science, Vol. 2583, Springer, pp 270–284

Chapter 28
Network Traffic Screening Using Frequent Sequential Patterns

Hisashi Tsuruta, Takayoshi Shoudai, and Jun'ichi Takeuchi

Abstract Darknet monitoring is very important for understanding various botnet activities for early detection and defense the threats on the Internet caused by the botnets. However, common illegal accesses by ordinary malware make such detection difficult. To remove such accesses by ordinary malware from the results of network monitoring, we propose a data screening method based on finding frequent sequential patterns that appear in given traffic data. We applied our method to traffic data observed in the darknet and report the results.

Keywords Incident detection • Frequent pattern mining • Sequential pattern • Data screening • Darknet monitoring

1 Introduction

The rapid growth of the high-speed Internet access service and mass storage media brings not only benefits to society but also harm. A notorious example is the social damage caused by various computer viruses. The bot worm is a typical example of a computer virus. It is Internet software controlled by a bot herder, and its infection spreads in a computer network. A bot herder can control the infected computers as a network, which is called a botnet, and cause many incidents, such as DDos attacks and sending of spam mail. To construct efficient countermeasures against these incidents, many effective methods for early detection of the tends of such incidents are being investigated in many projects, e.g., [3, 6, 7].

Our purpose is to discover information enabling signs of incidents to be detected. If we could prevent such incidents, the safety and the confidence of the Internet would be increased. Many researchers extract the trends of particular senders

H. Tsuruta • T. Shoudai (✉) • J. Takeuchi
Department of Informatics, Kyushu University, Fukuoka 819-0395, Japan
e-mail: hisashi.tsuruta@inf.kyushu-u.ac.jp; shoudai@inf.kyushu-u.ac.jp; tak@inf.kyushu-u.ac.jp

S.I. Ao et al. (eds.), *Intelligent Control and Innovative Computing*, Lecture Notes in Electrical Engineering 110, DOI 10.1007/978-1-4614-1695-1_28,

or patterns and analyze them to detect evidence of new attacks. For example, a flow-based method for detecting abnormal traffic was proposed by Kim et al. [4], and by focusing on the average number of packets sent by a source address and its frequency of appearances, a method for detecting subtle attacks was proposed by Fukushima et al. [2].

However, illegal packets caused by well-known malware make anomaly detection harder. Thus, we focus on finding the attack patterns of well-known malware. The majority can often be detected easily even by people unknowledgeable about network incidents. Therefore, we define attack patterns as a class of time-span sequence patterns that is easily detected by computers. Furthermore, we introduce a method to discover a set of frequent patterns that appear in the darknet observation data and to delete packets caused by those patterns. Frequent pattern discovery from structured data, such as web data and chemical compounds, has been extensively studied by many researchers: Miyahara et al. [5] and Yamasaki et al. [9].

We used data observed in the darknet, which is a network that cannot be accessed through conventional means. Most packets that the darknet receives are illegal, so they could be considered as traces of malicious attacks. Therefore, we might detect attack patterns of malware in the darknet access records. We think of each packet as a three tuple (transmitter's address, transmitter's port, and receiver's port), and call it an *event*. An event with its received time is called an *incident*. The observation data, called an *incident database*, is a set of incidents. We regard each item of an event as a string and propose a pattern class of sequences of string patterns with time delays. The following pattern is an example of incident patterns, called *informative event pattern delay (IEPD) sequences* in this paper.

$$(??.???. * .*, *, 445) \xrightarrow{0.50} (??. * . * .*, ?345, ?????) \xrightarrow{0.50} (??. * . * .*, ?345, ?????)$$

In this example, '?' stands for any one constant symbol, and '$*$' stands for any string whose length is at least 1. Each of the three items in a set of parentheses respectively denotes the transmitter's address, transmitter's port number, and receiver's port number. The real number over a rightwards arrow means the maximum time delay between the first pattern and the second pattern (or the second pattern and the third pattern).

First, we introduce a class of IEPD sequences to represent common illegal packets and then formally discuss a computational problem of finding IEPD sequences in a given incident database. Next, we present an effective heuristic algorithm for discovering IEPD sequences in an incident database. Last, we propose an automatic data screening method and report experimental results on the darknet traffic data.

2 Preliminaries

Let X be a set of distinct *events*. For an event $e \in X$, let t be a time when the event e occurs. We call a pair (t, e) an *incident* and a set of incidents an *incident database*.

Definition 28.1 (ED sequences). Let r be a positive integer and $T_{\max}$ a positive real number. Let $a_1, a_2, \ldots, a_r$ be r events (not necessarily distinct), and let $\tau_1, \tau_2, \ldots, \tau_{r-1}$ be $r-1$ positive real numbers that are less than or equal to $T_{\max}$. We then call $\pi = (a_1, \tau_1, a_2, \tau_2, \ldots, \tau_{r-1}, a_r)$ an *$(r, T_{\max})$-event delay sequence* (abbreviated to *$(r, T_{\max})$-ED sequence*).

Definition 28.2 (Matching of ED sequences). Let D be an incident database and $\pi = (a_1, \tau_1, a_2, \tau_2, \ldots, \tau_{r-1}, a_r)$ an $(r, T_{\max})$-ED sequence. For a subset D' of D with $|D'| = r$, we say that π *matches* D' if the following conditions hold: let $((t_1, e_1), (t_2, e_2), \ldots, (t_r, e_r))$ be a sorted sequence of the incidents in D' with respect to t_i $(1 \le i \le r)$, i.e., $t_1 \le t_2 \le \cdots \le t_r$. Then,

(a) for all i $(1 \le i \le r)$, $e_i = a_i$, and
(b) for all i $(1 \le i \le r-1)$, $t_{i+1} - t_i \le \tau_i$.

Definition 28.3 (Cover rate of ED sequences). Let D be an incident database and π an $(r, T_{\max})$-ED sequence. We denote by $D(\pi)$ the union of all subsets of D that are matched by π, i.e.,

$$D(\pi) = \bigcup_{D' \subseteq D \text{ s.t. } \pi \text{ matches } D'} D'.$$

The *cover rate* of π for D is defined as $cover_D(\pi) = |D(\pi)|/|D|$. Let P be a set of $(r, T_{\max})$-ED sequence and $D(P) = \bigcup_{\pi \in P} D(\pi)$. The *cover rate* of P for D is defined as $cover_D(P) = |D(P)|/|D|$.

First, we consider the following computational problem, which plays an important role in this paper.

R-EVENT DELAY SEQUENCE COVER (*R*-EC)

INSTANCE: Incident database D, cover rate σ $(0 \le \sigma \le 1)$, maximum time delay $T_{\max}$, and positive integer K.
QUESTION: Is there a set P of $(R, T_{\max})$-ED sequences such that $|P| \le K$ and the cover rate of P for D is at least σ?

We show the following theorem to address this problem.

Theorem 28.4. *3-EC is NP-complete.*

Proof. It is easy to see that 3-EC is in NP. We construct a reduction from the following well-known NP-complete problem X3C.

EXACT COVER BY 3-SETS (X3C)

INSTANCE: A set X with $|X| = 3q$ and a collection C of 3-element subsets of X.
QUESTION: Does C contain an exact cover for X, i.e., a subcollection $C' \subseteq C$ such that every element of X occurs in exactly one member of C'?

Let $X = \{e_1, e_2, \ldots, e_n\}$ where $n = 3q$ and $C = \{c_1, c_2, \ldots, c_m\}$ where $c_i = \{a_i^{(1)}, a_i^{(2)}, a_i^{(3)}\}$ $(1 \le i \le m)$. We construct an incident database D as follows.

Let $D' = \{((i-1)(n+2)+j, a_i^{(j)}) \mid 1 \le i \le m \text{ and } j = 1,2,3\}$ and $D'' = \{(m(n+2)+k, e_k) \mid 1 \le k \le n\}$. Let $D = D' \cup D''$, $\sigma = (3q+n)/(3m+n)$, $T_{\max} = n-1$, and $K = q$.

If X3C returns *true*, there is an exact cover $C' = \{c_{i_1}, c_{i_2} \dots, c_{i_q}\}$ $(1 \le i_1 < i_2 < \dots < i_q \le m)$. Let $P = \{(a_{i_\ell}^{(1)}, T_{\max}, a_{i_\ell}^{(2)}, T_{\max}, a_{i_\ell}^{(3)}) \mid 1 \le \ell \le q\}$. It is easy to see that $|\bigcup_{\pi \in P} D'(\pi)| = 3q$ and $|\bigcup_{\pi \in P} D''(\pi)| = n$. The cover rate of P is then equal to $\sigma = (3q+n)/(3m+n)$. Therefore 3-EC returns *true*.

Conversely, if 3-EC returns *true*, there is a set P of $(3, T_{\max})$-ED sequences such that $|P| \le q$. Let $P = \{(\alpha_\ell^{(1)}, \tau_\ell, \alpha_\ell^{(2)}, \tau'_\ell, \alpha_\ell^{(3)}) \mid 1 \le \ell \le q\}$ where $\alpha_\ell^{(1)}, \alpha_\ell^{(2)}, \alpha_\ell^{(3)} \in \{e_1, e_2, \dots, e_n\}$ and $\tau_\ell, \tau'_\ell \le T_{\max}$. Since $|\bigcup_{\pi \in P} D(\pi)| \ge 3q+n$ and $|D''| = n$, $|\bigcup_{\pi \in P} D'(\pi)| \ge 3q$. Therefore, for each $(3, T_{\max})$-ED sequence $(\alpha_\ell^{(1)}, \tau_\ell, \alpha_\ell^{(2)}, \tau'_\ell, \alpha_\ell^{(3)})$, there is an index $f(\ell)$ $(1 \le f(\ell) \le m)$ such that 3 continuous time events $\{((f(\ell)-1)(n+2)+1, a_{f(\ell)}^{(1)}), ((f(\ell)-1)(n+2)+2, a_{f(\ell)}^{(2)}), ((f(\ell)-1)(n+2)+3, a_{f(\ell)}^{(3)})\}$ are matched by $(\alpha_i^{(1)}, \tau_i, \alpha_i^{(2)}, \tau'_i, \alpha_i^{(3)})$. It is easy to see that $C' = \{\{\alpha_{f(\ell)}^{(1)}, \alpha_{f(\ell)}^{(2)}, \alpha_{f(\ell)}^{(3)}\} \mid 1 \le \ell \le q\}$ is an exact cover for X. □

In the next section, we give an effective heuristic algorithm for computing one of the more generalized problems within the framework of 3-EC.

3 Informative Event Pattern Delay Sequences

A typical Internet access log includes the time, transmitter's IP address, transmitter's port number, receiver's IP address, and receiver's port number. In this section, we formally define an algorithmic problem that deals with events consisting of several strings of variable length like Internet access logs. Moreover, we define sequential event patterns to represent frequent Internet access patterns

Let '?' and '$*$' be two special symbols. Let Σ be a finite alphabet ($|\Sigma| \ge 2$) that includes neither '?' nor '$*$' (i.e., $\Sigma \cap \{`?\text{'}, `*\text{'}\} = \emptyset$). Let N and $k_1, k_2, \dots, k_N$ be positive integers. An *informative event*, abbreviated to $\mathcal{I}$*-event*, is an event that consists of N strings $w_1, w_2, \dots, w_N$ in Σ^*, each of whose length is at most k_i $(1 \le i \le N)$. We denote an $\mathcal{I}$-event by $e = (w_1, w_2, \dots, w_N) \in \Sigma^{k_1} \times \Sigma^{k_2} \times \dots \times \Sigma^{k_N}$.

An *atom pattern* is a string in $\omega \in (\Sigma \cup \{`?\text{'}\})^+ \cup \{`*\text{'}\}$. For any atom pattern ω, we denote by $|\omega|$ the length of ω. For an atom pattern ω, if $\omega = `*\text{'}$, we can replace '$*$' with an atom pattern $\omega' \in (\Sigma \cup \{`?\text{'}\})^+$. If ω includes '?', we can replace it with a symbol in Σ. We call a set of such replacements a *substitution*. Let θ be a substitution. We denote by $\omega\theta$ the atom pattern that is obtained from ω by applying all replacements in θ to ω. We denote by $\mathcal{P}$ the set of atom patterns. For an integer k, we denote by $\mathcal{P}^{[k]}$ the set of atom patterns in $\mathcal{P}$ of length at most k. For two atom patterns ω and ω', we write $\omega' \preceq \omega$ if there is a substitution θ such that $\omega' = \omega\theta$.

Definition 28.5 ($\mathcal{I}$-Event patterns). We call a sequence of atom patterns $p = (\omega_1, \omega_2, \dots, \omega_N) \in \mathcal{P}^{[k_1]} \times \mathcal{P}^{[k_2]} \times \cdots \times \mathcal{P}^{[k_N]}$ an $\mathcal{I}$-event pattern. For two $\mathcal{I}$-event patterns $p = (\omega_1, \omega_2, \dots, \omega_N)$ and $p' = (\omega'_1, \omega'_2, \dots, \omega'_N)$, we write $p \preceq p'$ if for all i $(1 \le i \le N)$, $\omega_i \preceq \omega'_i$.

In a similar way to Definition 28.1, we define an IEPD sequence as follows.

Definition 28.6 (IEPD sequences). Let r be an integer ($r > 0$) and $T_{\max}$ a real number ($T_{\max} > 0$). Let $p_1, p_2, \dots, p_r$ be r $\mathcal{I}$-event patterns, and let $\tau_1, \tau_2, \dots, \tau_{r-1}$ be $r - 1$ positive real numbers that are less than or equal to $T_{\max}$. We then call $\pi = (p_1, \tau_1, p_2, \tau_2, \dots, \tau_{r-1}, p_r)$ an $(r, T_{\max})$-*informative event pattern delay sequence (abbreviated to* $(r, T_{\max})$*-IEPD sequence)*.

Definition 28.7 (Matching of IEPD sequences). Let D be an incident database and $\pi = (p_1, \tau_1, p_2, \tau_2, \dots, \tau_{r-1}, p_r)$ an $(r, T_{\max})$-IEPD sequence. For a subset D' of D with $|D'| = r$, we say that π *matches* D' if the following conditions hold: let $((t_1, e_1), (t_2, e_2), \dots, (t_r, e_r))$ be a sorted sequence of the incidents in D' with respect to t_i $(1 \le i \le r)$, i.e., $t_1 \le t_2 \le \cdots \le t_r$. Then,

(a) for all i $(1 \le i \le r)$, $e_i \preceq p_i$, and
(b) for all i $(1 \le i \le r - 1)$, $t_{i+1} - t_i \le \tau_i$.

Let D be an incident database. In a similar way to Definition 28.3, we define the cover rate of an IEPD sequence π for D and the cover rate of a set P of IEPD sequences for D.

To consider a similar computational problem to 3-EC, we have to define the order of IEPD sequences. The most generalized IEPD sequence is $\pi_0 = (p_0, T_{\max}, p_0, \dots, T_{\max}, p_0)$, where $p_0 = \underbrace{(*, *, \dots, *)}_{N}$. All incidents in any incident database can be covered by π_0.

Definition 28.8 (Size of delay sequences). For an $\mathcal{I}$-event pattern $p \in \mathcal{P}^{[k_1]} \times \mathcal{P}^{[k_2]} \times \cdots \times \mathcal{P}^{[k_N]}$, we denote by $n_\Sigma(p)$ the number of symbols in Σ that appear in p, and by $n_{\text{`?'}}(p)$ the number of '?' that appear in $p$. We define the size of an $\mathcal{I}$-event pattern $p$ as $size(p) = n_\Sigma(p) \times (Q + 1) + n_{\text{`?'}}(p)$, where $Q = \sum_{i=1}^{N} k_i$.

In a similar way to the definition of R-EC, we define R-INFORMATIVE EVENT PATTERN DELAY SEQUENCE COVER (R-IEPC) as follows.

R-INFORMATIVE EVENT PATTERN DELAY SEQUENCE COVER (R-IEPC)

INSTANCE: An incident database D and the following four parameters:

(a) K: maximum number of IEPD sequences,
(b) σ: minimum cover rate $(0 \le \sigma \le 1)$,
(c) S: minimum size of $\mathcal{I}$-event patterns, and
(d) $T_{\max}$: maximum time delay.

QUESTION: Is there a set P of $(R, T_{\max})$-IEPD sequences that satisfies the following conditions: for all $\pi = (p_1, \tau_1, \ldots, \tau_{R-1}, p_R) \in P$,

(a) $|P| \le K$,
(b) $cover_D(P) \ge \sigma$,
(c) $size(p_i) \ge S$ $(1 \le i \le R)$, and
(d) $\tau_i \le T_{\max}$ $(1 \le i \le R-1)$.

We have the following theorem. It is shown by reduction from 3-EC.

Theorem 28.9. *3-IEPC is NP-complete.*

4 Heuristic Algorithms for *R*-IEPC

4.1 Apriori-Like Method

Let D be an incident database. We denote by D_i the set of all the i-th strings w_i of the incidents $(t, (w_1, w_2, \ldots, w_N))$ in D. For any i $(1 \le i \le N)$ and atom pattern $\omega \in \mathcal{P}^{[k_i]}$, let $freq_{D_i}(\omega) = |\{w_i \in D_i \mid w_i \preceq \omega\}|/|D_i|$. For any $\mathcal{I}$-event pattern $p \in \mathcal{P}^{[k_1]} \times \mathcal{P}^{[k_2]} \times \cdots \times \mathcal{P}^{[k_N]}$, let $freq_D(p) = |\{e \mid \exists t \text{ s.t. } (t, e) \in D \text{ and } e \preceq p\}|/|D|$.

In Fig. 28.1, we give an algorithm FIND_IEPD_SEQUENCES (FIES) for computing R-IEPC by using an Apriori-like method twice. Figures 28.2 and 28.3 describe those Apriori-like methods used in Algorithm FIES. Let δ be a positive real number less than or equal to a minimum cover rate σ. This parameter plays an important role in Algorithm FIES to produce a good set of $(R, T_{\max})$-IEPD sequences.

Algorithm FIND_IEPD_SEQUENCES (FIES);
input: Incident database $D = \{(t, e) \mid t > 0 \text{ and } e \in \Sigma^{k_1} \times \cdots \times \Sigma^{k_N}\}$,
minimum cover rate σ, real number δ $(0 < \delta \le \sigma)$, minimum size S of IEPD sequences, integer R $(R > 0)$, and real number $T_{\max}$ $(T_{\max} > 0)$;
begin
$D_i^a := \{w_i \mid (t, (w_1, \ldots, w_i, \ldots, w_N)) \in D\}$ $(1 \le i \le N)$;
forall i $(1 \le i \le N)$ **do** $A_i := \{\omega \in \mathcal{P}^{[k_i]} \mid freq_{D_i^a}(\omega) \ge \delta\}$;
$F :=$FREQ_$\mathcal{I}$-EVENT_PATTERNS$(D, \delta, S, \{A_i \mid 1 \le i \le N\})$ (Fig. 1.2);
$P :=$FREQ_PATTERN_SEQUENCES$(D, \sigma, F, R, T_{\max})$ (Fig. 1.3);
foreach $\pi = (p_1, \tau_1, \ldots, \tau_{R-1}, p_R) \in P$ **do begin**
    Decrease each time delay τ_j $(1 \le j \le R-1)$ while $cover_D(P) \ge \sigma$
end;
output P
end.

Fig. 28.1 Algorithm FIND_IEPD_SEQUENCES (FIES)

```
Procedure FREQ_I-EVENT_PATTERNS(D, δ, S, {A_i | 1 ≤ i ≤ N});
begin
  F_1 := ∪_{1≤i≤N} {(*, ..., *, ω, *, ..., *) | ω ∈ A_i (ω is the i-th entry)};
  for k := 2 to N do begin
    F_k := ∅;
    foreach p, p' ∈ F_{k-1} do begin
      Let p = (ω_1, ω_2, ..., ω_N) and p' = (ω'_1, ω'_2, ..., ω'_N);
      if there are two indices i and j (i < j) s.t.
        (a) ω_ℓ = ω'_ℓ (1 ≤ ℓ ≤ N, ℓ ≠ i, and ℓ ≠ j),
        (b) ω_i ≠ '*' and ω_ℓ = '*' (i + 1 ≤ ℓ ≤ N), and
        (c) ω'_j ≠ '*' and ω'_ℓ = '*' (i ≤ ℓ ≤ j − 1, j + 1 ≤ ℓ ≤ N)
      then begin
        Let p'' = (ω''_1, ..., ω''_ℓ, ..., ω''_N) be the I-event pattern s.t.
          if ℓ = j then ω''_ℓ = ω'_ℓ else ω''_ℓ = ω_ℓ;
        if freq_D(p'') ≥ δ then F_k := F_k ∪ {p''}
      end
    end
  end;
  F := {p ∈ ∪_{1≤k≤N} F_k | size(p) ≥ S};
  foreach p ∈ F do
    if there is an I-event pattern p' ∈ F (p' ≠ p) s.t. p ⪯ p' then
      F := F \ {p};
  return F
end;
```

Fig. 28.2 Procedure FREQ_$\mathcal{I}$-EVENT_PATTERNS used at line 3 in Algorithm FIES

```
Procedure FREQ_PATTERN_SEQUENCES(D, σ, F, R, T_max);
begin
  P_1 := F;
  for r := 2 to R do begin
    P_r := ∅;
    foreach π, π' ∈ P_{r-1} do begin
      Let π = (p_1, T_max, ..., T_max, p_{r-1}) and
      π' = (p'_1, T_max, ..., T_max, p'_{r-1});
      if π and π' satisfy that
        for all i (1 ≤ i ≤ r − 2), p_i = p'_i, and p_{r-1} ≠ p'_{r-1} then
      begin
        Let π'' := (p_1, T_max, ..., T_max, p_{r-1}, T_max, p'_{r-1});
        if cover_D(π'') ≥ σ then P_r := P_r ∪ {π''}
      end
    end
  end;
  return P_R
end;
```

Fig. 28.3 Procedure FREQ_PATTERN_SEQUENCES used at line 4 in Algorithm FIES

4.2 Machine Learning Method

Here we define a partial order of IEPD sequences for a given incident database. Then we propose an algorithm for solving our problem.

Let $\pi = (p_1, \tau_1, p_2, \tau_2, \ldots, \tau_{r-1}, p_r)$ be an $(r, T_{\max})$-IEPD sequence and $\pi' = (p'_1, \tau'_1, p'_2, \tau'_2, \ldots, \tau'_{r'-1}, p'_{r'})$ an $(r', T_{\max})$-IEPD sequence. We write $\pi' \preceq \pi$ if $r' \geq r$ and there are r integers $j_1, j_2, \ldots, j_r$ with $1 \leq j_1 < j_2 < \cdots < j_r \leq r'$ such that for all i $(1 \leq i \leq r)$,

$$p'_{j_i} \preceq p_i \text{ and } \sum_{\ell=j_i}^{j_{i+1}-1} t'_\ell \leq t_i.$$

Definition 28.10 (Maximal IEPD sequences). Let D be an incident database and σ a minimum cover rate $(0 \leq \sigma \leq 1)$. We say that π is a *maximal* $(r, T_{\max})$-*IEPD sequence* with respect to D and σ if there is no $(r, T_{\max})$-IEPD sequence π' $(\pi' \neq \pi)$ such that $\pi' \preceq \pi$ and $cover_D(\pi') \geq \sigma$.

Let P be a set of $(r, T_{\max})$-IEPD sequence. We say that P is a *maximal K-set of $(r, T_{\max})$-IEPD sequences with respect to D and σ* if the following four conditions hold:

1. $|P| \leq K$,
2. $cover_D(P) \geq \sigma$,
3. For any pair $\pi, \pi' \in P$ $(\pi \neq \pi')$, neither $\pi \preceq \pi'$ nor $\pi' \preceq \pi$ hold, and
4. For any $\pi \in P$, there is no $(r, T_{\max})$-IEPD sequence $\pi' \notin P$ such that $\pi' \preceq \pi$ and $cover_D(P - \{\pi\} \cup \{\pi'\}) \geq \sigma$.

We define a formal algorithmic problem as follows.

R-MAXIMAL INFORMATIVE EVENT PATTERN DELAY SEQUENCE COVER (R-MIEPC)

INSTANCE: An incident database D and the following three parameters:
K: maximum number of IEPD sequences, σ: minimum cover rate $(0 \leq \sigma \leq 1)$, $T_{\max}$: maximum time delay.
QUESTION: Is there a maximal K-set P of $(R, T_{\max})$-IEPD sequences that satisfies the following conditions: (i) $cover_D(P) \geq \sigma$ and (ii) for all $\pi = (p_1, \tau_1, \ldots, \tau_{R-1}, p_R) \in P$, $\tau_i \leq T_{\max}$.

Definition 28.11 (Size of IEPD sequences). For an $(r, T_{\max})$-IEPD sequence $\pi = (p_1, \tau_1, p_2, \tau_2, \ldots, \tau_{r-1}, p_r)$, we define the size of π as follows:

$$size(\pi) = \sum_{i=1}^{r} size(p_i) + \sum_{i=1}^{r-1} (T_{\max} - \tau_i).$$

Lemma 28.12. *Let π and π' be $(r, T_{\max})$-IEPD sequence and $(r', T_{\max})$-IEPD sequence, respectively. Then, if $\pi' \preceq \pi$ then $size(\pi') \geq size(\pi)$.*

We can solve R-MIEPC by specializing IEPD sequences step by step with a machine learning method proposed in Arimura et al. [1]. We omit the details of the algorithm. From Lemma 28.12, a maximal K-set of $(R, T_{\max})$-IEPD sequences has a local optimal solution with respect to its size. In the next section, we use Algorithm FIES rather than a strategy based on maximalities in order to obtain a set of IEPD sequences that has a sufficiently large size.

5 A Screening Method on Network Traffic Data

In this section, we propose a screening method for removing irregular packets that are supposed to be caused by well-known malware and discuss its effectiveness by experiments on darknet observation data.

In Fig. 28.4, we give a formal description of our screening method. First, Algorithm SCREENING finds a small set of IEPD sequences of large size that covers a certain rate of the incidents by Algorithm FIND_IEPD_SEQUENCES, and then removes those incidents. We suppose that those incidents are caused by well-known malware.

We call the time span between the first received packet and the last received packet in an incident database the *attack period* of the database. Algorithm SCREENING spends exponentially increasing time depending on the size of a given incident database D. To overcome this difficulty, we divide a given database into smaller databases whose attack periods are specified previously. In Fig. 28.5, we give Algorithm LARGE_SCREENING_BY_ATTACK_PERIOD for a very large database.

5.1 *Experiments on Internet Access Logs*

In our previous work [8], we conducted some preliminary experiments with a prototype of our system. In this paper, we show the effectiveness of our system by sufficient experiments on real network data.

Algorithm SCREENING;
input: Incident database $D = \{(t, e) \mid t > 0 \text{ and } e \in \Sigma^{k_1} \times \cdots \times \Sigma^{k_N}\}$,
  minimum cover rate σ, real number δ $(0 < \delta \leq \sigma)$, minimum size S of
  IEPD sequences, integer R $(R > 0)$, and real number $T_{\max}$ $(T_{\max} > 0)$;
begin
  P :=FIND_IEPD_SEQUENCES$(D, \sigma, \delta, S, R, T_{\max})$;
  output $D \setminus D(P)$
end.

Fig. 28.4 Algorithm SCREENING

```
Algorithm LARGE_SCREENING_BY_ATTACK_PERIOD;
input: Incident database D = {(t, e) | t > 0 and e ∈ Σ^{k_1} × ··· × Σ^{k_N}},
   minimum cover rate σ, real number δ (0 < δ ≤ σ), minimum size S of
   IEPD sequences, integer R (R > 0), real number T_max (T_max > 0), and
   attack period s (s > 0) of divided data;
begin
   Let ((t_1, e_1), ..., (t_|D|, e_|D|)) be the time-sorted sequence of D,
      i.e., t_1 ≤ t_2 ≤ ··· ≤ t_|D|;
   i := 1; t := t_i; k := 1; D_k := ∅;
   while i ≤ |D| begin
      if t_i < t + s then begin D_k := D_k ∪ {(t_i, e_i)}; i := i + 1 end
      else begin t := t + s; k := k + 1; D_k := ∅ end
   end;
   foreach D_j (1 ≤ j ≤ k) do D'_j :=SCREENING(D_j, σ, δ, S, R, T_max);
   output ⋃_{1≤j≤k} D'_j
end.
```

Fig. 28.5 Algorithm LARGE_SCREENING_BY_ATTACK_PERIOD

A darknet observation data includes the time, transmitter's IP address, transmitter's port number, and receiver's port number. The receiver's IP address is not included. This is due to the concealment of the darknet. Let $\Sigma = \{0, 1, 2, \ldots, 9\}$. We divide the transmitter's address into four parts according to the form of the IP address, and those parts are referred to as *address*-1, *address*-2, *address*-3, and *address*-4 from the head, respectively. In addition to those addresses, we have two port numbers, which are referred to as *port*-1 and *port*-2. Then, we have a set of atom data D_i^a $(1 \leq i \leq 6)$ (defined in Algorithm FIES), which includes all atom data about *address*-*i* $(1 \leq i \leq 4)$ and *port*-*i* $(i = 1, 2)$. Each *address*-*i* is a string in $\{0, 1, \ldots, 255\}$, and *port*-*i* is a string in $\{0, 1, \ldots, 65535\}$. An $\mathcal{I}$-event then consists of 6 strings whose lengths are at most 3, 3, 3, 3, 5, 5, respectively. In the experiments, we consider only $(3, T_{\max})$-IEPD sequences with the following user specified parameters.

- Experimental Data
  The observation data and the quantity of packets that we used in each experiment are shown in Table 28.1.
- Parameters for screening

Minimum cover rate:	$\sigma = 0.1$,
Maximum time delay:	$T_{\max} = 0.5$ (seconds),
Minimum size of $\mathcal{I}$-event patterns:	$S = 69$ (This is the size of $\mathcal{I}$-event patterns that consists of three symbols in Σ and four $*$'s.),
Attack period of each divided data:	$s = 600$ (seconds) (= 10 (minutes)).

Table 28.1 Quantities of received and remaining packets after screening and their cover rates

Observation day	Received packets	Remaining packets	Cover rate (%)
8 Nov, 2006	4,039,197	2,006,190	50.33
10 Sep, 2008	2,864,698	1,736,558	39.38
23 Jan, 2009	2,884,381	1,775,265	38.45
2 Apr, 2009	10,261,071	7,336,516	28.50
24 May, 2009	6,358,187	4,258,470	33.02
25 May, 2009	4,022,849	3,039,185	24.45
21 Jun, 2009	2,275,630	1,797,655	21.00
12 Jul, 2009	6,691,671	3,109,089	53.54

We show the relationship between the quantities of screened databases and the unit time in the line graphs of Fig. 28.6 for continuous two days data: 24-25th May, 2009, and one day data: 21st June, 2009, which are typical results of our experiments. Those graphs show the transitions of quantities of packets that are received every 10 min in the time sequence of each day and the quantities of packets that are left after screening the packets. The horizontal and vertical axes indicate the time sequence and the quantities of packets, respectively. The solid and dotted lines describe the quantities of packets before screening and after screening, respectively. A peak in a solid line shows a mass of packets is received during a short period. In the graph for 24-25th May, 2009, there are no significant peaks of the dotted line. Thus, we can conclude that attacks indicated by peaks in the solid line are caused by well-known malware and that our screening method could detective them. In other observation data, we found some peaks in the solid line still remain after screening. For example, the line graph of 21st June, 2009, 10-min intervals around 10:30 and 22:30 are such cases. There are also many packets caused by complex attack patterns which cannot be detected by our method. We show the quantities of all remaining packets obtained from databases in Table 28.1 with their cover rates. The average cover rate for all databases is 36.39%.

Next we show the relationship between the increasing rates of peaks and their corresponding cover rates. We call a 10-min interval an *m-peak* if the quantity of packets received during the interval is more than m times the average of two 10-min intervals before and after it. Moreover, for $m < n$, we call a 10-min interval an *(m, n)-peak* if the interval is an *m-peak* but not an *n-peak*. We show in Fig. 28.7 the distribution of the number of (m, n)-peaks against their cover rates for some m and n $(m < n)$. These distributions were obtained from the experiments on all data in Table 28.1. For example, the number of $(1.2, 1.6)$-peaks whose cover rate is greater than or equal to 15% and less than 20% is 8.

From these experimental results, we can say that our method succeeds to detect attacks by malware that has simple attack patterns even if the quantities of them are large. Furthermore, we can conclude that the higher the peak, the larger the cover rate. Thus, in most darknet observation data, most of peaks contain a large number of packets by well-known malware.

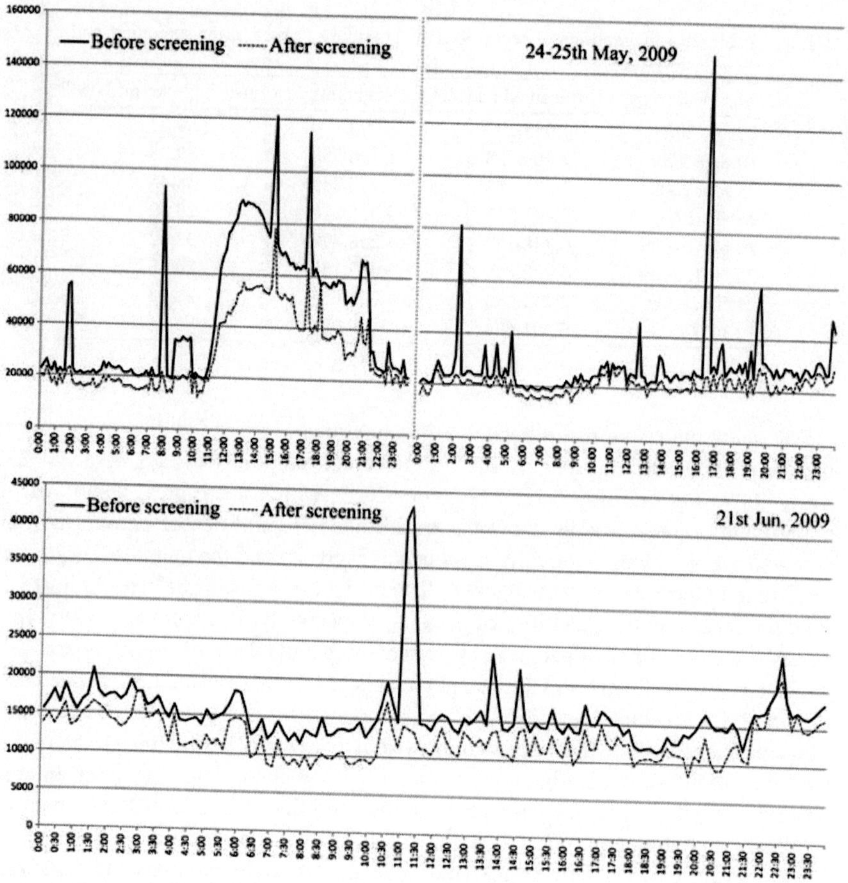

Fig. 28.6 In the line graphs, *solid line* (resp. *dotted line*) describes the number of packets before (resp. after) screening every 10 min. The total numbers of packets before and after screening are shown in Table 28.1

6 Conclusions

We proposed a screening method taking advantage of time-span sequential patterns. We applied our proposed method to darknet observation data and showed its effectiveness for identifying packets caused by well-known malware. In many cases, the reduction rate before and after screening was proportional to the frequency of received packets. However, experiments showed there are some cases in which the reduction rate is low despite concentrated attacks.

As future work, we should investigate the relation between the reduction rate and the frequency of received packets on tuning appropriate establishments. Furthermore, to attain online screening, we are considering efficient algorithms to find a set of time-span sequential patterns.

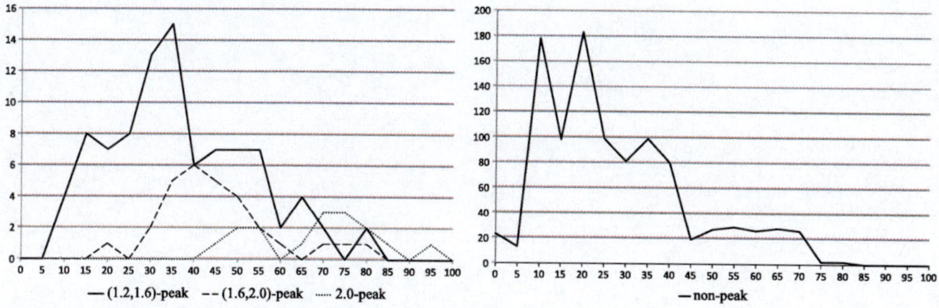

Fig. 28.7 The horizontal and vertical axes mean the cover rates and the numbers of peaks, respectively. The average cover rates are 29.90%, 38.44%, 47.73%, and 70.72% for the (0, 1.2)-peaks (*non-peaks*), (1.2, 1.6)-peaks, (1.6, 2.0)-peaks, and 2.0-peaks, respectively

Acknowledgments This research is supported by the National Institute of Information and Communications Technology (NICT) of Japan, entitled "Research and Development for Widespread High-speed Incident Analysis".

References

1. Arimura H, Shinohara T, Otsuki S (1994) Finding minimal generalizations for unions of pattern languages and its application to inductive inference from positive data. Proceedings of The 11th Symposium on Theoretical Aspects of Computer Science, Springer, Lecture Notes in Computer Science 775:649–660
2. Fukushima Y, Hori Y, Sakurai K (2009) A consideration of feature extraction for attacks on darknet. IEICE technical report, 109(285):37–42, in Japanese
3. JPCERT/CC. http://www.jpcert.or.jp/english/
4. Kim M, Kang H, Hong S, Chung S, Hong JW (2004) A flow-based method for abnormal network traffic detection. Proc IEEE/IFIP Network Oper Manag Sym (1):599–612
5. Miyahara T, Suzuki Y, Shoudai T, Uchida T, Takahashi K, Ueda H (2002) Discovery of frequent tag tree patterns in semistructured web documents. Proceedings of The 5th Pacific-Asia Conference on Knowledge Discovery and Data Mining, Springer, Lecture Notes in Artificial Intelligence, 2336:341–355
6. Nicter project. http://www.nict.go.jp/en/nsri/index.html
7. SANS Internet Storm Center. http://isc.sans.org/
8. Tsuruta H, Shoudai T, Takeuchi J (2011) Frequent sequential pattern discovery for data screening, Lecture notes in engineering and computer science: Proceedings of the international multiConference of engineers and computer scientists, IMECS 2011, 16–18 March, 2011, Hong Kong, pp 315–322
9. Yamasaki H, Sasaki Y, Shoudai T, Uchida T, Suzuki Y (2009) Learning block-preserving graph patterns and its application to data mining. Mach Learn 76(1):137–173

Chapter 29
Informative Gene Discovery in DNA Microarray Data Using Statistical Approach

Kentaro Fukuta and Yoshifumi Okada

Abstract Preventing, diagnosing, and treating disease is greatly facilitated by the availability of biomarkers. Recent improvements in bioinformatics technology have facilitated large-scale screening of DNA microarrays for candidate biomarkers. Here we discuss a gene selection method, which is called *LEAve-one-out Forward selection method* (LEAF), for discovering informative genes embedded in gene expression data, and propose an additional algorithm for extending LEAF's capabilities. LEAF is an iterative forward selection method incorporating the concept of leave-one-out cross validation (LOOCV) and provides a discrimination power score (DPS) for genes, which is a criterion for selecting the candidate of informative genes. We show that LEAF identifies genes that are practically used as biomarkers. Our method should be useful bioinformatics tool for biomedical, clinical, and pharmaceutical researchers.

Keywords Biomarkers • Data mining • Gene expression profiles • Cancer classification

1 Introduction

Recent progress in bioinformatics technology has facilitated large-scale screening for candidate biomarkers [7]. A biomarker, as the name implies, is a cell-derived substance such as a gene, protein or enzyme that can be used to elucidate

K. Fukuta
Satellite Venture Business Laboratory, Muroran Institute of Technology,
27-1, Mizumoto, Muroran, Hokkaido 050-8585, Japan
e-mail: fukuta@mail.svbl.muroran-it.ac.jp

Y. Okada (✉)
College of Information and Systems, Muroran Institute of Technology,
27-1, Mizumoto, Muroran, Hokkaido 050-8585, Japan
e-mail: okada@epsilon2.csse.muroran-it.ac.jp

S.I. Ao et al. (eds.), *Intelligent Control and Innovative Computing*, Lecture Notes in Electrical Engineering 110, DOI 10.1007/978-1-4614-1695-1_29,

physiological or pathological process [2]. In our previous study, we have proposed a novel method called LEAve-one-out Forward selection method (LEAF) for analysis of gene expression data [4, 5]. This method enabled us to construct a ranking system of informative genes using a parameter reflecting the efficiency of the class discriminant designated the Discriminant Power Score (DPS). We applied LEAF to four public leukemia datasets (ALL/AML, ALL/MLL, and MLL/AML) [1]. The results showed that our method yields a stable discriminant result with 100% accuracy using a three-gene set. Furthermore, some genes with high DPS values are cancer-related genes (top-h genes), as clarified by research in recent years.

Nevertheless, two problems remain to be resolved, namely: (1) We have not selected a criterion for defining the h-value. (2) The candidate list of associated genes is insufficient to assign a discrete biological function (correlation and causal relation between genes).

Here we briefly introduce LEAF and then propose a solution to address these problems. Thus, using public gene function database, we propose a simple and straightforward method for determining the top-h genes (h-value) and conduct a biological functional analysis of the genes. Moreover, we investigate the usefulness of the method by applying it to other datasets. Subsequently, we conduct a biological functional analysis of the genes, using public gene function database.

2 Methodology

2.1 Datasets

We used six well-known cancer datasets provided by Armstrong et al., Golub et al., Gordon et al. and Singh et al. These datasets are available at the Broad Institute [3]. Details of the datasets are summarized in Fig. 29.1a.

Figure 29.1b presents two datasets are arranged in the form of a data matrix. The matrix size is $CN \times TG$, where CN denotes $CN1 + CN2$. Furthermore, $CN1$ and $CN2$, respectively, represent the number of samples in Class 1 and Class 2, and g_k ($k = 1, 2, \ldots, TG$) corresponds to a gene expression value, and TG signifies the total number of genes.

2.1.1 Leukemia Dataset 1 (1a, 1b, 1c)

These datasets are about subtype of leukemia (ALL: Acute lymphocytic leukemia, MLL: Mixed lineage leukemia, and AML: Acute myelogenous leukemia) [1]. The total of training samples is 57 samples (20 ALL, 17 MLL, and 20 AML), while the total of testing samples is 15 samples (4 ALL, 3 MLL, and 8 AML). The total number of genes is 12,582. We build the three data matrix using two kinds of these datasets.

a Preparation of gene expression dataset

Dataset name	# Genes	Class	# Training Samples	# Test samples
Leukemia 1a		ALL/AML	43 (20/23)	9 (4/5)
Leukemia 1b	12582	ALL/MLL	37 (20/17)	7 (4/3)
Leukemia 1c		MLL/AML	40 (17/23)	8 (3/5)
Leukemia 2	7129	ALL/AML	38 (27/11)	34 (20/14)
Lung Cancer	12533	MPM/ADCA	32 (16/16)	149 (15/134)
Prostate Cancer	12600	Tumor/Normal	102 (52/50)	34 (25/9)

b Data matrix

			Gene index number					
	Sample ID	Sample Gene	1	2	⋯	k	⋯	*TG*
Class 1	1	s_1^1	g_1	g_2	⋯	g_k	⋯	$g_{\#Genes}$
	2	s_2^1						
	⋮	⋮						
	CN1	s_{CN1}^1						
Class 2	*CN1* + 1	s_{CN2}^2						
	⋮	⋮						
	CN = *CN1* + *CN2*	s_{CN}^2						

Fig. 29.1 Preparation of dataset

2.1.2 Leukemia Dataset 2

This dataset is the well-known leukemia data studied by Golub et al. [7]. In this dataset, there are 72 tissue samples (47 ALL, 25 AML). The training set contains 38 samples (27 ALL, 11 AML); the remaining 34 samples are used for testing (20 ALL, 14 AML). The total number of genes is 7,129.

2.1.3 Lung Cancer Dataset

This dataset is concerned with the malignant pleural mesothelioma (MPM) vs. adenocarcinoma (ADCA) of the lung [8]. There are 180 tissue samples (31 MPM, 150 ADCA). The training set contains 32 samples (16 MPM, 16 ADCA); the remaining 149 samples are used for testing (15 MPM, 134 ADCA). The total number of genes is 12,533.

2.1.4 Prostate Cancer Dataset

This dataset is concerned with prostate tumor and normal tissue [11]. The training set contains 102 samples (52 tumor, 50 normal) and testing set contains 34 samples (25 tumor, 9 normal). The total number of genes is 12,600.

2.2 LEAF: LEAve-One-Out Forward Selection Method

We have proposed a robust and accurate gene selection method based on forward selection called forward selection method (FSM) [10]. To satisfy a maximal variance ratio (F-value) between two disease classes by using Mahalanobis distance, FSM cumulatively selects gene one-by-one and ultimately identifies a set of genes (a gene ranking) that is informative for disease classification.

The flow of the FSM algorithm is described as follows:

1. Calculation of the F-value (F_1) for all genes and selection of a gene having the maximum F_1 as the first gene.
2. For k ($\geq$2)th gene, we pick up a k-th gene from the rest of genes, and add it into the set of $k - 1$ genes.
3. Step 2 is repeated for all the genes in the rest set, and k-th genes is determined by selecting the gene with the maximum F_k.
4. Step 2 and Step 3 are repeated for $k \leq (Class1\_N + Class2\_N) - 2$ till the ranking of the genes is accomplished.

In fact, LEAF is an iterative FSM inspired by leave-one-out cross validation (LOOCV) [9]. Details of the algorithm have been published [4]. Figure 29.2a outlines the method. First, one test sample is taken from the dataset. Then the remaining samples are used as a learning set. Subsequently, we apply FSM to the learning set and obtain a gene ranking. These steps are repeated for every test sample. Finally, we extract a highly robust set of genes in a classification based on discriminant power, called DPS. DPS is a parameter of the class discriminant ability defined for all genes. $DPS(k)$ $(1 \leqq k \leqq TG)$ represents the DPS value of the gene with the k-th gene-index-number.

The discrimination power of selected genes greatly depends on the genes that compose the ranking. In the method of this paper, the genes with low discrimination power are excluded by using threshold parameter $Thrs1$. In this research, we adopt $Thrs1 = 2.0$, and use the genes which satisfy $F_1 > 2.0$ for selecting the k ($\geq$2)th gene.

2.3 Determination Method of h-Value (Top-h Genes)

Because previous work [4] has not provided any criterion (cut-off threshold) for obtaining a set of discriminative genes, here we introduce an interactive method for extracting the top-h genes that are used to generate a final discriminant function. The identification method of the h-value is illustrated in Fig. 29.2b. The h-value is calculated by the following steps:

1. Descending sort of DPS (Fig. 29.2b.A).
2. Decision of h-value.

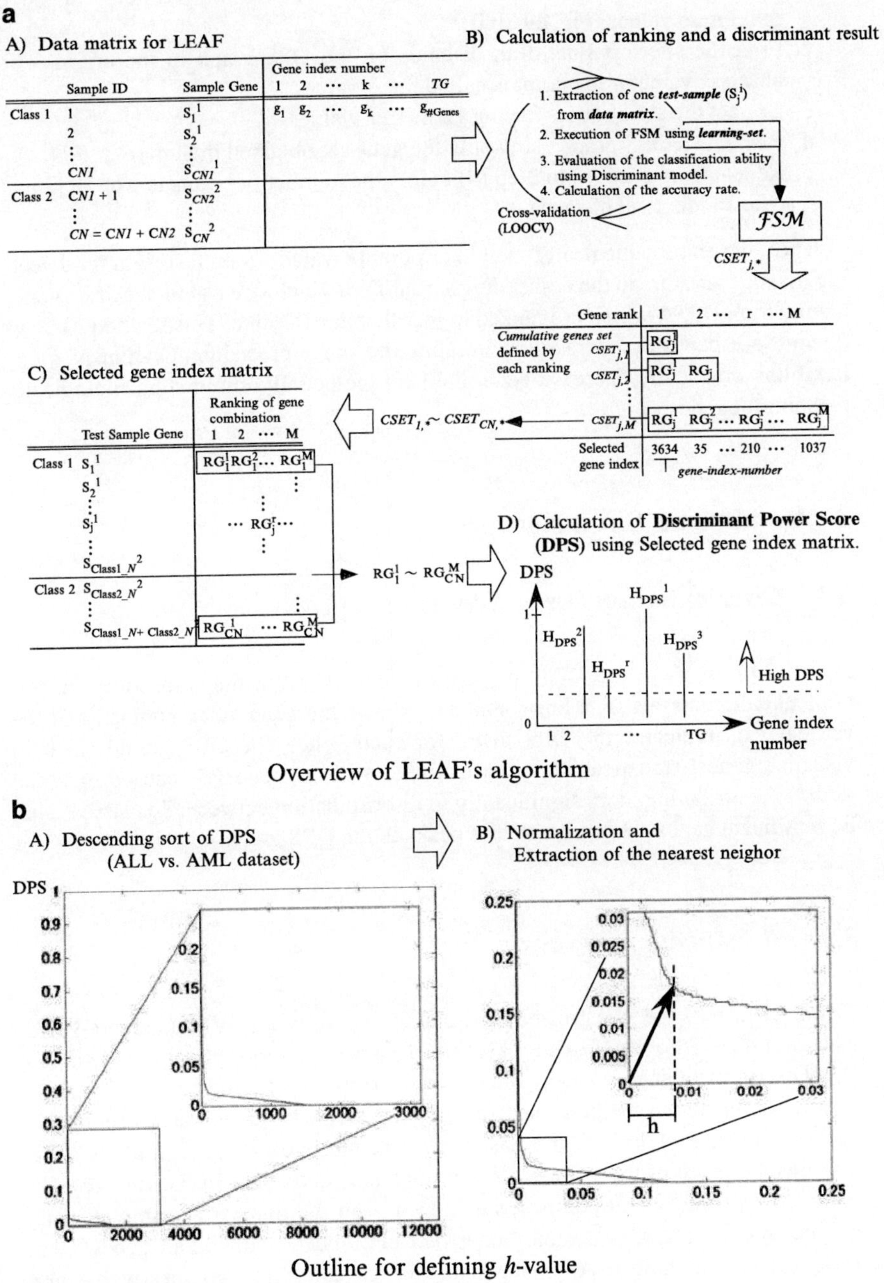

Fig. 29.2 Overview of our methodology. (**a**) Overview of LEAF's algorithm. (**b**) Outline for defining h-value

a. Normalize the horizontal and vertical axes by dividing by their respective maximum values (Fig. 29.2b.B).
b. Find the shortest Euclidean distance on the DPS graph to the origin. The abscissa value of the point is called the h-value.
c. Extract the set of genes having DPSs $\geqq h$-value.
d. Recreate a DPS graph using only the gene set obtained in Step (c).
e. Repeat from Step (a) to Step (d) unless the number of points is 1 or all points take an identical distance.

Thus, we employ the nearest neighbor point (h-value) from the origin for detecting drastic curvature in the descending sorted-DPS graph. We can then extract genes having high DPSs, which are ranked higher than the h-value. This method narrows down top-h-genes by interactively iterating the above procedure. Obviously, many iterations drastically decrease gene numbers, potentially eliminating biologically meaningful genes.

3 Results and Discussion

3.1 Discrimination Power Score

Figure 29.3 displays the DPSs of genes calculated from the respective pairs of the leukemia datasets. The horizontal axis shows the gene index number, and the vertical axis indicates the DPS given for each gene. The DPS graph can help visualize genes' statistical importance. Genes with higher DPSs can be regarded as those contributing more significantly to discrimination between the classes. That is, significant genes are represented as peaks in the DPS graph.

3.2 Classification Results

In this research, we constructed five classifiers, such as SVM (support vector machine), Naive Bayes classifier, kNN (K Nearest Neighbor classifier), C4.5 and DA (Discriminant analysis using Mahalanobis' distance), for the high DPS gene set selected by using our method, and evaluated the accuracy rate to the test set in each data set. Figure 29.4 displays the classification results in each test set, and Fig. 29.4a contains the result of the multi-class identification (Leukemia 1a, 1b, and 1c). The number of variables that compose DA depends on the number of samples of the training set. Therefore, we cannot construct the model including all genes. Here, we construct the mold (DA + Non-selection) by using the variables with higher Mahalanobis distance of each gene.

In case of all, the accuracy rate of the classifier with our method is higher than the accuracy of Non-selection.

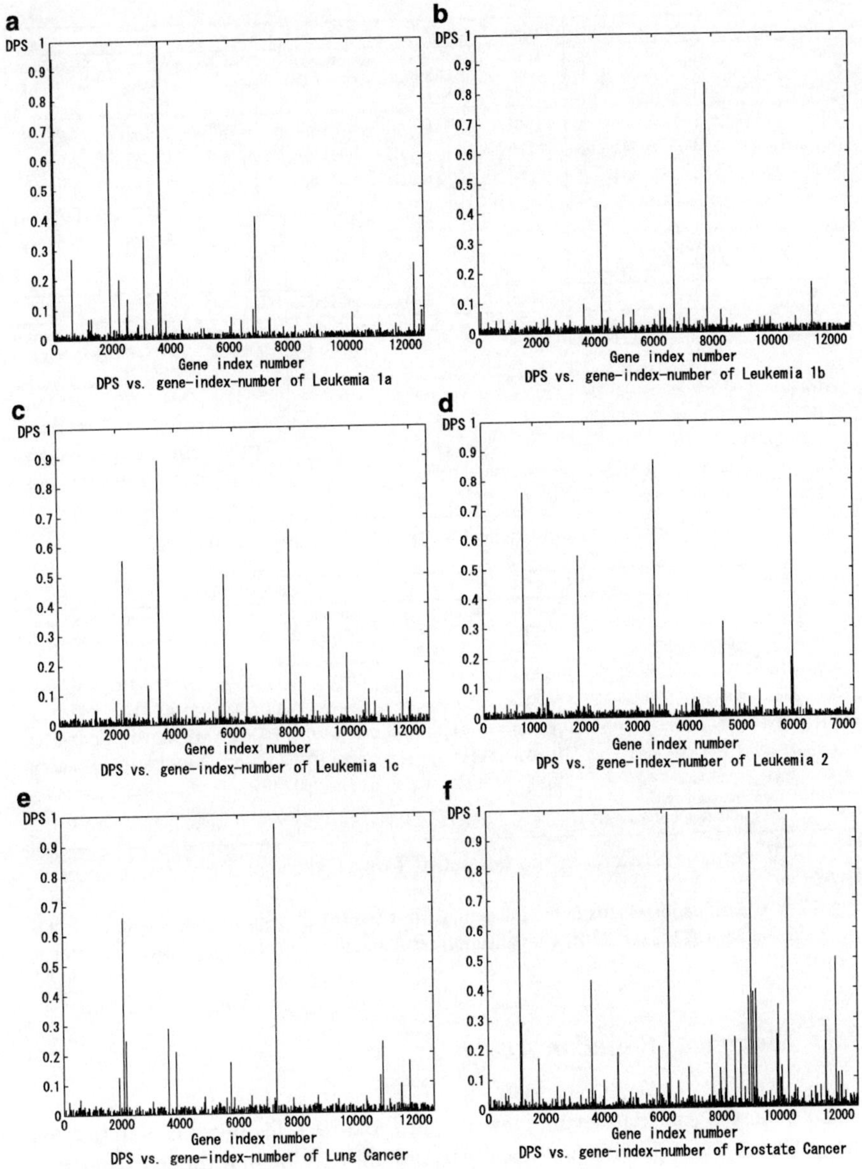

Fig. 29.3 DPS vs. gene-index-number of cancer dataset

a

Dataset name	Leukemia 1a			Leukemia 1b		
Classifier + Feature selection	Used genes	# Correctly classified samples	Maximum accuracy (%)	Used genes	# Correctly classified samples	Maximum accuracy (%)
SVM + LEAF [+ Non-selection]	1 [12582]	12/12 [12/12]	1.0 [1.0]	1 [12582]	7/7 [7/7]	1.0 [1.0]
Naive Bayes + LEAF [+ Non-selection]	1 [12582]	12/12 [12/12]	1.0 [1.0]	1 [12582]	7/7 [7/7]	1.0 [1.0]
kNN + LEAF [+ Non-selection]	1 [12582]	12/12 [12/12]	1.0 [1.0]	1 [12582]	7/7 [4/7]	1.0 [1.0]
C4.5 + LEAF [+ Non-selection]	1 [12582]	12/12 [11/12]	1.0 [0.92]	1 [12582]	7/7 [7/7]	1.0 [0.57]
DA + LEAF [+ Non-selection] (* DA: Discriminant analysis using Mahalanobis' distance)	1 [38]	12/12 [11/12]	1.0 [0.92]	1 [35]	7/7 [5/7]	1.0 [0.71]

Leukemia 1c			Leukemia 1a, 1b, 1c (One-vs-one classifier)			Leukemia 2		
Used genes	# Correctly classified samples	Maximum accuracy (%)	Used genes	# Correctly classified samples	Maximum accuracy (%)	Used genes	# Correctly classified samples	Maximum accuracy (%)
2 [12582]	11/11 [11/11]	1.0 [1.0]	9	15/15	1.0	13 [7129]	33/34 [31/34]	0.97 [0.91]
2 [12582]	11/11 [8/11]	1.0 [0.73]	4	15/15	1.0	4 [7129]	32/34 [32/34]	0.94 [0.94]
2 [12582]	11/11 [11/11]	1.0 [1.0]	5	15/15	1.0	3 [7129]	31/34 [24/34]	0.91 [0.71]
3 [12582]	11/11 [8/11]	1.0 [0.73]	6	15/15	1.0	2 [7129]	31/34 [31/34]	0.91 [0.91]
2 [35]	11/11 [10/11]	1.0 [0.91]	9	15/15	1.0	29 [36]	33/34 [24/34]	0.97 [0.71]

Classification results in the test set of Leukemia 1a, 1b, 1c, and 2

b

Dataset name	Lung Cancer			Prostate Cancer		
Classifier + Feature selection	Used genes	# Correctly classified samples	Maximum accuracy (%)	Used genes	# Correctly classified samples	Maximum accuracy (%)
SVM + LEAF [+ Non-selection]	1 [12533]	149/149 [148/149]	1.0 [1.0]	1 [12600]	33/34 [31/34]	0.97 [0.91]
Naive Bayes + LEAF [+ Non-selection]	1 [12533]	148/149 [140/149]	0.99 [0.94]	1 [12600]	33/34 [9/34]	0.97 [0.26]
kNN + LEAF [+ Non-selection]	1 [12533]	148/149 [146/149]	0.99 [0.98]	1 [12600]	33/34 [21/34]	0.97 [0.61]
C4.5 + LEAF [+ Non-selection]	1 [12533]	135/149 [122/149]	0.91 [0.82]	1 [12600]	32/34 [28/34]	0.97 [0.82]
DA + LEAF [+ Non-selection] (* DA: Discriminant analysis using Mahalanobis' distance)	1 [30]	145/149 [91/149]	0.97 [0.61]	1 [100]	33/34 [26/34]	0.97 [0.76]

Classification results in the test set of Lung Cancer and Prostate Cancer

Fig. 29.4 Classification results in the Leukemia 1 test set. (**a**) Classification results in the test set of Leukemia 1a, 1b, 1c, and 2. (**b**) Classification results in the test set of lung cancer and prostate cancer

3.3 *Biological Function Analysis*

The *h*-values of each dataset are presented in Table 29.1. Ideally, it is preferred that the extracted genes provide biologically useful information in addition to imparting high discriminatory power to different classes. We conducted a biological function analysis of gene group in reference to the Gene ontology tool [6] and the University of Washington's L2L microarray analysis tool [12]. Below we focus on the top-10 genes' biological function.

Tables 29.2 and 29.3 summarize the primary functions of the top-10 genes obtained using Gene ontology. As expected, genes related to leukemia in addition

Table 29.1 $h1$, $h2$ and $h3$ values

Dataset name	$h1$	DPS	$h2$	DPS	$h3$	DPS
Leukemia 1a	103	0.019	11	0.096	4	0.405
Leukemia 1b	145	0.021	6	0.078	4	0.162
Leukemia 1c	157	0.019	14	0.081	6	0.231
Leukemia 2	91	0.022	8	0.105	5	0.319
Lung Cancer	149	0.021	12	0.059	3	0.288
Prostate Cancer	219	0.022	27	0.098	7	0.396

Table 29.2 Summary of the top $h2$ genes ranked by DPS (Leukemia 1 and 2)

Leukemia 1a	
DPS ranking: probe name / Gene name	- Description
1: 39318_AT / TCL1A	- T-cell leukemia/lymphoma 1A
2: 34168_AT / DNTT	- Deoxynucleotidyltransferase, terminal
3: 266_S_AT / CD24	- CD24 molecule
4: 37988_AT / CD79B	- CD79b molecule, immunoglobulin-associated beta
5: 36122_AT / PSMA6	- proteasome (prosome, macropain) subunit, alpha type, 6
6: 1096_G_AT / CD19	- CD19 molecule
7: 38017_AT / CD79A	- CD79a molecule, immunoglobulin-associated alpha
8: 36571_AT / TOP2B	- topoisomerase (DNA) II beta 180kDa
9: 39598_AT / GJB1	- gap junction protein, beta 1, 32kDa
10: 39707_AT / MTMR4	- myotubularin related protein 4
Leukemia 1b	
DPS ranking: probe name / Gene name	- Description
1: 33412_AT / LGALS1	- lectin, galactoside-binding, soluble, 1
2: 39081_AT / MT2A	- metallothionein 2A
3: 41058_G_AT / THEM2	- thioesterase superfamily member 2
4: 1389_AT / MME	- membrane metallo-endopeptidase
5: 39257_AT / KLF12	- Kruppel-like factor 12
6: 38266_AT / RBBP6	- retinoblastoma binding protein 6
7: 40763_AT / MEIS1	- Meis homeobox 1
8: 31319_AT / IGKV2OR22-4	- immunoglobulin kappa variable 2/OR22-4 pseudogene
9: 34699_AT / CD2AP	- CD2-associat
10: 35315_AT / ORM1	- orosomucoid 1
Leukemia 1c	
DPS ranking: probe name / Gene name	- Description
1: 38869_AT / PLCH1	- phospholipase C, eta 1
2: 33910_AT / PTPRD	- protein tyrosine phosphatase, receptor type, D
3: 34525_AT / TCL1B	- T-cell leukemia/lymphoma 1B

(continued)

Table 29.2 (continued)

Leukemia 1a	
DPS ranking: probe name / Gene name	- Description
4: 36048_AT / ZNF318	- zinc finger protein 318
5: 38459_G_AT / CYB5A	- cytochrome b5 type A (microsomal)
6: 40272_AT / CRMP1	- collapsin response mediator protein 1
7: 38360_AT / RASGRP2	- RAS guanyl releasing protein 2 (calcium and DAG-regulated)
8: 992_AT / ERCC2	- excision repair cross-complementing rodent repair deficiency, complementation group 2
9: 35350_AT / GALNAC4S-6ST	- B cell RAG associated protein
10: 37528_AT / ASPH	- aspartate beta-hydroxylase
Leukemia 2	
DPS ranking: probe name / Gene name	- Description
1: U50136_RNA1_AT / LTC4S	- Leukotriene C4 synthase
2: Y00339_S_AT / CA2	- carbonic anhydrase II
3: D90097_AT / AMY2B	- amylase, alpha 2B (pancreatic)
4: M23197_AT / CD33	- CD33 molecule
5: X78992_AT / ZFP36L2	- zinc finger protein 36, C3H type-like 2
6: J00105_S_AT / B2M	- beta-2-microglobulin
7: J05096_RNA1_AT / ATP1A2	- ATPase, Na + /K+ transporting, alpha 2(+) polypeptide
8: U60519_AT / CASP10	- caspase 10, apoptosis-related cysteine peptidase
9: X95735_AT / ZYX	- zyxin
10: X76648_AT / GLRX	- glutaredoxin (thioltransferase)

to leucocyte communication, such as TCL1A (Leukemia 1a: rank 1), LGALS1 (Leukemia 1b: rank 1), TCL1B (Leukemia 1c: rank 3) and LTC4S (Leukemia 2: rank 1), are selected from every leukemia dataset pair. These genes are peculiar disease associated-genes of leukemia. In the L2L program, a p value for the significance of overlap between the given list and the function list of the databases is calculated by using the binomial distribution. Tables 29.4–29.6 summarize dominant top-12 functions in the L2L results. In the each datasets, we can observe that functions related to human cancer and tumor, such as colon carcinoma, glioma and breast cancer, exhibit statistical significance.

4 Conclusion

LEAF is an iterative FSM incorporating the concept of LOOCV; it also provides a DPS of genes. Moreover, we can determine the top-h according to the distribution of DPS value for each dataset using a simple algorithm for determining h-values.

Table 29.3 Summary of the top $h2$ genes ranked by DPS (Lung Cancer and Prostate Cancer)

Lung Cancer	
DPS ranking: probe name / Gene name	- Description
1: 37205_AT / FBXL7	- F-box and leucine-rich repeat protein 7
2: 32046_AT / PRKCD	- protein kinase C, delta
3: 33569_AT / CLEC10A	- C-type lectin domain family 10, member A
4: 32128_AT / CCL18	- chemokine (C-C motif) ligand 18 (pulmonary and activation-regulated)
5: 40765_AT / PDXDC1	- pyridoxal-dependent decarboxylase domain containing 1
6: 33833_AT / SPTAN1	- spectrin, alpha, non-erythrocytic 1 (alpha-fodrin)
7: 35672_AT / AHCTF1	- AT hook containing transcription factor 1
8: 41657_AT / STK11	- serine/threonine kinase 11
9: 31888_S_AT / PHLDA2	- pleckstrin homology-like domain, family A, member 2
10: 40681_AT / ITGBL1	- integrin, beta-like 1 (with EGF-like repeat domains)
Prostate Cancer	
DPS ranking: probe name / Gene name	- Description
1: 37639_AT / HPN	- hepsin (transmembrane protease, serine 1)
2: 37720_AT / HSPD1	- heat shock 60kDa protein 1 (chaperonin)
3: 41504_S_AT / MAF	- v-maf musculoaponeurotic fibrosarcoma oncogene homolog (avian)
4: 34138_AT / GNAO1	- guanine nucleotide binding protein (G protein), alpha activating activity polypeptide O
5: 863_G_AT / SERPINB5	- serpin peptidase inhibitor, clade B (ovalbumin), member 5
6: 39206_S_AT / ACAN	- aggrecan
7: 38127_AT / SDC1	- syndecan 1
8: 38028_AT / LMO3	- LIM domain only 3 (rhombotin-like 2)
9: 37394_AT / C7	- complement component 7
10: 40562_AT / GNA11	- guanine nucleotide binding protein (G protein), alpha 11 (Gq class)

The h-values can be used as criteria for identifying candidate or informative genes. Our method shows that high classification accuracy and the biological functions of extracted genes correspond well with those reported in the literature. Finally, we propose a gene analysis method for using LEAF for basic biomedical research and drug discovery. From these results, we expect that our method will become a powerful tool for exploring biomarker candidates or for various diagnosing practices (such as cancer, carcinoma tumor, etc).

We plan to investigate the effect of threshold parameter ($Thrs1$) on the result of function analysis to evaluate the usefulness of the method by applying it to other datasets.

Table 29.4 Function enrichment analysis (L2L) for the top-10 genes of Leukemia 2

Function name	*p*-Value	Discription
as3_fibro_c3	0.000	Upregulated late by sodium arsenite in fibroblasts (Cluster 3)
as3_fibro_up	0.000	Upregulated by sodium arsenite in fibroblasts (Clusters 1, 2, and 3)
hdaci_colon_but2hrs_up	0.001	Upregulated by butyrate at 2 hrs in SW260 colon carcinoma cells
il2_pbmc_4hr_dn	0.002	Down-regulated at 4 hr following treatment of peripheral blood mononuclear cells (PBMC) with recombinant IL-2.
hdaci_colon_but_up	0.009	Upregulated by butyrate at any timepoint up to 48 hrs in SW260 colon carcinoma cells
gh_exogenous_early_dn	0.010	Down-regulated at early time points (1–4 hours) following treatment of mammary
hdaci_three_up	0.010	carcinoma cells (MCF-7) with exogenous human growth hormone Upregulated in several cell lines by all of TSA, SAHA, and MS-275
uvb_nhek4_24hrs_up	0.014	Upregulated at 24 hours by UV-B light in normal human epidermal keratinocytes
bleo_human_lymph_high_24hrs_up	0.015	Up-regulated at 24 hours following treatment of human lymphocytes (TK6) with a high dose of bleomycin
bay_pbmc_4hr_dn	0.016	Down-regulated at 4 hr following treatment of peripheral blood mononuclear cells (PBMC) with BAY 50-4798, an IL-2 receptor agonist.
sarcomas_gistromal_up	0.016	Top 20 positive significant genes associated with GI stromal tumors, versus other soft-tissue tumors.
bay_pbmc_2hr_dn	0.019	Down-regulated at 2 hr following treatment of peripheral blood mononuclear cells (PBMC) with BAY 50-4798, an IL-2 receptor agonist.

Table 29.5 Function enrichment analysis (L2L) for the top-10 genes of Leukemia 1

Function name	*p*-Value	Discription
hdaci_colon_tsa48hrs_dn	0.005	Downregulated by TSA at 48 hrs in SW260 colon carcinoma cells
tnfalpha_tgz_adip_up	0.008	Upregulated in mature, differentiated adipocytes following treatment with TNFalpha
brcalko_mef_up	0.010	Up-regulated in mouse embryonic fibroblasts following targeted deletion of BRCA1 (exon 11) compared to wild-type MEFs
elongina_ko_dn	0.011	Downregulated in MES cells from elongin-A knockout mice
tnfalpha_tgz_adip_up	0.015	Upregulated in mature, differentiated adipocytes following simultaneous treatment with troglitazone and TNFalpha
insulin_adip_insens_dn	0.016	Down-regulated by insulin in murine adipocytes, and continue to respond following induction of insulin-resistance with TNFalpha treatment
adipocyte_pparg_up	0.017	Adipocyte genes induced by both PPARgamma and rosiglitazone
hdaci_colon_but_dn	0.023	Downregulated by butyrate at any timepoint up to 48 hrs in SW260 colon carcinoma cells
aged_mouse_retina_any_up	0.024	Up-regulated in the retina of 16-month aged mice from any of four strains (S8, S10, SR1, B6J), versus 3-month young controls
human_tissue_thymus	0.028	Genes expressed specifically in human thymus tissue
il2_pbmc_4hr_up	0.033	Up-regulated at 4 hr following treatment of peripheral blood mononuclear cells (PBMC) with recombinant IL-2.
brcal_sw480_up	0.034	Up-regulated by infection of human colon adenocarcinoma cells (SW480) with Ad-BRCA1, versus Ad-LacZ control

a) Function enrichment analysis (L2L) for the top-10 genes of Leukemia 1a

(continued)

Table 29.5 (continued)

Function name	*p*-Value	Discription
hdaci_colon_cur12hrs_dn	0.004	Downregulated by curcumin at 12 hrs in SW260 colon carcinoma cells
fsh_rat_granulosa_up	0.008	Up-regulated by FSH in immortalized rat granulosa cells (rFSH-17)
hdaci_colon_cur16hrs_dn	0.010	Downregulated by curcumin at 16 hrs in SW260 colon carcinoma cells
hdaci_colon_cursul_dn	0.010	Downregulated by both curcumin and sulindac at any timepoint up to 48 hrs in SW260 colon carcinoma cells
5azac_hepg2_up	0.012	Up-regulated in human hepatoma cells (HepG2) following 48 hrs treatment with 2.5 microM 5-aza-2-deoxycytidine (5azaC).
as3_hek293_up	0.012	Upregulated in HEK293 cells by treatment with arsenite
dsrna_dn	0.013	Downregulated by dsRNA (polyI:C) in IFN-null GRE cells
cmv_hcmv_timecourse_6hrs_up	0.017	Up-regulated in fibroblasts following infection with human cytomegal ovirus (at least 3-fold, with Affymetrix change call, in at least two consectutive timepoints), with maximum change at 6 hours
fsh_human_granulosa_up	0.017	Up-regulated by FSH in primary human granulosa cells (rFSH-17)
5azac-tsa_hepg2_up	0.020	Up-regulated in human hepatoma cells (HepG2) following 24 hrs treatment with 2.5 microM 5-aza-2-deoxycytidine (5azaC) and 24 hrs treatment with both 5azaC and 500 nM trichostatin A (TSA).
oxstress_breastca_up	0.021	Upregulated by H2O2, Menadione and t-BH in breast cancer cells
cmv_hcmv_timecourse_10hrs_up	0.022	Up-regulated in fibroblasts following infection with human cytomegalovirus (at least 3-fold, with Affymetrix change call, in at least two consectutive timepoints), with maximum change at 10 hours

b) Function enrichment analysis (L2L) for the top-10 genes of Leukemia 1b

human_tissue_pancreas	0.002	Genes expressed specifically in human pancreas tissue
uvc_high_d1_dn	0.023	Down-regulated at 6 hours following treatment of WS1 human skin fibroblasts with UVC at a high dose (50 J/m^2) (cluster d1)
cited1_ko_het_dn	0.026	Down-regulated in pubertal mammary glands from CITED1 knockout mice, which display disturbed mammary development, versus heterozygotes
bay_pbmc_6hr_dn	0.027	Down-regulated at 6 hr following treatment of peripheral blood mononuclear cells (PBMC) with BAY 50-4798, an IL-2 receptor agonist.
il2_pbmc_24hr_up	0.028	Up-regulated at 24 hr following treatment of peripheral blood mononuclear cells (PBMC) with recombinant IL-2.
il2_pbmc_4hr_dn	0.034	Down-regulated at 4 hr following treatment of peripheral blood mononuclear cells (PBMC) with recombinant IL-2.
il2_pbmc_6hr_dn	0.034	Down-regulated at 6hr following treatment of peripheral blood mononuclear cells (PBMC) with recombinant IL-2.

c) Function enrichment analysis (L2L) for the top-10 genes of Leukemia 1c

Table 29.6 Function enrichment analysis (L2L) for the top-10 genes of Lung Cancer and Prostate Cancer

Function name	*p*-Value	Discription
dixoin_pbmnc_up	0.007	Up-regulated in peripheral blood mononuclear cells of subjects with high exposure to dioxin, compared to low-exposure controls.
bws_imprinted	0.010	Imprinted genes on Chr. 11p15 that are dysregulated in Beckwith-Wiedmann syndrome
hbx_nl_dn	0.018	Downregulated by expression of Hepatitis B virus HBx protein in normal hepatocytes (Hhep)
dex_keratinocyte_4hr_up	0.019	Up-regulated in primary human keratinocytes at 4 hours following treatment with 0.1 microM dexamethasone
uvb_nhek3_c3	0.021	Regulated by UV-B light in normal human epidermal keratinocytes, cluster 3
pituitary_fetal_up	0.022	Up-regulated in human fetal pituitary tissue, compared to adult pituitary tissue
but_tsa_up	0.027	Upregulated by the combination of TSA and butyrate in HT-29 cells
uvb_nhek1_c2	0.029	Upregulated by UV-B light in normal human epidermal keratinocytes, cluster 2
hbx_hcc_dn	0.035	Downregulated by expression of Hepatitis B virus HBx protein in SK-Hep-1 hepatocellular carcinoma cells
human_tissue_placenta	0.035	Genes expressed specifically in human placental tissue
vegf_huvec_2hrs_up	0.038	Up-regulated 2 hours after VEGF treatment in human umbilical vein endothelial cells
bcrabl_h160_cdna_dn	0.040	Down-regulated by expression of p210(BCR-ABL) in human leukemia (HL-60) cells; detected by spotted cDNA arrays

a) Function enrichment analysis (L2L) for the top-10 genes of Lung Cancer

hdaci_colon_tsa12hrs_dn	0.009	Downregulated by TSA at 12 hrs in SW260 colon carcinoma cells
uvb_nhek4_1hr_dn	0.011	Downregulated at 1 hour by UV-B light in normal human epidermal keratinocytes
uvb_nhek4_24hrs_up	0.015	Upregulated at 24 hours by UV-B light in normal human epidermal keratinocytes
sarcomas_synovial_dn	0.020	Top 20 negative significant genes associated with synovial sarcomas, versus other soft-tissue tumors.
uvb_nhek4_6hrs_dn	0.023	Downregulated at 6 hours by UV-B light in normal human epidermal keratinocytes
hdaci_colon_tsa48hrs_up	0.024	Upregulated by TSA at 48 hrs in SW260 colon carcinoma cells
ros_mouse_aorta_up	0.028	Up-regulated in mouse aorta by chronic treatment with PPARgamma agonist rosiglitazone
bay-il2_pbmc_middle_dn	0.030	Down-regulated at 2–6 hours following treatment of peripheral blood mononuclear cells (PBMC) with BAY 50-4798, an IL-2 receptor agonist, compared to recombinant IL-2.
gh_exogenous_all_up	0.034	Up-regulated consistently at all time points (1–24 hours) following treatment of mammary carcinoma cells (MCF-7) with exogenous human growth hormone
il6_scar_fibro_up	0.036	Upregulated following IL-6 treatment in hypertrophic scar fibroblasts
myc_targets	0.045	Myc-responsive genes reported in multiple systems
hdaci_colon_cur24hrs_up	0.046	Upregulated by curcumin at 24 hrs in SW260 colon carcinoma cells

b) Function enrichment analysis (L2L) for the top-10 genes of Prostate Cancer

Acknowledgements A part of this work was supported by Promotion for Young Research Talent and Network from Northern Advancement Center for Science & Technology (NOASTEC Japan) and Grant-in-Aid for Young Scientists (B) No.21700233 from MEXT Japan.

References

1. Armstrong SA, Staunton JE, Silverman LB, et al (2001) Mll translocations specify a distinct gene expression profile that distinguishes a unique leukemia. BioinNature Genet 30(1):41–47
2. Turck, Chris (Ed.) (2009) Biomarkers for Psychiatric Disorders. Springer, U.S.
3. Kevin Ahern, Ph.D. (2009) GEN Best of the Web. Genetic Engineering & Biotechnology News 29(8):66
4. Fukuta K, Nagashima T, Okada Y (2010) Leaf: leave-one-out forward selection method for cancer classification using gene expression data. Proceedings of The 9th IEEE/ACIS International Conference on Computer and Information Science, ICIS 2010, 18–20, August, pages 31–36
5. Fukuta K, Okada Y (2011) Leaf: Leave-one-out forward selection method for gene selection in dna microarray data. Lecture Notes in Engineering and ComputerScience: Proceedings of The International MultiConference of Engineers and Computer Scientists 2011, IMECS 2011, 16–18 March, Hong Kong, pages 175–180
6. Gene Ontology Consortium (2000a) Gene ontology: tool for the unification of biology. Nature Genet 25:25–29
7. Golub T-R, Slonim D-K, Tamayo P, Huard C, Gaasenbeek M, Mesirov J-P, Coller H, Loh M-L, Downing J-R, Caligiuri M-A, et al (1999) Molecular classification of cancer: class discovery and class prediction by gene expression monitoring. Science 286:531–537
8. Gordon GJ, Jensen RV, Hsiao LL, Gullans SR, Blumenstock JE, Ramaswamy S, Richards WG, Sugarbaker DJ, Bueno R (2002) Translation of microarray data into clinically relevant cancer diagnostic tests using gene expression ratios in lung cancer and mesothelioma. Cancer Res 62:4963–4967
9. Lachenbruch PA (1979) Discriminant analysis. Gendai-Sugakusha, Kyoto
10. Mitsubayashi H, Aso S, Nagashima T, Okada Y (2008) Accurate and robust gene selection for disease classification using a simple statistic. Bioinformation 3(2):68–71
11. Singh D, Febbo PG, Ross K, Jackson DG, Manola J, Ladd C, Tamayo P, Renshaw AA, D'Amico AV, Richie JP, Lander ES, Loda M, Kantoff PW, Golub TR, Sellers WR8 (2002) Gene expression correlates of clinical prostate cancer behavior. Cancer Cell 1:203–209
12. Newman JC, Weiner AM (2005) L2L: a simple tool for discovering the hidden significance in microarray expression data. Genome Biol 6(9):R81

Chapter 30
High Capacity Image Steganography Based on Genetic Algorithm and Wavelet Transform

Elham Ghasemi, Jamshid Shanbehzadeh, and Nima Fassihi

Abstract This paper presents the application of wavelet transform and genetic algorithm (GA) in a novel steganography scheme. We employ a GA based mapping function to embed data in discrete wavelet transform coefficients in 4 × 4 blocks on the cover image. The optimal pixel adjustment process (OPAP) is applied after embedding the message. We utilize the frequency domain to improve the robustness of steganography and, we implement GA and OPAP to obtain an optimal mapping function to reduce the difference error between the cover and the stego-image, therefore improving the hiding capacity with low distortions. Our simulation results reveal that the novel scheme outperforms adaptive steganography technique based on wavelet transform in terms of peak signal to noise ratio and capacity, 39.94 dB and 50% respectively.

Keywords Steganography • Discrete wavelet transform • Genetic algorithm • Optimal pixel adjustment process • Image processing

1 Introduction

Steganography is the art and science to hide data in a cover that it can be text, audio, image, video, etc. Data hiding techniques are generally divided in two groups: spatial and frequency domain [1–3]. The first group is based on embedding message in the least significant bit (LSB) of image pixel. The basic LSB method has a

E. Ghasemi (✉) • N. Fassihi
Department of Computer Engineering at Science and Research branch
Islamic Azad University, Tehran, Iran
e-mail: elham_ghasemi@yahoo.com; nimafassihi@gmail.com

J. Shanbehzadeh
Department of Computer Engineering, Tarbiat Moallem University Tehran, Iran
e-mail: jamshid@tmu.ac.ir

S.I. Ao et al. (eds.), *Intelligent Control and Innovative Computing*, Lecture Notes in Electrical Engineering 110, DOI 10.1007/978-1-4614-1695-1_30,

simple implementation and high capacity [4]. However it has low robustness versus some attacks such as low-pass filtering and compression [5]. There are two types of LSB insertion methods, fixed-size and variable-size. In the fixed-size methods are embedded the same number of message bits in each pixel of cover image [6, 7], and the variable-size embeds, the variant number of LSBs in each pixel used for message embedding depends on the image characteristics [8, 9]. A variant of LSB method can be found in [7] that it proposes an optimal pixel adjustment process (OPAP) in which imperceptibility of the stego-image can be improved. Furthermore, this hiding method improved the sensitivity and imperceptibility problem found in the spatial domain.

The second group embeds the messages in the frequency coefficients of images. These hiding methods overcome the problem related to robustness and imperceptibility found in the spatial domain. JPEG, a standard image compression technique, employs discrete cosine transform (DCT). Several steganography techniques for data hiding in JPEG have been proposed; such as JSteg [2], JP Hide&Seek [2] and OutGuess [11]. Most recent researches apply discrete wavelet transform (DWT) due to its wide application in the new image compression standard, JPEG2000. An example is the employment of an adaptive data embedding technique with the use of OPAP to hide data in the integer wavelet coefficients of the cover image [12]. Raja et al. [13] presented a genetic algorithm (GA) based steganography in discrete cosine transforms (GASDCT) domain and, GA based steganography using discrete wavelet transforms (GASDWT). GASDWT has an improvement in bit rate error compared to GASDCT.

The application of GA in steganography can increase the capacity or imperceptibility [14, 15]. Fard, Akbarzadeh and Varasteh [14] proposed a GA evolutionary process to make secure steganography encoding on the JPEG images. Elshafie, Kharma and Ward [15] introduced a parameter optimization using GA that maximizes the quality of the watermarked image. This paper proposes a method to embed data in DWT coefficients using a mapping function based on GA in 4×4 blocks on the cover image and, it applies the OPAP after embedding the message to maximize the Pick signal to noise ratio (PSNR).

This paper is organized as follows: Sect. 2 introduces the proposed algorithm in detail. Section 3 discusses the achieved results and compares the proposed scheme with the state of the art algorithms. Section IV concludes the paper.

2 The Steganography Method

The proposed method embeds the message in DWT coefficients based on GA and OPAP algorithm and then applies on the obtained embedded image. This section describes this method, and embedding and extracting algorithms in detail.

Fig. 30.1 The image Lena after one Haar wavelet transform [17]

2.1 Haar Discrete Wavelet Transform

Wavelet transform has the capability to present data information in time and frequency simultaneously. This transform passes the time domain data through low-pass and high-pass filters to extract low and high frequency information respectively. This process is repeated for several times and each time a section of the signal is drawn out.

DWT analysis divides the discrete signal into two segments (i.e. approximation and detail) by signal decomposition for various frequency bands and scales. DWT utilizes two function sets: scaling and wavelet which associate with low and high pass filters. Such a decomposition manner bisects time separability. In other words, only half of the samples in a signal are sufficient to represent the whole signal, doubling the frequency separability.

Haar wavelet operates on data by calculating the sums and differences of adjacent elements. This wavelet operates first on adjacent horizontal elements and then on adjacent vertical elements. One nice feature of Haar wavelet transform is that the transform is equal to its inverse. Figure 30.1 shows image Lena after one Haar wavelet transform.

After each transformation, the size of the square that contains the most important information is reduced by 4. For detail information on DWT, we can see [16].

2.2 OPAP Algorithm

The main idea of applying OPAP is to minimize the error between the cover and the stego image. For example if the pixel number of the cover is 10,000 (decimal number 16) and the message vector for 4 bits is 1,111, then the pixel number will change to 11,111 (decimal number 31) and the embedding error will be 15, while

Fig. 30.2 A simple chromosome with 16 genes

3	10	4	8	12	7

after applying OPAP algorithm the fifth bit will be changed from 1 to 0, and the embedding error is reduced to 1.

The OPAP algorithm can be described as follows:

Case 1 $(2^{k-1} < \delta_i < 2^k)$: if $p_i' \geq 2^k$, then $p_i'' = p_i' - 2^k$ otherwise $p_i'' = p_i'$;

Case 2 $(-2^{k-1} < \delta_i < 2^{k-1}) : p_i'' = p_i'$;

Case 3 $(-2^k < \delta_i < -2^{k-1})$: if $p_i' < 256 - 2^k$, then $p_i'' = p_i' + 2^k$; otherwise $p_i'' = p_i'$;

P_i, P_i' and P_i'' are the corresponding pixel values of the ith pixel in the three images; cover, stego and the obtained image by the simple LSB method, respectively. $(\delta_i = P_i'' - P_i)$ is the embedding error between P_i and P_i'' [7]. Therefore after embedding k-LSBs of P_i with k message bits, δ_i will be as follows:

$$-2^K < \delta_t < 2^K \tag{30.1}$$

2.3 *Genetic Algorithm*

GA is a technique which mimics the genetic evolution as its model to solve problems. The given problem is considered as input and the solutions are coded according to a pattern. The *fitness* function evaluates every candidate solution most of which are chosen randomly. Evolution begins from a completely random set of entities and is repeated in subsequent generations. The most suitable, and not the bests, are picked out in every generation. Our GA aims to improve the image quality. Pick signal to noise ratio (PSNR) can be an appropriate evaluation test. Thus the definition of fitness function will be:

$$PSNR = 10\log_{10}\frac{M \times N \times 255^2}{\sum_{ij}(y_{ij} - x_{ij})^2}. \tag{30.2}$$

Where M and N are the image sizes and, x *and* y are the image intensity values before and after embedding respectively.

A solution to the problem is translated into a list of parameters known as chromosomes. These chromosomes are usually displayed as simple strings of data. At the first step, several characteristics are generated for the pioneer generation randomly and the relevant proportionality value is measured by the fitness function. A chromosome is encoded as an array of 16 genes containing permutations 1 to 16 that point to pixel numbers in each block. Each chromosome produces a mapping function as shown in "Fig. 30.2".

The next step associates with the formation of the second generation of the society which is based on selection processes via genetic operators in accordance with the formerly set characteristics. A pair of parents is selected for every individual. Selections are devised so that to find the most appropriate component. In this way, even the weakest components enjoy their own chance of being selected and local solutions are bypassed. This paper employs Tournament method.

The contents of the two chromosomes which enter the generation process are interacted to produce two newborn chromosomes. In this approach two of the bests are mixed to give a superb one. In addition, during each process, it is likely for a series of chromosomes to undergo mutations and breed a succeeding generation of different characteristics.

2.4 *Embedding Algorithm [17]*

The following steps explain the embedding process:

Step1. Divide the cover image into 4×4 blocks.

Step2. Find the frequency domain representation of blocks by 2D Haar DWT and get four subbands LL1, HL1, LH1, and HH1.

Step3. Generate 16 genes containing the pixels numbers of each 4×4 blocks as the mapping function.

Step4. Embed the message bits in k-LSBs DWT coefficients each pixel according to mapping function. For selecting value of k, images are evaluated from $k = 3$ to 6. k equal to 1 or 2, provide low hiding capacity with high visual quality of the stego image and k equal to 7 or 8, provide low visual quality versus high hiding capacity.

Step5. Fitness evaluation is performed to select the best mapping function.

Step6. Apply OPAP on the image.

Step7. Calculate inverse 2D-HDWT on each 4×4 block.

2.5 *Extraction Algorithm [17]*

The extraction algorithm consists of four steps as follows:

Step1. Divide the cover image into 4×4 blocks.

Step2. Extract the transform domain coefficient by 2D HDWT of each 4×4 block.

Step3. Employ the obtained function in the embedding phase and find the pixel sequences for extracting.

Step4. Extract k-LSBs in each pixel.

Table 30.1 Comparison of PSNR of images for variant values of k [17]

Cover image	PSNR $k = 3$	$k = 4$	$k = 5$	$k = 6$
Lena	46.83	39.94	32.04	24.69
Jet	51.88	45.20	37.45	29.31
Boat	**48.41**	**40.44**	**31.17**	**23.60**
Baboon	**47.32**	**40.34**	**32.79**	**24.80**

3 Experimental Results [17]

The proposed method is applied on 512×512 8-bit grayscale images "Jet", "Boat", "Baboon" and "Lena". The messages are generated randomly with the same length as the maximum hiding capacity. Table 30.1 shows the stego image quality by PSNR as described in (30.2). Human visual system is unable to distinguish the grayscale images with PSNR more than 36 dB [12]. This paper embedded the messages in the k-LSBs, from $k = 3$ to $k = 6$ and received a reasonable PSNR. Table 30.1 shows PSNR for variant values of k.

This paper embedded the messages in the 4-LSBs and received a reasonable PSNR.

Table 30.1 presents the results and we can see that for k equal to 4 or 5, we obtain the highest hiding capacity and reasonable visual quality. Therefore, we take k equal to 4 as the number of bits per pixel.

Figure 30.3 shows the original cover images along with their histogram analyze which will be used later to compare it with the ones of the resulting stego image to test for imperceptibility.

Figure 30.4 shows images for k equal to 4 that there is no significant change in stego image histogram for 4-LSBs images, thus it is robust against some statistical attacks.

4 Conclusions

This paper introduced a novel steganography technique to increase the capacity and the imperceptibility of the image after embedding. GA employed to obtain an optimal mapping function to lessen the error difference between the cover and the stego image and use the block mapping method to preserve the local image properties. Also we applied the OPAP to increase the hiding capacity of the algorithm in comparison to other systems. However, the computational complexity of the new algorithm is high. The simulation results showed that capacity and imperceptibility of image had increased simultaneity. Also, we can select the best block size to reduce the computation cost and to increase the PSNR using optimization algorithms such as GA (Table 30.2).

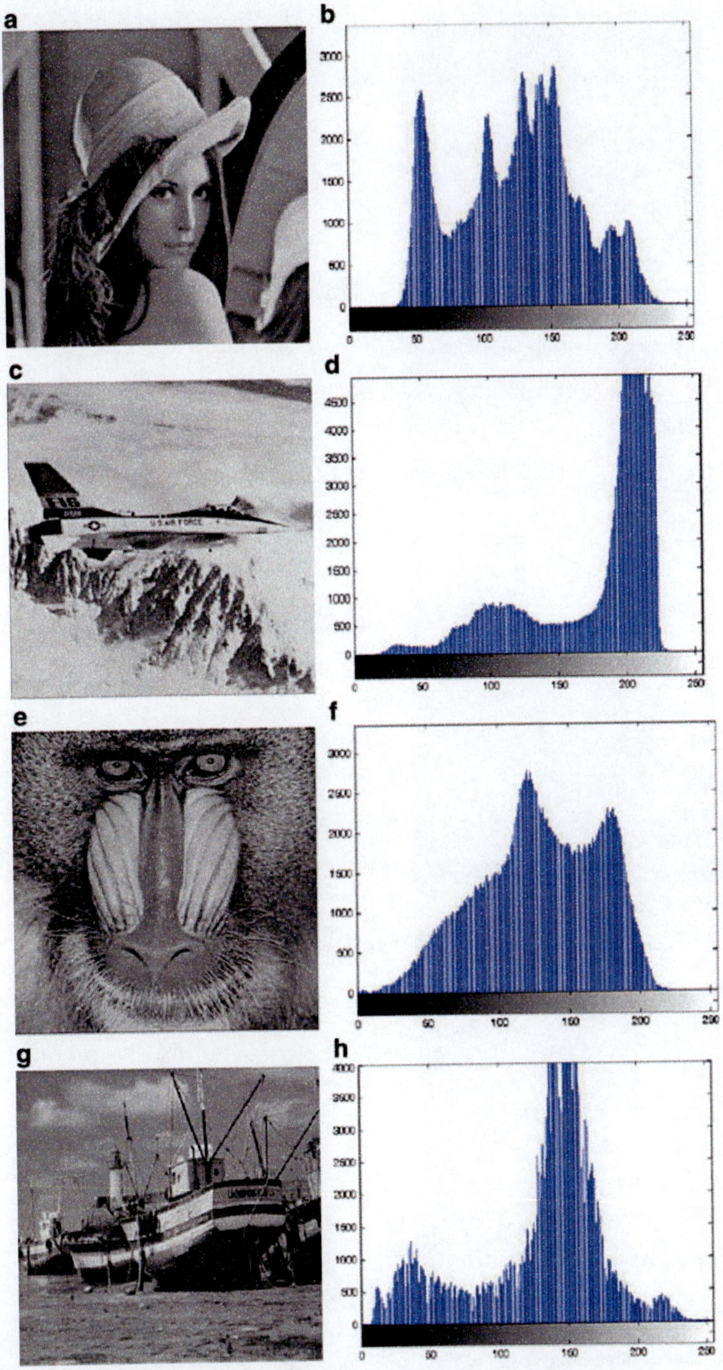

Fig. 30.3 Four cover images used in system simulation and their corresponding histogram: (**a**) cover image Lena; (**b**) Lena histogram; (**c**) cover image Jet; (**d**) Jet histogram; (**e**) cover image Baboon; (**f**) Baboon histogram; (**g**) cover image boat; (**h**) boat histogram [17]

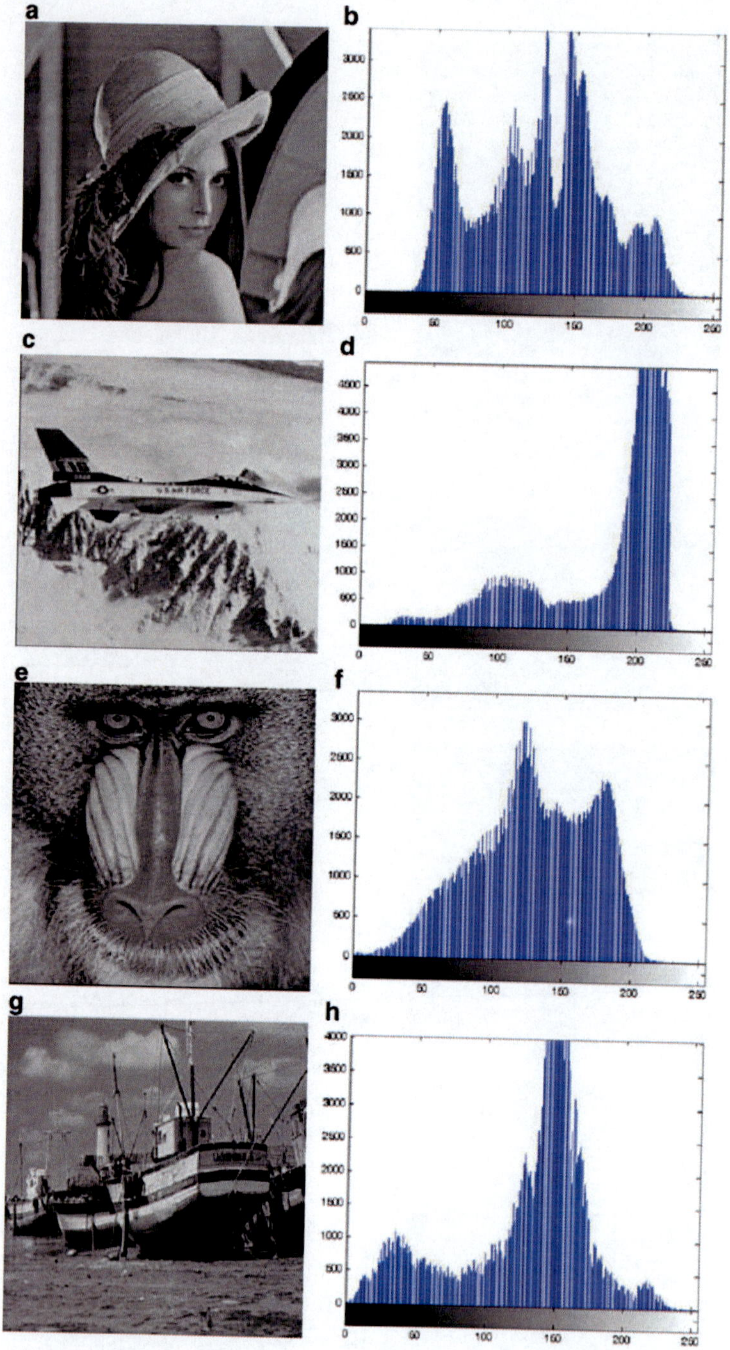

Fig. 30.4 Output stego image of $k = 4$ for embedding data and their corresponding histograms: (**a**) Lena image; (**b**) Lena histogram; (**c**) Jet image; (**d**) Jet histogram; (**e**) Baboon image; (**f**) Baboon histogram; (**g**) Boat image; (**h**) Boat histogram [17]

Table 30.2 Comparison of hiding capacity achieved and the obtained PSNR between our proposed method and methods in [10, 12] and [16, 17]

Cover image	Method	Hiding capacity (bit)	Hiding capacity (%)	PSNR (DB)
Lena	Proposed method	1,048,576	50%	39.94
	adaptive [12]	986,408	47%	31.8
	HDWT [10]	801,842	38%	33.58
	DWT [16]	573,550	27.34%	44.90
Baboon	Proposed method	1,048,576	50%	40.34
	adaptive [12]	1,008,593	48%	30.89
	HDWT [10]	883,220	42%	32.69
	DWT [16]	573,392	27.34%	44.96
Jet	Proposed method	1,048,576	50%	45.20
	DWT [16]	573,206	27.33%	44.76
Boat	Proposed method	1,048,576	50%	40.44
	DWT [16]	573,318	27.33%	44.92

References

1. Inoue H, Miyazaki A, Katsura T (2002) An image watermarking method based on the wavelet transform. In: International Conference on Image Processing, IEEE ICIP, Kobe, vol 1, pp 296–300
2. Provos N, Honeyman P (2003) Hide and seek: an introduction to steganography. In: IEEE Journal Computer Society, ISSN: 1540-7993, 11 June 2003, pp 32–44
3. Provos N (2001) Defending against statistical steganalysis. In: SSYM'01 Proceedings of the 10th conference on USENIX Security Symposium, USENIX Association Berkeley, CA, USA, vol 10, pp 323–335
4. Wu N, Hwang M (2007) Data hiding: current status and key issues. Int J Network Security 4(1):1–9
5. Chen W (2003) A comparative study of information hiding schemes using amplitude, frequency and phase embedding, PhD thesis, National Cheng Kung University, Taiwan
6. Chi-kwong C, Cheng LM (2002) Improved hiding data in images by optimal moderately significant-bit replacement. IEE Electron 37:1017–1018
7. Chan CK, Chang LM (2004) Hiding data in images by simple LSB substitution. Pattern Recogn 37:469–474
8. Chang K-C, Chang C-P, Huang PS, Te-Ming T (2008) A novel image steganographic method using tri-way pixel-value differencing.J Multimedia 3(2):37–44
9. Lee YK, Chen LH (2000) High capacity image steganographic model. IEE Proc Image Signal Process 147:288–294
10. Lai B, Chang L (2006) Adaptive data hiding for images based on haar discrete wavelet transform. In: Lecture Notes in Computer Science, Springer-verlag Berlin Heidelberg, vol 4319, pp 1085–1093
11. Provos N (2001) Defending against statistical steganalysis. In: Proceedings of 10th usenix security symposium, Usenix Assoc, pp 323–335
12. El Safy RO, Zayed HH, El Dessouki A (2009) An adaptive steganography technique based on integer wavelet transform. In: ICNM international conference on networking and media convergence, Cairo 24–25 March, pp 111–117

13. Raja KB, Kiran KK, Satish KN, Lashmi MS, Preeti H, Venugopal KR, Patnaik LM (2007) Genetic algorithm based steganography using wavelets. In: International conference on information system security, ICISS, Springer-Verlag Berlin Heidelberg, vol 4812, pp 51–63
14. Fard AM, Akbarzadeh MR, Varasteh AF (2006) A new genetic algorithm approach for secure JPEG steganography. International conference on engineering of intelligence systems, ICEIS Islamabad, 18 September, pp 1–6
15. ElShafie DR, Kharma N, Ward R (2008) Parameter optimization of an embedded watermark using a genetic algorithm. In: International symposium on communications, control and signal processing, ISCCSP, St Julians12–14 March, pp 1263–1267
16. Chen P, Lin H (2006) A DWT based approach for image steganography. Int J Appl Sci Eng 4(3):275–290
17. Ghasemi E, Shanbehzadeh J, Fassihi N High capacity image steganography using wavelet transform and genetic algorithm. In: Lecture notes in engineering and computer science: proceedings of the international multiconference of engineers and computer scientists 2011, IMECS 2011, Hong Kong, 16–18 March 2011, pp 495–498

Chapter 31
Reduce Measurement Uncertainty of Ball Crater Method by Using Image Processing Techniques

Jungang Huang, Panos Liatsis, and Kevin Cooke

Abstract A method of reducing the uncertainty in measuring coating thickness using the ball crater method is proposed in this research. A crucial step of the ball crater method is measurement of the radii of the crater circles from a microscope image. While traditional methods of measuring the radii involve human operators, who inevitably introduce measurement uncertainty, the proposed approach measures a radius by fitting a circle (or an ellipse) to all edge points of a crater circle edge, which is extracted from the microscope image of the crater by using image processing techniques. This eliminates the subjectiveness introduced by human operators and reduces the measurement uncertainty. Experimental results confirm the feasibility of our method and its potential in reducing measurement uncertainty and increasing measurement accuracy of the ball crater method. The use of all edge points in the estimation of the radius in the proposed method also enables accurate determination of coating thickness from an image of a crater taken from a direction other than the normal to the coating surface.

Keywords Digital image processing • Measurement uncertainty • Thickness measurement • Coatings

J.G. Huang (✉) • P. Liatsis
School of Engineering and Mathematical Sciences, City University London, Northampton Square, London EC1V 0HB, UK
e-mail: j.huang.1@city.ac.uk; p.liatsis@city.ac.uk

K. Cooke
Teer Coatings Ltd., Berry Hill Industrial Estate, Droitwich, Worcestershire WR9 9AS, UK
e-mail: kevin.cooke@miba.com

S.I. Ao et al. (eds.), *Intelligent Control and Innovative Computing*, Lecture Notes in Electrical Engineering 110, DOI 10.1007/978-1-4614-1695-1_31,

1 Introduction

The ball crater method is a low cost method of determining coating thickness and thus it is widely used in both the laboratory and industry [1]. Figure 31.1 illustrates the method in measuring a coating layer on a substrate. A depression into both the coating and the substrate is first created by rotating a ball with a known radius R in the presence of an abrasive paste. The depression is then observed under an optical microscope for measuring the radii of the two circles R_1 and R_2. The thickness d of the coating can be obtained as:

$$d = \sqrt{R^2 - R_1^2} - \sqrt{R^2 - R_2^2} \tag{31.1}$$

Thus, a crucial step in this method is to determine the radii of the crater circles from its microscopic image. The accuracy of the thickness value measurement is mostly defined by the accuracy of radius measurements. Traditional methods of measuring the radii of the circles are subjective as they involve the use of human operators. For example, one method (two point method) is to move the sample stage by a micrometer drive in a direction normal to a line on the microscope eyepiece reticles and record the readings of the micrometer at two positions, where the line becomes the tangent of the circle [2]. The subtraction of the two readings provides the diameter of the circle. These two positions are defined by the operator. The other method (three point method) is to calculate the radius by using the coordinates of three points on the circle [2]. In the method, the locations of the three points are chosen by the operator.

The involvement of human operators inevitably introduces measurement uncertainty in the ball crater method since people perceive an image subjectively and even the same person may perceive the same image differently at different times. Hence,

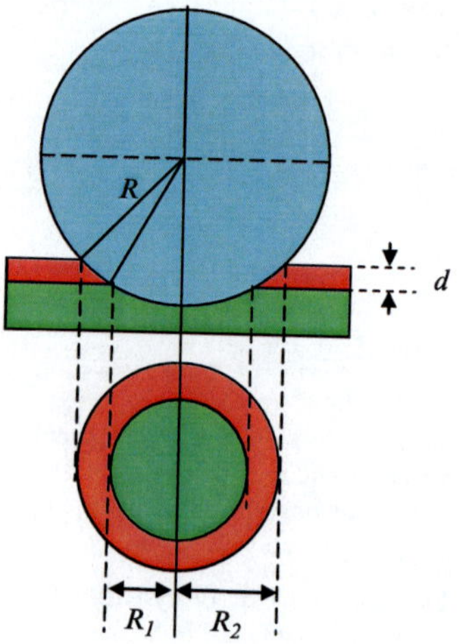

Fig. 31.1 Schematic illustration of ball crater method for coating thickness measurements

traditional approaches introduce randomness through the subjective selection of a limited number of points on the circle edge and subsequently the estimation of the radius, which is based on the point coordinates. It should be noted that the circle edges of a real crater do not resemble a perfect circle, as shown in Fig. 31.2a. The random nature of the points selection, thus increases measurement uncertainty. This measurement uncertainty becomes more pronounced when the crater has to be observed at a direction other than the normal of the coating surface because the crater circles appear to be ellipses in the microscope image.

In this work, we propose a method of measuring the radii with the aid of image processing. This method measures a radius by fitting a circle (or an ellipse) to all the edge points of a crater edge, which is extracted from the microscope image of the crater by using image processing techniques. The proposed method will always result to a specific thickness value for a given crater image. Therefore, it improves the measurement repeatability of the ball crater method [3]. Furthermore, the use of all edge points in the estimation of the radius increases measurement accuracy and enables accurate determination of coating thickness from an image of a crater taken from a direction other than the normal of the coating surface.

2 Image Processing and Thickness Estimation

Figure 31.2a shows the image of a crater taken under a microscope with 10× magnification. The crater is created by rotating a 30 mm chromium steel ball with the presence of 0.25 μm diamond abrasive agent on top of a flat coated sample. The coated sample is composed of a titanium alloy substrate, a pure, metallic interlayer and a multi-element alloy topcoat. We now illustrate the method by providing the details of the process of extracting the radius value of the top circle edge, i.e., the edge separating the crater from the top surface (The result is shown in Fig. 31.2b).

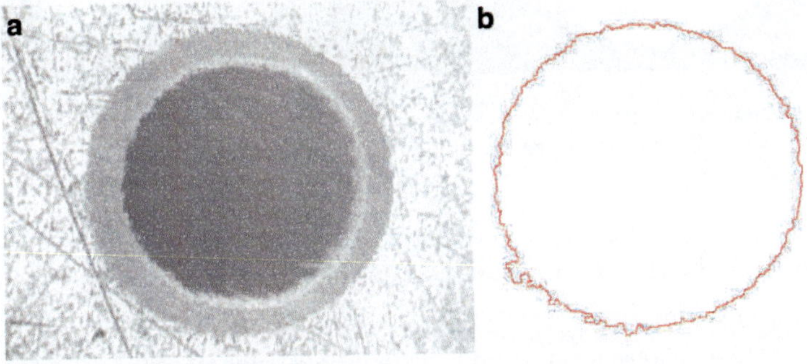

Fig. 31.2 (**a**) Crater image and (**b**) top circle edge of a coating sample (a titanium alloy substrate, a pure, metallic interlayer and a multi-element alloy topcoat) viewed under a microscope with 10× magnification

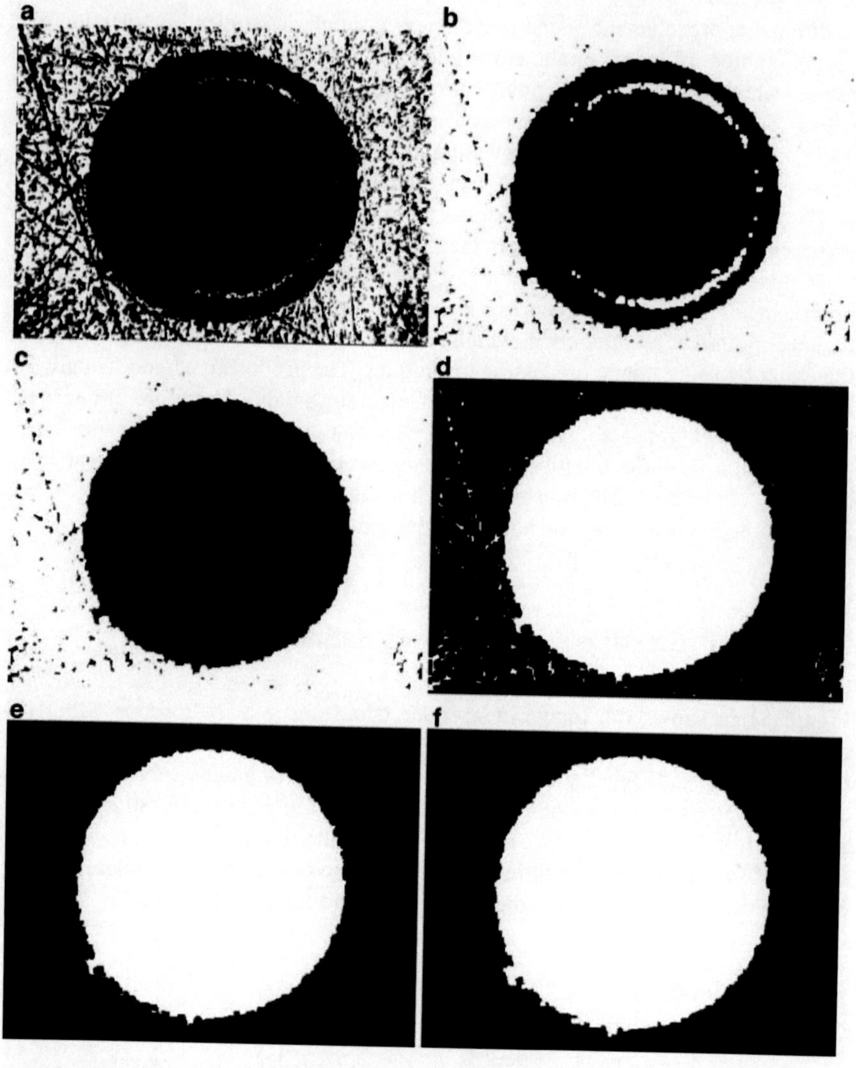

Fig. 31.3 Intermediate stage images when processing Fig. 31.2a to obtain Fig. 31.2b: (**a**) thresholding; (**b**) dilation; (**c**) low pass filtering; (**d**) image inversion; (**e**) low pass filtering; (**f**) dilation

To calculate the radius of the top circle edge, we first need to extract all the edge points from the image. This can be divided into six steps:

1. Thresholding [4] (Fig. 31.3a): the top surface area and the crater area are segmented by choosing an appropriate threshold value. Figure 31.3a shows the binary image obtained at this step. It is obvious that thresholding does not lead to perfect segmentation, rather there are noisy pixels originating from both the top surface and the crater area.

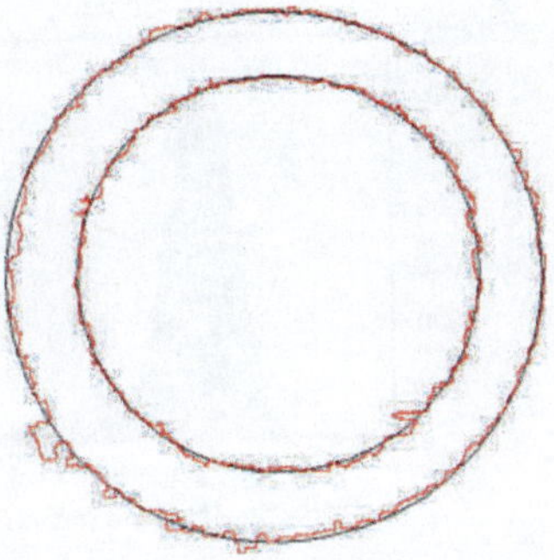

Fig. 31.4 Best fit circles on the two edges of the crater shown in Fig. 31.2a

2. Dilation [4] (Fig. 31.3b): the mathematical morphology operator of dilation is used to connect the pixels of the top surface area into one component so that it becomes the largest object in the image.
3. Low pass filtering [4] (see Fig. 31.3c): all small objects inside the crater area are removed by a low pass filter.
4. Image inversion [4] (Fig. 31.3d): the image is inverted so that the crater area now becomes the largest object in the image.
5. Low pass filtering (Fig. 31.3e): all small objects on the top surface area corresponding to scratches and defects are removed by a low pass filter.
6. Dilation (Fig. 31.3f): dilation is used to reverse the effect of dilation in step (2) on the circle edge. Since the image was inverted in step (4), this dilation now corresponds to erosion [4].

Following these six steps, the edge points of the top circle (Fig. 31.2b) can be obtained by a simple edge detection scheme. Once the coordinates of the edge points are known, the best fit circle can be obtained using the Levenberg–Marquardt least squares fitting [5]. We can consider the radius value of the best fit circle as the radius value of the circle edge.

According to (31.1), thickness estimation requires two radius values, i.e., R_1 and R_2. Here, we take the measurement of the total thickness of the two layers of coatings as an example. For this purpose, we need the radius values of the top circle edge and the bottom circle edge (the edge separating the substrate from the coatings). Figure 31.4 shows the edge points and the best fit circles for both circle edges. The bottom circle edge can be obtained by the procedure described above.

In theory, the two boundaries are co-circular. In real measurements, however, the coordinates of the circle centres given by the two best fit circles vary. We choose the one from the circle with the smallest fitting residue as the common centre for both circular boundaries. The average distance from this common centre to all edge points of the boundary is then taken as the radius value of the circular boundary. The radii of the bottom and the top circular boundaries measured using our method are 299.29 μm and 400.69 μm, respectively. By inserting these values into (31.1), we obtain the total thickness of the two layers to be 2.02 μm.

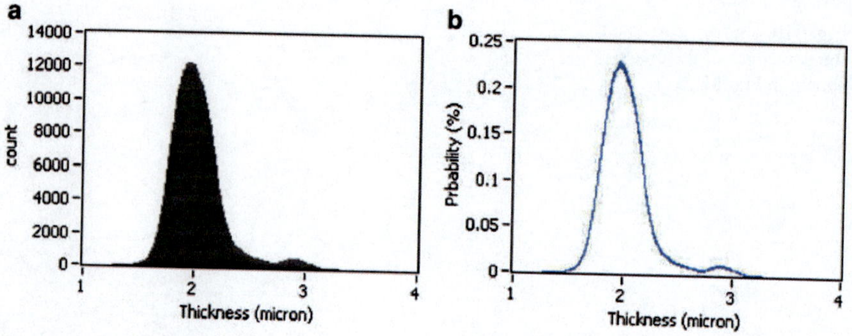

Fig. 31.5 (**a**) Histogram and (**b**) probability distribution of thickness values obtained by a traditional method (estimated by using the 5,319,006 combinations of all points from the top and bottom circular boundaries in Fig. 31.2a)

3 Estimating Measurement Uncertainty

By using the coordinates of the edge points and the previously obtained boundary centre, we can estimate measurement uncertainty introduced by interactive methods. We assume that the distance between the boundary centre and an edge point is a possible radius value obtained from random measurements, each with equal probability of occurrence. There are 2,233 and 2,382 points, respectively, in the bottom and top circular boundaries, which give 5,319,006 combinations, equivalent to the number of random measurements. The thickness value calculated for each combination has a 1/5319006 of probability of occurrence in a measurement.

The histogram of these 5,319,006 thickness values is shown in Fig. 31.5a, and b shows the probability density distribution of the thickness values. It is evident that it resembles a Gaussian distribution. The mean value is 2.02 μm and the standard deviation is 0.23 μm. There is direct correspondence between the above procedure of radii estimation and the one utilising human operators, and thus the obtained results are highly relevant.

While state-of-the-art methods result to a standard deviation of 0.23 μm in radius estimation, which reflects the degree of uncertainty, the proposed technique always results to the mean radius value. As such, it reduces measurement uncertainty. Furthermore, because it utilizes all edge points to estimate the radius value, it has the potential to increase measurement accuracy.

4 Thickness Measurement of Coatings on Tilted Surfaces

The effectiveness of the improved method can be demonstrated by using it to measure the thickness of a coating on a tilted surface. Here the tilted surface refers to a surface that is not perpendicular to the optical axis of the microscope.

In most case, the crater is observed at a plane that is perpendicular to the optical axis of the microscope so that the crater edges appear to be circles. However, there

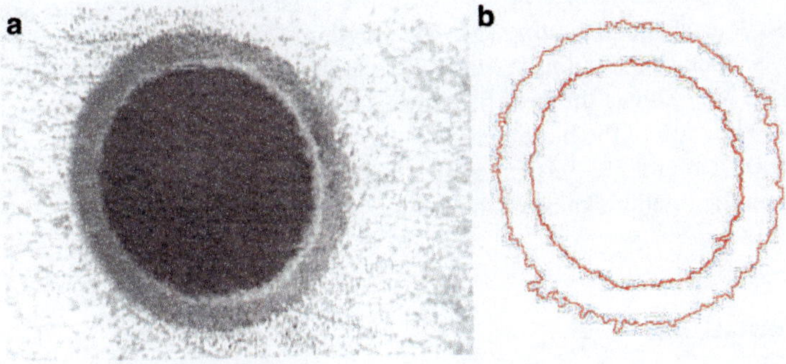

Fig. 31.6 (**a**) Crater image and (**b**) boundaries of a coating sample (a titanium alloy substrate, a pure, metallic interlayer and a multi-element alloy topcoat) viewed at a tilted angle under a microscope with 10× magnification

Fig. 31.7 Best fit ellipses on the two edges of the crater shown in Fig. 31.6a

are some circumstances, where the crater cannot be physically put on such a plane. For instance, the crater is on a heavy part, which cannot be rotated and thus there is not sufficient space to place the microscope in a direction that is perpendicular to the crater. In these circumstances, the crater edges appear to have elliptical shapes in the microscope image and the radius values of the crater edges are equal to the lengths of the semi-major axis of the ellipses. This is because the projection of a circle on a plane that is not parallel to the plane of the circle is an ellipse and the length of the semi-major axis of the ellipse is equal to the radius of the circle. Figure 31.6a shows the image of a crater on a tilted surface taken under a microscope with 10× magnification. The edges of the crater clearly exhibit elliptical.

Traditional methods have difficulty in obtaining an accurate thickness value of a coating on a tilted surface because it is complex to accurately determine the major axis of a non-smooth elliptical shape.

On the contrary, the proposed method can handle this situation quite easily. The elliptical edges of the crater image [see Fig. 31.6b] are extracted using the image processing technique, described in Sect. 2. Once the coordinates of all edge points on the edge is known, the best fit ellipse can be obtained using the Levenberg–Marquardt least squares fitting [5]. The length of the semi-major axis of the best fit ellipse is taken as the radius value of the edge.

Figure 31.7 shows the best fit ellipse for the top edge and bottom edge of the sample. The radii of the bottom and the top edges obtained from the best fit ellipses are 301.01 μm and 397.93 μm, respectively. By inserting these values into (31.1), we obtain the total thickness of the two layers to be 1.04 μm.

5 Conclusion

A method of measuring the radii of crater circles by using image processing techniques was described. It measures a radius of a crater boundary by fitting a circle (or an ellipse) to all the edge points of a crater circle edge, which is extracted from the microscope image of the crater by using image processing techniques.

Compared with the traditional ball crater radius measurement methods, which use a limited number of points on a circle edge to calculate the radius value, the proposed method can reduce the measurement uncertainty introduced by human operators and, therefore, improves the measurement repeatability and potentially increases measurement accuracy of the thickness measurement of ball crater method. The improved method also enables accurate determination of coating thickness from an image of a crater taken from a direction other than the normal of the coating surface.

Acknowledgment The authors would like to acknowledge the financial support of TSB/Advantage W. Midlands under KTP Partnership No. 006504.

References

1. Gee MG, Gant A, Hutchings I, Bethke R, Schiffman K, Van Acker K, Poulat S, Gachon Y, von Stebut J (2003) Progress towards standardisation of ball cratering. Wear 255:1–13
2. Teer BC-2 (2009) Ball crater device specification and user guide. Teer Coatings Ltd, Droitwich
3. Huang JG, Liatsis P, Cooke K (2011) Improving measurement repeatability of ball crater method by using image processing techniques, lecture notes in engineering and computer science: proceedings of the international multiconference of engineers and computer scientists 2011, IMECS 2011, Hong Kong, 16–18 March 2011
4. Relf CG (2004) Image acquisition and processing with LabVIEW. CRC, Boca Raton
5. Levenberg K (1944) A method for the solution of certain non-linear problems in least squares. Quart Appl Math 2:164–168

Chapter 32
Decision Level Fusion Framework for Face Authentication System

V. Vaidehi, Teena Mary Treesa, N.T. Naresh Babu, A. Annis Fathima, S. Vasuhi, P. Balamurali, and Girish Chandra

Abstract In this paper, multiple algorithm and score-level fusion for enhancing the performance of the face based biometric person authentication system is proposed. Though many algorithms are conferred, several crucial issues are still involved in the face authentication. Most traditional algorithms are based on certain assumptions failing which the system will not give appropriate results. Due to the inherent variations in face with time and space, it is a big challenge to formulate a single algorithm based on the face biometric that works well under all variations. This paper addresses the problem of illumination and pose variations, by using three different algorithms for face recognition: Block Independent Component Analysis (B-ICA), Discrete Cosine Transform (DCT) and Kalman filter. The weighted average based score level fusion is performed to improve the results obtained by the system. An intensive analysis of the various algorithms has been performed and the results indicate an increase in accuracy of the proposed system.

Keywords Biometric • B-ICA • DCT • Empirical mode decomposition • Face detection • Kalman filtering • Score level fusion

---

V. Vaidehi (✉)
Head of the Department of Computer Technology, MIT, Anna University, Chennai, India
e-mail: vaidehi@annauniv.edu

T.M. Treesa • N.T.N. Babu • A.A. Fathima • S. Vasuhi
Department of Electronics Engineering, MIT, Anna University, Chennai, India
e-mail: teenamarytreesa@yahoo.com; ntnareshbabu@gmail.com; annisfathima@yahoo.com; vasuhisrinivasan@yahoo.com

P. Balamurali • G. Chandra
Innovations lab, TCS Bangalore, India
e-mail: balamurali.p@tcs.com; m.gchandra@tcs.com

S.I. Ao et al. (eds.), *Intelligent Control and Innovative Computing*, Lecture Notes in Electrical Engineering 110, DOI 10.1007/978-1-4614-1695-1_32,

1 Introduction

Biometric authentication systems evolved in the wake of intense concerns about security and advancements in networking and computer vision systems. Biometrics is described as recognizing an individual based on physiological or behavioral qualities. Biometric systems relying on physiological traits are more secured and free of spoof attacks. Face is considered as the appropriate physiological feature for authentication since it does not demand a strict posture for the person and the biometric can be captured and analyzed even without the person's knowledge. The biometric algorithms developed are based on assumptions about the input images and operate on assumed ideal input conditions. But in any real time system, enforcing constraints is time consuming, user-hostile and undependable. Real time systems have to be regulated and adapted to the changes in the input introduced due to environmental conditions, time and other factors.

The algorithms developed for face recognition are under the assumption of ideal conditions and addresses any one or two face variations. In the face recognition system, the variation due to pose, orientation, illumination and occlusion affects the identification of an individual. In the literature, there is no face recognition theory that is invariant to all face variations. The complementarities of established face recognition algorithms can be utilized to develop a robust system for real-time and can be able to overcome the variations. Thus focusing the objective for developing multi-algorithm based face recognition system.

The multi-algorithmic authentication methods for face recognition proposed in [1, 2] have not handled on the challenges of occlusion, pose and illumination altogether. So, based on these approaches the person may be falsely rejected. For successful and input invariant recognition, extracted features should be invariant to the changes caused due to the factors mentioned above.

The multi-algorithmic biometric system considered in this work is the combination of three independent face recognition algorithms. Each independent module is capable of giving a decision of the person's identity claim independently. The aim of decision level fusion has been to combine the decisions of each of the independent modules into a unanimous decision on the person's identity. To make the system invariant to input image quality, quality estimates have also been incorporated into the fusion schemes.

Thus this paper proposes a face recognition architecture which uses three different feature extraction algorithms to address the problems of illumination, pose variation and incorporates sensor fusion for handling occlusion along with a score level fusion in an intelligent way of assigning weights based on the analysis of the input quality. Rest of the paper is organized as follows. Section 2 describes the proposed face authentication system. Section 3 provides the results of the implementation and analysis of the decision level fusion. Section 4 gives the conclusion.

2 Face Authentication System

A typical biometric system is comprised of five integrated components; sensor module, image processing module, data storage, classifier, and decision making. The sensor module, image sensor captures the image and transmits it for further processing. Image processing module processes the captured image to extract the region of interest (ROI) and to obtain the distinct features. The feature templates are stored in database. The classifier compares the test feature template to one or more templates kept in data storage and percentage of matching is obtained. Finally, decision making module with the percentage of matching makes the final decision with score-level fusion.

The proposed PAS is a multi-algorithmic, multi-sensor biometric system for face recognition. The sensors used are visible camera and IR camera. Visible and thermal IR sensors capture complementary information of reflectance and radiation from the face. Fusion of visible and thermal images is done for robust face recognition regardless of illumination conditions and occlusion. An image fusion technique, utilizing Empirical Mode Decomposition (EMD) [3, 4], is used for improved face recognition. The complete frame work of the proposed PAS is shown in Fig. 32.1.

Face detection is an important step in face recognition. Since algorithms respond differently to differing backgrounds, it is advisable to extract the ROI from the

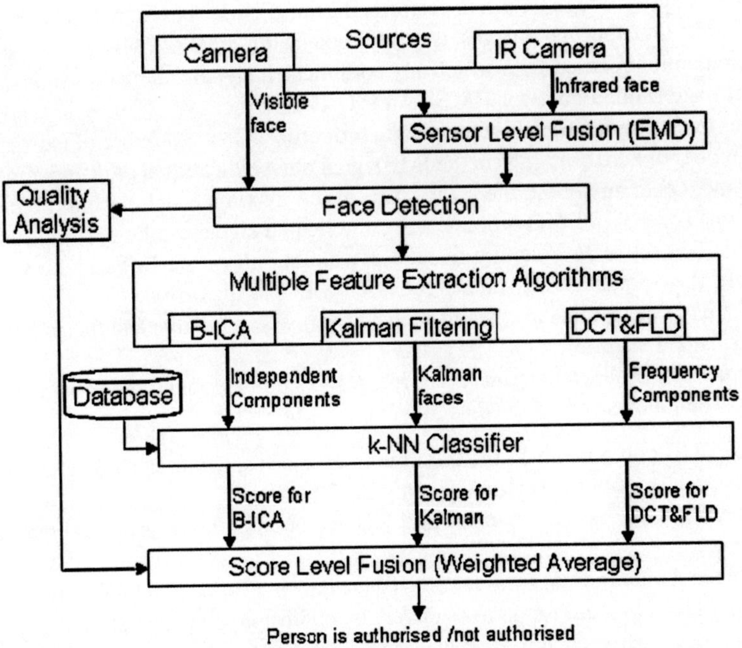

Fig. 32.1 Block diagram of Face recognition system

captured image. In this work ROI is achieved using Haar based face detection. Block Independent Component Analysis (B-ICA), Kalman filtering and Discrete Cosine Transform (DCT) have been chosen as appropriate feature extraction algorithms. The choice of these features was promoted by the need to make the system invariant to illumination and pose. B-ICA has been found to perform well when subjected to images taken in poor lighting conditions. DCT requires a well illuminated image for good results. Kalman faces take care of the varying poses of a person's face.

The k-nearest neighbor (k-NN) classification is dependent on the inter-class and intra-class variance among features. The results are obtained for each algorithm with k-NN classifier, as it is the simplest form of classifier found in the literature on pattern analysis. Based on the face quality results, the weights for score of each algorithm are dynamically assigned. Then, the weighted average score level fusion is the strategy adapted to take the final decision on the person's authenticity.

2.1 Sensor Level Fusion

Fusion of visible and thermal images enhances the rate of recognition and efficiency under varying illumination conditions. Visible and thermal IR sensors capture complementary information of reflectance and radiation from the face [5]. A pair of visible and thermal IR face images must be spatially registered before any useful information is extracted from the fused images [3, 6].

In this system, registration is performed using Fourier based method and fusion of visible and thermal images is performed using EMD. EMD is an adaptive decomposition method with which any complicated signal can be decomposed into its Intrinsic Mode Functions (IMF). An IMF is an indicative of intrinsic oscillatory modes in a complex signal, with specific properties, (1) the number of extrema and the number of zero crossings in the IMF signal must either equal or differ at most by one; and, (2) at any point, the mean value of the maxima envelope and the minima envelope is zero. In EMD process, any complicated signal can be decomposed into a redundant set of IMFs and a residue. By adding all of the IMFs along with the residue, the original signal can be recovered with less distortion.

The basic steps that have been followed for fusion of visible and thermal images are explained as follows:

A pair of visible and thermal color images of same face captured in same view is taken. The images are obtained as 3-D data.

1. The RGB components in the both the images are separated. Three pairs of 2-D data are decomposed separately using EMD.
2. The 2-D matrices are converted into 1-D vector using lexicographical rearrangement.
3. The local maxima and local minima of the data are found and two envelopes viz. upper and lower are found respectively by cubic spline interpolation.
4. The mean between the envelopes are found and then preceded with sifting process to find all the IMFs using EMD as described above for all the six vectors.

5. For fusion of images, these IMFs of RGB components of two different modalities are added as

$$F_{(R,\,G,\,B)}(X,\,Y) = \sum_{j=1}^{k} \alpha_j V_j + \beta_j T_j \tag{32.1}$$

where $F(x, y)$ is the fused image of individual components, α_j and β_j are weighting factors for visible and thermal IMFs, V_j is the jth visible IMF, T_j is the jth thermal IMF, k is the number of IMFs. Here weighting factors are used to decrease the mutual information between visible and thermal [7].

2.2 *Face Detection*

Face detection module feeds the face recognition algorithms. Face recognition algorithms recognize faces based on the features extracted from the face. The captured image may encompass the surrounding objects and there is the need for getting the region of interest (i.e. the face) out from the captured image.

Haar based face detection algorithm is a sub-window based algorithm with a dense or overcomplete [8] feature set from which effective features are selected using Adaboost algorithm [9]. Haar wavelet features are suitable for rigid objects and are capable of extracting the structural content of an object [10]. Adaboost is the learning based algorithm; boost the performance of simple learning algorithm. The features obtained from a large set of face and non-face samples are trained and classified using a set of weak classifiers.

2.3 *Multiple Feature Extraction Algorithms*

Feature extraction algorithms are a special form of dimensionality reduction. Feature extraction involves simplifying the amount of resources required to describe a large set of data accurately. The different feature extraction algorithms that have been employed in this work are B-ICA (Block Independent Component Analysis), DCT (Discrete Cosine Transform) followed by FLD (Fisher's Linear Discriminant), and Kalman filtering.

Block Independent Component Analysis

The Independent Component Analysis (ICA) [11] based face recognizer captures the higher order statistics of the image overcoming the drawback of PCA based system which considers only the second order statistics. The flow of execution of BICA algorithm is to obtain independent components of a face image.

Given training images as input, BICA extracts significant features and store it in the database. In BICA the optimal discriminant features are calculated. The whole image is partitioned into many sub-images, i.e. blocks of the same size and then a common demixing matrix for all the blocks are calculated [12]. Compared with ICA, whose training vector is stretched from the whole image, B-ICA stretches only part of the face image as the training vector. B-ICA greatly dilutes the dimensionality dilemma of ICA.

DCT and FLD

The DCT [13] is applied on the entire face image and the frequency components of the face image are obtained. The DCT converts high-dimensional face images into low-dimensional spaces in which more significant facial features are maintained. The truncated DCT works like a band pass filter which inhibits high-frequency irrelevant information and low-frequency illumination variations.

In order to obtain the most salient and invariant feature of human faces, the FLD is applied in the truncated DCT domain. The FLD is applied after clustering such that the most discriminating facial feature can be effectively extracted. The discriminating feature vectors projected from the truncated DCT domain to the optimal subspace can be calculated as follows:

$$P = E_{\text{optimal}}^{T} \cdot X \qquad (32.2)$$

where X is the truncated DCT coefficient vectors and E_{optimal} is the FLD optimal projection matrix.

Kalman Filtering

Kalman filter [14] identifies significant traits in the face. "Kalmanfaces" [15] are compact visual model that represents the invariant proportions of the face classes. Each face class is represented in single feature vector. Image normalization converts the image into luminance metrics (gray scale) of the same size. During feature extraction, Kalman averaging is computed from normalized faces using Kalman filter to obtain the Kalman face. Only those Kalman faces that are sufficiently invariant with respect to different poses are considered as a feature. Kalman average of the image is computed by the following equation:

$$X_t = X_{t-1} + k_t(X_{t-1} - L_t) \qquad (32.3)$$

where, X_t = pixel average with the time t, k_t = Kalman weighing factor, L_t = luminance value.

$$k_t = \frac{\sigma_{t-1}}{\sigma_{t-1} + \sigma_t} \qquad (32.4)$$

Weight k_t, is the crucial factor in this simplified application of the Kalman filter. σ_t is the standard deviation of the considered face region at time t. k_t approaches zero if the variance increases, i.e. if the luminance of a pixel changes from sample to sample and k_t approaches if the variance decreases.

2.4 k-NN Classifier

k-NN classifier locates the k nearest samples to the query sample and determines its class by identifying the single most frequent class label. The relative distance between instances is determined by using a distance metric which minimizes the distance between two similarly classified instances, while maximizing the distance between instances of different classes.

The Euclidean distance [16] between any two vectors $X = (x_1, x_2, \ldots, x_n)$ and $Y = (y_1, y_2, \ldots, y_n)$ is defined by $d(x, y)$ as

$$d(x, y) = \sqrt{(x_1 - y_1)^2 + \ldots + (x_n - y_n)^2} \tag{32.5}$$

The percentage of match for each algorithm is determined using the formula:

$$C = \left(1 - \frac{\min(d(x, y))}{\max(d(x, y))}\right) * 100 \tag{32.6}$$

3 Decision Level Fusion

3.1 Face Quality Estimation

A quantitative estimate of image quality can be effective in predicting the performance of the recognition unit and subsequently compensating for the degradation due to unavoidable noisy input. The quality estimates are fed to the decision logic to account for the noise in the input and consequently to improve recognition performance. This work has implemented a face illumination analysis for the face quality estimation [17]. The pseudo code for face illumination analysis is as follows:

Start
{
Normalize the image to zero mean and unit variance;
Partition the image into 16 regions;
for (each of the 16 regions)
{
Calculate mean of a block and store in a matrix;
}

0.5095	0.4325	0.1198	0.0285	-0.0493	-0.3491	-0.5872	-0.7467	-0.8306	-0.9213

Fig. 32.2 Face quality analysis of images captured in lab

Generate a Gaussian weight matrix of the same size as the mean matrix;
Calculate the weighted sum of the mean matrix (W_{mi});

$$W_{mi} = \sum_{i=1}^{10} w_i * \overline{I_i} \tag{32.7}$$

and

$$\overline{I_i} = 1/(M * N) \sum_{x=1}^{N} \sum_{y=1}^{M} I(x, y)$$

}
Stop
W_i is the Gaussian weight factor, MxN is the size of the image and $I(x, y)$ is the pixel value at x, y location. The value of W_{mi} determines the illumination of the face image. The measure W_{mi} spans a range of values from -1 to $+1$, where -1 implies a very dark image and $+1$ for a very bright image. The result is shown in Fig. 32.2. The programs are executed using the OpenCV image processing library. OpenCV [18] is an open source Intel's computer vision library and MIT-India Database is used for execution.

3.2 *Score Level Fusion*

An intelligent way of fusion i.e. a correct decision on weighing the obtained score of the different algorithms, will provide the necessitate results and making the system invariant to pose, illumination and occlusion [19]. A weighted average strategy of fusion can easily accommodate varying performance of different algorithms. The matching result ranges are normalized so that the percentage of match for algorithm x is

$$P\_x : \sum_{x=1}^{3} P\_x = 1 \tag{32.8}$$

where algorithm 1 stands for DCT, 2 stands for B-ICA and 3 stands for Kalman. The initial weights given to the respective algorithm is:

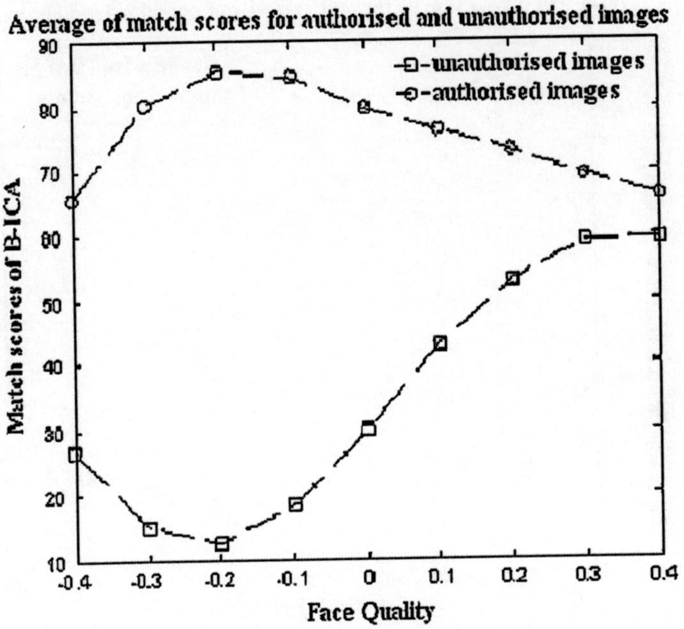

Fig. 32.3 Match scores vs Face Quality for B-ICA algorithm

$$W\_x : \sum_{x=1}^{3} W\_x = 1 \tag{32.9}$$

The initial weights for the algorithms are assigned by evaluating the performance of the algorithm during training and testing. The final score for the person is determined using the formula:

$$S = \sum_{x=1}^{3} W\_x * P\_x \tag{32.10}$$

The performance of the three different face recognition schemes (B-ICA, Kalman and DCT) with respect to the face quality is plotted in Figs. 32.3–32.5 respectively. As seen, the match scores of the B-ICA algorithm falls much more rapidly as compared to the Kalman algorithm for values of quality below −0.2 while the match scores of the Kalman algorithm falls rapidly as compared to the B-ICA algorithm for values of quality above −0.2. The DCT algorithm provides good scores for all values of quality but it is not reliable because it shows a good score for one person when compared with another person. It is reliable only for the template image it was trained with.

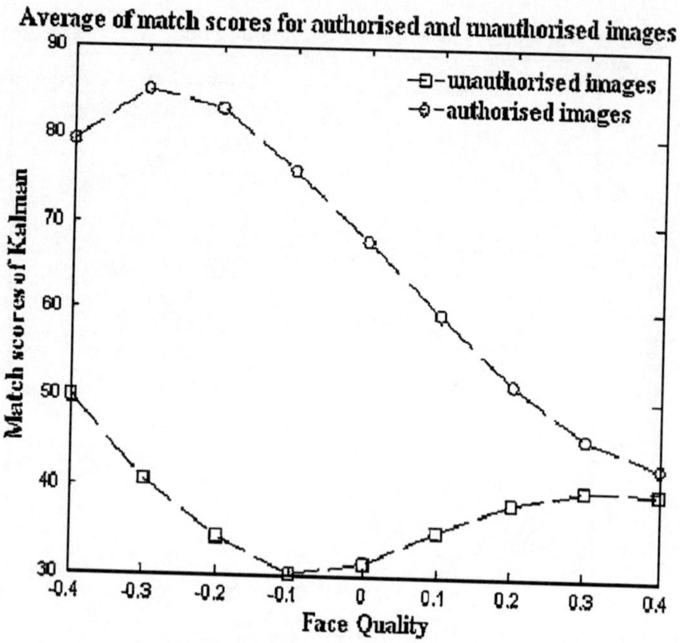

Fig. 32.4 Match scores vs Face Quality for Kalman algorithm

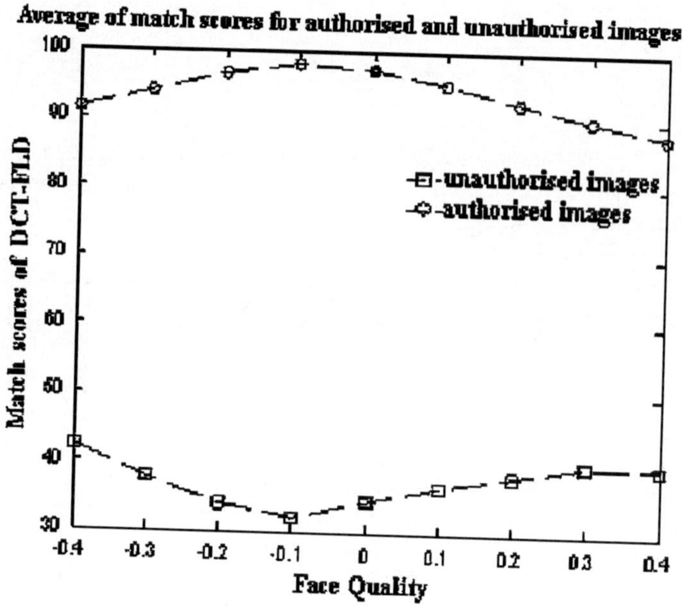

Fig. 32.5 Match scores vs Face Quality for DCT and FLD algorithm

Thus the weights to the three algorithms can be dynamically formulated using the following equations:

$$W_{BICA} = \begin{cases} 0.4 + W_{i/d}(Q_{face}), \forall -0.4 \leq Q_{face} \leq -0.2 \\ 0.8 + W_{i/d}(Q_{face}), \forall -0.2 < Q_{face} \leq 0.4, excluding 0 \\ 0.25 + W_{i/d}(Q_{face}), \forall Q_{face} = 0 \\ 0.2, abs(Q_{face}) > 0.4 \end{cases} \quad (32.11)$$

$$W_{Kalman} = \begin{cases} 0.6 - W_{i/d}(Q_{face}), \forall -0.4 \leq Q_{face} \leq -0.2 \\ 0.2 - W_{i/d}(Q_{face}), \forall -0.2 < Q_{face} \leq 0.4, excluding 0 \\ 0.25 - W_{i/d}(Q_{face}), \forall Q_{face} = 0 \\ 0.2, abs(Q_{face}) > 0.4 \end{cases} \quad (32.12)$$

$$W_{DCT} = \begin{cases} 0.5, \forall Q_{face} = 0 \\ 0, otherwise \end{cases} \quad (32.13)$$

$$W_{i/d}(Q) = [\Pi] * [Q^x] \quad (32.14)$$

$$[\Pi] = [16244 \quad -2847 \quad -4040 \quad 843 \quad 231 \quad -61 \quad 3 \quad -1 \quad 0 \quad 0]$$

$$[Q^x] = [Q^9 \quad Q^8 \quad Q^7 \quad Q^6 \quad Q^5 \quad Q^4 \quad Q^3 \quad Q^2 \quad Q^1 \quad Q^0]'$$

where W_{BICA}, W_{Kalman} and W_{DCT} are the quality dependent weight assigned to B-ICA, Kalman and DCT algorithm, $W_{i/d}$ is the quality dependent weight increment or decrement, $[\Pi]$ is the row vector containing the coefficients of the weight increment/decrement equation and $[Q^x]$ is the column vector containing the powers of the estimated quality.

The selection of quality range from –0.4 to +0.4 in the weight determination equation is due to the fact that for absolute values of quality above 0.4 the percentage of match for authorized as well as unauthorized images starts to increase or decrease, in an undistinguishable way, respectively as can be seen in Fig. 32.6.

The final score after score level fusion is given by the following equation:

$$P_{face} = W_{DCT} * P_{DCT} + W_{BICA} * P_{BICA} + W_{kalman} * P_{kalman} \quad (32.15)$$

where, P_{DCT}, P_{BICA}, P_{Kalman} are the individual match scores of the respective algorithms. For the face quality falling below –0.4 and rising above 0.4, the recognition is carried on with the fused visible and IR image. For the occluded image, the IR image and visible face image of the person is fused using EMD. The fusion of the two images gives more details and face recognition algorithms are performed on it. The final result after performing the score level fusion with three different algorithms considering the quality analysis is given in Fig. 32.7.

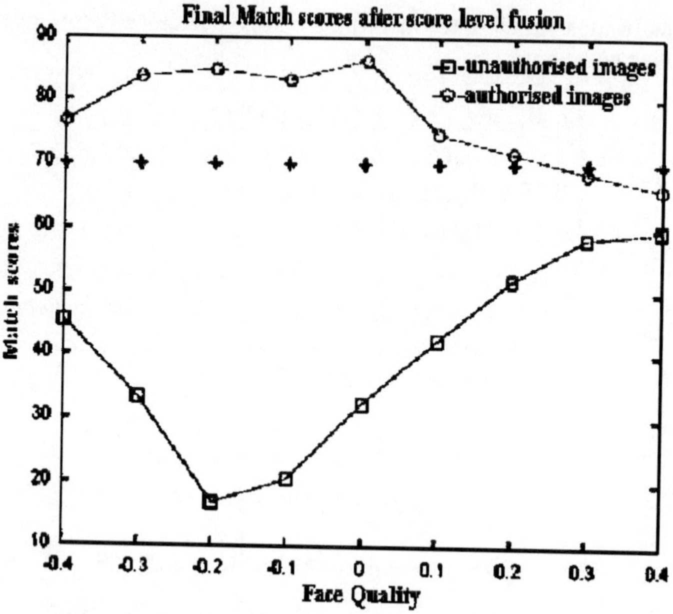

Fig. 32.6 Average match scores with respect to face illumination quality with score level fusion

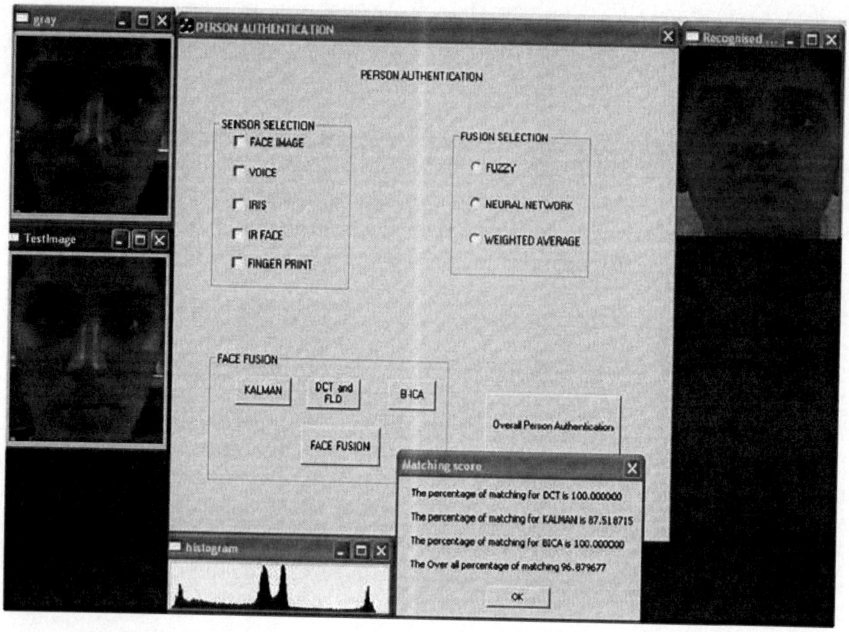

Fig. 32.7 Result of Face authentication with Score level fusion

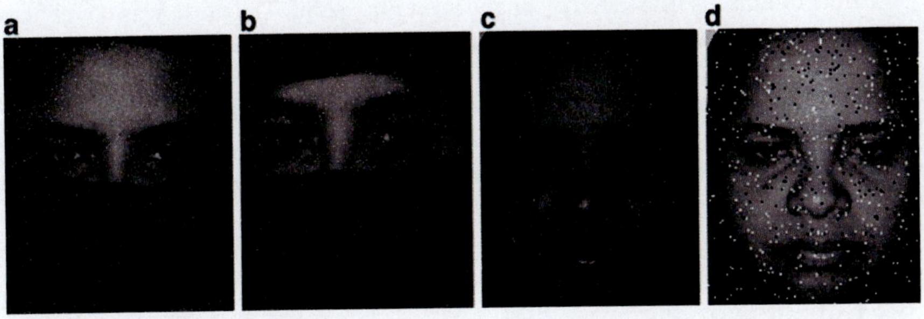

Fig. 32.8 (**a**) and (**b**) Test images with covering on face. (**c**) Test image wearing spectacles. (**d**) Noise added to original image

Table 32.1 Result analysis for modified test images

	Percentage matching			
Test image	BICA	DCT	Kalman	Overall
Fig. 32.8(a)	89.136147 (correct image was recognized)	75 (wrong image was recognized)	76.67993 (correct image was recognized)	82.488060 (correct image was recognized)
Fig. 32.8(b)	15.064096 (not authenticated)	58 (incorrect image was recognized)	41.83242401 (not authenticated)	32.491394 (not authenticated)
Fig. 32.8(c)	10.0000 (not authenticated)	99 (correct image was recognized)	75.790535 (correct image was recognized)	43.697632 (not authenticated)
Fig. 32.8(d)	99.990082 (correct image recognized)	99 (correct image recognized)	87.242401 (correct image recognized)	96.555641 (correct image recognized)

3.3 *Result Analysis*

Further an analysis was performed to judge the effect of some distortions on the face images. Some masked and noise added images of the person present in the database were taken as the test images for the analysis.

The effect of covering was analyzed using Fig. 32.8a and b. In Fig. 32.8a half of the face is covered, in Fig. 32.8b even forehead is also covered. The results are discussed in Table 32.1, it is observed that the fully covered face is not authenticated, but half covered is recognized correctly, though DCT fails to authenticate the image.

The Fig. 32.8c is of a person in the database, who originally does not wear spectacles, but in the test image with spectacles. This reveals that slight changes in the image leads to drastic recognition results. In the above analysis, DCT provides a 99% match whereas B-ICA shows no match.

Effect of noise is illustrated by analysis on Fig. 32.8d. The noise was not able to distort the figure, so that important features were retained, and the analysis did not deviate from the original image.

The analysis reveals that face images are affected in many ways which show their effect on the recognition. Effect of such changes can be accounted for by the introduction of the dynamic weight assignment strategy in the decision stage so that the system bears with the changes in the inputs.

4 Conclusion

This paper performed an analysis into the need for multiple algorithms in the recognition of a face biometric and the approach makes the system adaptive to variations in the image capture or due to environmental conditions. The effect of surroundings is analysed, which makes it necessary for implementing face detection before feature extraction. It is concluded that the proposed scheme outperforms other face recognition approaches. The proposed PAS is pose, illumination and orientation invariant. The system can be further enhanced to include much more biometrics in order to make it spoof-free.

Acknowledgment Authors acknowledge Tata Consultancy Service (TCS), Bangalore, INDIA, for supporting this work.

References

1. Abboud AJ, Sellahewa H, Jassim SA (2009) Quality based approach for adaptive face recognition. In: Proceedings of mobile multimedia/image processing, security, and applications, SPIE, vol 7351, 73510N
2. Ozayy N, Frederick YT, Wheeler W, Liu X (2009) Improving face recognition with a quality-based probabilistic framework. In: IEEE computer society conference on computer vision and pattern recognition workshops, (CVPR Biometrics Workshop 2009), Miami Beach, Florida, pp 134–141
3. Brown LG (1992) A survey of image registration techniques-ACM Comput Surv 24(4): 325–376
4. Kong SG, Heo J, Boughorbel F, Zheng Y, Abidi BR, Koschan A, Yi M, Abidi MA (2007) Multiscale fusion of visible and thermal IR images for illumination-invariant face recognition. Int J Comput Vis 71(2):215–233
5. Looney D, Danilo PM (2009) Multiscale image fusion using complex extensions of EMD, IEEE Trans Sig Process 57(4)
6. Blum RS, Liu Z (2006) Multi-sensor image fusion and its applications. CRC Taylor & Francis group, First edition
7. Goshtasby A (2005) Fusion of multi-exposure images, image and vision computing, 23:611–618. www.imagefusion.org.
8. Papageorgiou C, Poggio T (2000) A trainable system for object detection. Int J Comput Vision 38(1):15–33
9. Viola P, Jones MJ (2004) Robust real-time face detection, Int Jof Comput Vis 57(2):137–154

10. Lienhart R, Maydt J (2002) An extended set of Haar-like features for rapid object detection. IEEE ICIP 2002 1:900–903
11. Bartlett MS, Movellan JR, Sejnowski TJ (2002) Face recognition by independent component analysis. IEEE Trans Neural Networks 13(6):1450–1464
12. Zhang L, Gao Q, Zhang D (2007) Block independent component analysis for face recognition. 14th international conference on image analysis and processing (ICIAP), Modena, Italy 0-7695-2877-5/07
13. Matos FM, Batista LV, Poel 1 (2008) Face recognition using DCT coefficients selection. In: Proc. ACM symposium on applied computing, (SAC '08), Fortaleza, Ceara, Brazil, pp 1753–1757
14. Welch G, Bishop G An introduction to the kalman filter, Technical report, University of North Carolina at Chapel Hill
15. Eidenberger H (2006) Kalman filtering for robust identification of face images with varying expressions and lighting conditions. In: 18th international conference on pattern recognition (ICPR'06), pp 1073–1076
16. Rafael CG, Richard EW (2007) Digital image processing. 3rd edn Prentice Hall, Inc., (now known as Pearson education, Inc.), upper Saddle river, New Jersey 07458, USA
17. Abdel-Mottaleb M, Mahoor MH (2007) Algorithms for assessing the quality of facial images, application notes, IEEE computational intelligence magazine (CIM), 2(2):10–17
18. Bradski G, Kaehler (2008) A Learning OpenCV. First Edition O'Reilly Media, Inc., 1005 Gravenstein Highway North USA
19. Vaidehi V, Treesa TM, Babu NTN, Annis Fathima A, Balamurali P, Chandra G Multi-algorithmic face authentication system, lecture notes in engineering and computer science: proceedings of the international multiconference of engineers and computer scientists 2011, IMECS 2011, 16–18 March 2011

Index

A
Absolute supremum values, 14
Aggregate production planning (APP)
fuzzy programming approach, 192–193
harmony search algorithm (HSA)
hunting search element on the pitch adjustment rate (HuSHSA), 194–195
novel global best harmony search algorithm (NGHSA), 196–197
optimisation process, 193
pseudo code, 194, 195
variable neighbourhood search method (VHS), 196
linear membership functions, 197
multi-objective linear programming (MOLP) model, 190–192
nominal demand
algorithms, 198–199
individual objective optimisation, 197–198
speed of convergence, 199–200
Aggregation-based indexing, 205
Approximate linear quadratic (LQ) regulator problem
computation method, 55–57
evaluation function, 54
feedback matrix, 54–55
inverted pendulum, on cart, 58–63
optimal control problem, 53
Apriori-like method, 368–369
Artificial immune systems
cloning and memory construction, 294–295
dynamic weighted B-cells, 293
image restoration, 294
spike noise, 291
tracking, 291
Artificial intelligent techniques. *See* Fixed-structure robust control techniques
Association rule. *See* Itemset hiding algorithm
Asynchronous communication, 310
Augmented web search system
characteristics, 304
client side
basic operation, 312–313
highlighting function, 313
ranking, number of current accesses, 313
communication through page
asynchronous communication, 310
real-time communication, 310
text highlighting, 310–311
evaluation
communication effect, 315
ranking effect, 313–315
flow of finding expert, 305, 307
numbers of access users, 305–306
ranking method
based on hyperlinks, 308
based on social links, 309
recursive user communication, links, 305, 307
server side
communication servlet, 311–312
search servlet, 311
web page, communication window, 305–306

B
Ball crater method
illustration of, 406
image processing and thickness estimation

S.I. Ao et al. (eds.), *Intelligent Control and Innovative Computing*, Lecture Notes in Electrical Engineering 110, DOI 10.1007/978-1-4614-1695-1,

Ball crater method (*cont.*)
crater image, 407
dilation and image inversion, 408, 409
low pass filtering, 408, 409
thresholding, 408
measurement uncertainty, estimation of, 410
thickness measurement, coatings, 410–412
Biometric authentication systems, 414
Block independent component analysis (B-ICA), 417–418
match scores *vs.* face quality, 421
Boil-step, cooking, 126
Bot herder, 363

C

Capacitated location allocation (CLA) problem, 113–115
C-DenStream, 282
Central level fusion, 29–31
CIQ. *See* Control intelligence quotient (CIQ)
Closure operator
autarkies, conjunctive normal form (CNF) formulas, 332–334
and equivalences, 327–328
kernelization, 331
rectangular coverings, point sets, 331–332
structural parameters, 328–330
variant, shadow problem, 334–335
Clustering, 235
Clustering data streams. *See* Data streams
Combinatorial game. *See* N-player White-Black Cutthroat game
Combinatorial problems, closure structures. *See* Closure operator
Communication servlet, 311–312
Complex-valued network inversion, 139–140
Computer numerical control (CNC)
accuracy analyses, 48–49
feedrate analyses, 47–48
interpolator scheme, 41
inverse kinematics, 43–45
tool orientation, 42–43
tool trajectory, 42
Content based image retrieval (CBIR)
artificial immune systems
cloning and memory construction, 294–295
dynamic weighted B-cells, 293
image restoration, 294
spike noise, 291
tracking, 291
effective and fast image retrieval system (EFIRS), 290
noise filtering, 290
query image evaluation, 295–299
spatial fuzzy linking histogram and feature extraction, 290–291
Control intelligence quotient (CIQ), 183
Cooking model
cooking processes simulation
next cooking-step selection, cook and stove/microwave oven, 129–130
washing pot, 128–129
cooking-step scheduling algorithm
algorithm parameter and cooking model simulation stage, 132
dish order generation stage, 131–132
dish order update stage, 132
preprocessing stage, 131
cooking-step scheduling problem
constraint, 131
input, 130
objective functions, 131
output, 130
evaluation
algorithm parameter adjustment, 132–133
cooking time accuracy evaluation, 133–134
cooking time, change in, 134–135
kitchen layout, 125–126
state transition diagram
cook, 127
cutting board, stove, microwave oven and sink, 128
ingredient, 126–127
pot, 127–128
Cooperative particle swarm optimization, 150
Cost adjusted machine intelligence quotient (CAMIQ)
derivation of, 184–185
interpretations and example
annual budget, 187
cost-per-MIQ, 186
netting out of operating costs, 186
Cut-step, cooking, 125–126
Cutthroat game. *See* N-player White-Black Cutthroat game

D

Darknet observation data, 372
Data hiding, 249
Data streams
density-based clustering algorithms, 277–278

density micro-clustering algorithms
C-DenStream, 282
characteristic, 284, 285
core-micro-cluster, 279
Denstream, 279–281
outlier micro-cluster, 279
potential-micro-cluster, 279
rDenStream, 281–282
SDStream, 283
two-phase clustering, 278
micro-cluster definition, 278
Decision level fusion
face quality estimation, 419–420
score level fusion
face authentication, 424
match scores *vs.* face quality, 421–422
quality dependent weight, 423
Density micro-clustering algorithms
C-DenStream, 282
characteristic, 284, 285
core-micro-cluster, 279
Denstream, 279–281
outlier micro-cluster, 279
potential-micro-cluster, 279
rDenStream, 281–282
SDStream, 283
two-phase clustering, 278
Density-based clustering algorithms, 277–278
Denstream, 279–281
Depth-first unary degree sequence (DFUDS), 353
Discrete cosine transform (DCT), 418
match scores *vs.* face quality, 422
Discrete-time LTI systems. *See* Approximate linear quadratic (LQ) regulator problem
Discrimination power score (DPS), 382
vs. gene-index-number, 383
ranking, 385–387
Dish order, 128
generation stage, 131–132
update stage, 132
Distributed generation
description, 138
inverse estimation, 141–142
results, 145–146
simulation
network architecture and network parameters, 143
regularization coefficient, 144
Distributed indexing, 205
Distribution networks (DN)
capacitated location allocation (CLA) problem, 113–115
case study, 116, 117
HGA
application of, 118–120
structure of, 115
typical evolution, 118
implementation, 116–118
location-allocation problems, 110–111
optimisation of, 112–113
Diversive curiosity, 154
Dynamic weighted B-cells, 293

E

ED sequences, 365
Effective and fast image retrieval system (EFIRS), 290
Electro-hydraulic servo system, 66–69
Element-base indexing, 205
Empirical mode decomposition (EMD), 415
Evolutionary particle swarm optimizer with inertia weight (EPSOIW), 154
Exact cover by 3-sets, 365–366
Exponential Histogram of Cluster Feature (EHCF), 283
Extensible markup language (XML)
compression technique, 208–210
data model, 204–205
evaluations, 212
indexing strategies, 205
INEX collection tests, 211–212
non-queriable XML compressors, 206–207
queriable XML compressors, 207
retrieval model, 210–211
score sharing scheme, 212–214

F

Face authentication system
block diagram of, 415
decision level fusion (*see* Decision level fusion)
face detection, 417
k-NN classifier, 419
multiple feature extraction algorithms
BICA, 417–418
DCT and FLD, 418
Kalman filtering, 418–419
result analysis, 425–426
sensor level fusion, 416–417
Feedback linearizable computed torque controller (FLCTC)
orientation trajectories, 87
position trajectories, 86
tracking errors, 88

FIND_IEPD_SEQUENCES (FIES), 368
First-order-plus-dead-time (FOPDT), 94
Five-axis curve machining
  advantages, 40
  computer numerical control (*see also* Computer numerical control (CNC))
    interpolator scheme, 41
    inverse kinematics, 43–45
    tool orientation, 42–43
    tool trajectory, 42
  motion code
    g-code format, 46
    rational Bezier curve, 45
  off-line approach, 40
Fixed-parameter tractable (FPT), 326
Fixed-structure robust control techniques
  electro-hydraulic servo system, 66–69
  three-phase power inverter
    error, output current, 75
    model of, 72
    multiplicative weight function, 73
    objective functions, 74
    parameters of, 72
    simulation of, 74, 75
    transformation matrix, 70
    weighted sensitivity function, 74
Four-dimensional sentiment vectors, 219
Freight train, 163–164
FREQ_PATTERN_SEQUENCES, 369
Frequent itemset hiding. *See* Itemset hiding algorithm
Fry-step, cooking, 126
Function enrichment analysis
  leukemia 1, 389–391
  leukemia 2, 388
  lung cancer and prostate cancer, 392–393
Fuzzy aggregate production planning. *See* Aggregate production planning (APP)
Fuzzy linking histogram, 290–291
Fuzzy ontology, 267
Fuzzy prefixspan algorithm
  prefix, 265
  projected database, 266–267
  suffix, 265–266
Fuzzy sequence mining
  algorithm, 268
  collection and data set, 267
  frequent pattern mining, 262–263
  instant of items, similar mental concept, 268–269
  membership functions, 264
  nomenclature, 267–268
  ontology, 267
  output sequences, 269–271
  prefixspan algorithm
    prefix, 265
    projected database, 266–267
    suffix, 265–266
  sequential patterns, 263
  SPADE, 262
  time series, 262
  transactional dataset, 268–269

H

Haar discrete wavelet transform, 397
Hard clustering, 236
Harmony search algorithm (HSA)
  evolutionary elements
    hunting search element on the pitch adjustment rate (HuSHSA), 194–195
    novel global best harmony search algorithm (NGHSA), 196–197
    variable neighbourhood search method (VHS), 196
  optimisation process, 193
  pseudo code, 194, 195
Heuristic algorithms, *R*-IEPC
  apriori-like method, 368–369
  machine learning method, 370
HGA. *See* Hybrid genetic algorithms (HGA)
H infinity optimal control, 66
HIQ. *See* Human intelligence quotient (HIQ)
Homemade cooking, 124
Human intelligence quotient (HIQ), 183–184
Hybrid genetic algorithms (HGA)
  application of, 118–120
  structure of, 115
  typical evolution, 118
Hybrid search, 153

I

Impulsive noise, 291
Incident database, 364
Informative event pattern delay (IEPD) sequences
  atom pattern, 366
  definition, 364
  informative event, 366
  matching of, 367
Informative gene discovery, DNA microarray data
  biological function analysis
    discrimination power score ranking, 385–387

function enrichment analysis, 388–393
h-values, dataset, 384, 385
primary functions, genes, 384, 385
classification results, 382, 385
determination method, h-value, 380–382
discrimination power score (DPS) vs. gene-index-number, 382, 383
LEAve-one-out forward selection method, 380 (*see also* LEAve-one-out forward selection method (LEAF))
methodology
leukemia dataset 1, 378
leukemia dataset 2, 379
lung and prostate cancer dataset, 379
Integral-absolute-error (IAE), 94
Internet access logs, 371–374
Intrinsic mode functions (IMF), 416–417
Inverse kinematics, 43–45
Inverse problems and network inversion
complex-valued network inversion, 139–140
definition, 138
forward training phase, 139
ill-posedness and regularization, 140–141
inverse estimation phase, 139
iterative update, of input, 138, 139
Itemset hiding algorithm
integrated sanitization
database, 253
disadvantages, 250–251
Ds, MFI, STE and TList, 252–253
lost itemset, 253
overall process, 251
schematic representation, 250–251
performance evaluation
cases, 253
databases, 257
integrating post-mining, 256–257
number of, sensitive itemsets, 254–255
simulation environment, 254
support, sensitive itemset, 255–256

K

Kalman filtering, 418–419, 422
Kitchen layout, 125–126

L

Leaf-only indexing, 205
Learning system, marshaling
computer simulations, 173–175
desirable layouts, in main track, 166–167
direct rearrangement, 167, 168
learning algorithm
car removal, 171
γ calculation, 171–172
rearrange order and removal destination, 170
problem description
freight yard, 165
position index and yard state, 166
rearrangement process
locomotive, transfer distance, 169, 170
operations, 167–168
Least significant bit (LSB) method, 395–396
LEAve-one-out forward selection method (LEAF), 380
algorithm and data matrix, 381
Localized random search (LRS), 153–154
Location-allocation problems, 110–111
Locomotive, transfer distance of, 169, 170
LP/MIP, 115
LRS. *See* Localized random search (LRS)

M

Machine intelligence quotient (MIQ)
CAMIQ (*see also* Cost adjusted machine intelligence quotient (CAMIQ))
derivation of, 184–185
interpretations and example, 186–187
CIQ, 183
description, 180
HIQ, 183–184
and HIQ and CIQ relationship, 184
household heating system, 180–181
intelligence task graph, 181–182
Machine learning method, 370
Main-cook, 125
Mamdani type fuzzy SMC (FSMC)
block diagram of, 83
tracked paths, WMR, 86
Man-machine cooperative system
components of, 179
functions, 178
Matrix-Apriori algorithm, 250
Maximum upper bounds on delays (MUBDs), 6–8
Measurement uncertainty, 410
Membership problem, tree patterns, 355–357
MEXIR retrieval system, 211
Micro-cluster, 278
MinPts, 278
Mixed-sensitivity approach, 65
Mix-step, cooking, 126
MPV plus extracts, 219

Multi-dimensional sentiment value
sentiment vector
comparison, 227
co-occurrence frequency, 224–226
co-occurrence ratio, 226–227
users experiment, 228
ten-dimensional sentiment vector, 223
Multiple particle swarm optimizers with inertia weight with diversive curiosity (MPSOIWα/DC)
algorithms
flowchart, 152–153
internal indicator, 154
LRS, 153–154
computational time, sphere problem, 160
computer experiments
benchmark optimization problems, 154, 155
LRS effect, 157
vs. MPSOIW*α/DC, 157–159
optimized swarms *vs.* original swarms, 157
parameters, 155
reinitialization frequencies, 156
features, 150
mean and standard deviation, 160
Multi-sensor architectures
central level fusion, 29–31
data processing, 26, 27
implementation
measurement vector, 31
merge measurement, 31–32
tracks and observations, distribution of, 32
sensor level fusion, 26–29
simulations
radars, parameters for, 33
scenario 1, 33–34
scenario 2, 34–35
scenario 3, 35–37
Multi-tape finite automaton, 339

N
Natural immune system, 290
ε-Neighbourhood, 278
Networked control systems
delayed systems, 2, 6
description, 2–3
examples, 6–8
robust stability analysis
Lyapunov-Krasovskii functional, 4
S-procedure, 5
Network traffic screening
algorithm LARGE SCREENING BY ATTACK PERIOD, 371, 372
algorithm SCREENING, 371
darknet observation data, 372
observation data and quantity of packets, 372, 373
parameters for, 372
Non-queriable XML compressors, 206–207
Novel global best harmony search algorithm (NGHSA), 196–197
N-player White-Black Cutthroat game
complexity, 320–323
queer game, 320
winning strategy, 319
Nuke-step, cooking, 126

O
Off-line approach, five-axis curve machining, 40
Optimal pixel adjustment process (OPAP) algorithm, 397–398

P
Particle swarm optimization (PSO), 150, 196–197
algorithms, 151
DC motor speed control plant, 67, 68
parameters and controller parameter ranges, 69
response, system outputs, 69, 71
stability margin *vs.* iterations, 69, 70
Particle swarm optimizers with inertia weight (PSOIW)
algorithms, 151–152
computational time, sphere problem, 160
execution, 153
mean and standard deviation, 160
Pattern matching algorithm, 357–360
Phrase rank, 210
Pick signal to noise ratio (PSNR)
definition, 398
variant k values, 400
Positive/negative value
multiple sentences, 222
one sentence, 221
teaching data extraction, 220
word calculation, 220–221
Premature convergence, 152
Privacy preserving data mining (PPDM)
approaches, 249
fake transactions, 250

input privacy, 249
Matrix-Apriori algorithm, 250
output privacy (*see* Itemset hiding algorithm)
Proportional-integrative-derivative (PID) controllers
complete tuning case, 103–104
design
balanced PID tuning, 101–102
performance optimality index, 97–99
robustness increase measure, 99–100
robustness/performance balance, 100–101
materials and methods
control system configuration, 93–94
performance, 94
robustness, 95
tuning rules, 95–97
particular process case, 104–105
regulatory-control, 92
servo-control operation, 92

Q
Q-Learning, 164
Quantified Boolean formulas (QBF), 320–321
Queer game, 320
Queriable XML compressors, 207

R
Ranking method
based on hyperlinks, 308
based on social links, 309
Rational Bezier curve, 45
Rational word relations
alphabet, 338
inclusion problem, 337–338
multi-tape finite automaton, 339
prefix-closure, 343–347
recognizable sets, 339
submonoids, 340–343
rDenStream, 281–282
Real-time communication, 310
Real-time interpolator, 41
Rearrangement process, cars
locomotive, transfer distance, 169, 170
operations, 167–168
Reciprocally convex combination, 2, 8–9
R-event delay sequence cover (*R*-EC), 365
R-informative event pattern delay sequence cover (*R*-IEPC), 367–368
apriori-like method, 368–369
machine learning method, 370
R-maximal informative event pattern delay sequence cover (*R*-MIEPC), 370–371
Robust stability analysis
Lyapunov-Krasovskii functional, 4
S-procedure, 5
Rule hiding, 249

S
Salt and pepper noise, 291
SDStream, 283
Search servlet, 311
Selective indexing, 205
Sensor level tracking system, 26, 27
Sentiment models, 219
Sentiment search. *See* Multi-dimensional sentiment value
Sentiment vector dimensions
comparison, 227
co-occurrence frequency, 224–226
co-occurrence ratio, 226–227
users experiment, 228
Sequence mining. *See* Fuzzy sequence mining
Sequential data, 261
Shadow problem, 334–335
Soft clustering, 236–237
Space robot
equations of motion, 13–14
model of, 12
phase I, 17
control force, 19
control input torque, 19
joint angles, 20
joint torque input, 20
robot attitude angle, 19
robot position, 18
phase II
attitude angle velocity, 22
joint angles and joint velocities, 23
robot attitude angle, 22
robot position and velocity, 21
space robot parameters, 18
supremum value, 17
torque control input, 22
torque input, 24
translational control input, 21
sliding mode control, 15–16
SPADE, 262
Spike noise, 291
Start cost average (SA), 118–119
State transition diagram
cook, 127
cutting board, stove, microwave oven and sink, 128

State transition diagram (*cont.*)
ingredient, 126–127
pot, 127–128
Statistical rank, 210
Steganography method
embedding algorithm, 399
experimental results, 400–403
extraction algorithm, 399
genetic algorithm, 398–399
Haar discrete wavelet transform, 397
OPAP algorithm, 397–398
Succinct data structures, 352–354

T
Ten-dimensional sentiment vector, 223
Three-phase power inverter
error, output current, 75
model of, 72
multiplicative weight function, 73
objective functions, 74
parameters of, 72
simulation of, 74, 75
transformation matrix, 70
weighted sensitivity function, 74
Three point method, 406
Tikhonov regularization, 141
Time series, 262
Translational control force, 14
Tree patterns
algorithm, 355–357
operations, 354–355
ordered term tree, 351–352
pattern matching algorithm, 357–360
succinct data structures, 352–354
Tree-structured data, 350
Two-dimensional sentiment vectors, 219
Two point method, 406

U
UML. *See* Unified modeling language (UML)
Unified modeling language (UML), 238

V
Variable neighbourhood search element on HSA (VHSA), 196

W
Wash-step, cooking, 126
Web search. *See* Augmented web search system
Wheeled mobile robots (WMRs)
fuzzy control system, 83–85
input torques *vs.* time, 87, 88
kinematic model, 79–80
orientation errors *vs.* time, 87, 88
position trajectory tracking errors, 86
sliding mode controller, 81–83
tracked paths, 86, 87
tracking errors, 88
White-Black Cutthroat game. *See* N-player White-Black Cutthroat game
WMRs. *See* Wheeled mobile robots (WMRs)
Words-of-wisdom search
calculation of, sentiment value, 229
experiments, 229–230
flow of extraction, 218
MPV plus, 219
by multi-dimensional sentiment value
dimensions, sentiment vector, 224–228
ten-dimensional sentiment vector, 223
by positive/negative value
multiple sentences, 222
one sentence, 221
teaching data extraction, 220
word calculation, 220–221
sentiment models, 219
value of, sentiment words, 228–229

X
XML. *See* Extensible markup language (XML)
XML mining application
methodology and design
CDATA section, 235
conceptual model, 237, 238
expectation maximum (EM) algorithm, 237
hard clustering, 236
nine-step model, 234–235
term frequency-inverse document frequency, 237
token, 236
software description
class diagram, 238–239
INEX collection, 239
modularity, 237
package, 237
PostgreSQL, 240
pseudocode, 241
relational model, 240
SAX and DOM, 241
Tree Tagger, 242
UML, 237